Section 10

Transport and Storage of Fluids

Meherwan P. Boyce, Ph.D., P.E. *Chairman and Principal Consultant, The Boyce Consultancy Group, LLC; Fellow, American Society of Mechanical Engineers; Registered Professional Engineer (Texas) (Section Editor, Measurement of Flow, Pumps and Compressors)*

Victor H. Edwards, Ph.D., P.E. *Process Director, Aker Kvaerner, Inc.; Fellow, American Institute of Chemical Engineers; Member, American Association for the Advancement of Science, American Chemical Society, National Society of Professional Engineers; Life Member, New York Academy of Sciences; Registered Professional Engineer (Texas) (Section Editor, Measurement of Flow)*

Terry W. Cowley, B.S., M.A. *Consultant, DuPont Engineering; Member, American Society of Mechanical Engineers, American Welding Society, National Association of Corrosion Engineers (Polymeric Materials)*

Timothy Fan, P.E., M.Sc. *Chief Project Engineer, Foster Wheeler USA; Member, American Society of Mechanical Engineers, Registered Professional Engineer (Massachusetts and Texas) (Piping)*

Hugh D. Kaiser, P.E., B.S., MBA *Principal Engineer, PB Energy Storage Services, Inc.; Senior Member, American Institute of Chemical Engineers; Registered Professional Engineer (Texas) (Underground Storage of Liquids and Gases, Cost of Storage Facilities, Bulk Transport of Fluids)*

Wayne B. Geyer, P.E. *Executive Vice President, Steel Tank Institute and Steel Plate Fabricators Association; Registered Professional Engineer (Atmospheric Tanks)*

David Nadel, P.E., M.S. *Senior Principal Mechanical Engineer, Aker Kvaerner, Inc.; Registered Professional Engineer (Pressure Vessels)*

Larry Skoda, P.E. *Principal Piping Engineer, Aker Kvaerner, Inc.; Registered Professional Engineer (Texas) (Piping)*

Shawn Testone *Product Manager, De Dietrich Process Systems (Glass Piping and Glass-Lined Piping)*

Kenneth L. Walter, Ph.D. *Process Manager—Technology, Aker Kvaerner, Inc.; Senior Member, American Institute of Chemical Engineers, Sigma Xi, Tau Beta Pi (Storage and Process Vessels)*

MEASUREMENT OF FLOW

PUMPS AND COMPRESSORS

PROCESS PLANT PIPING

STORAGE AND PROCESS VESSELS

Nomenclature and Units

In this listing, symbols used in the section are defined in a general way and appropriate SI and U.S. customary units are given. Specific definitions, as denoted by subscripts, are stated at the place of application in the section. Some specialized symbols used in the section are defined only at the place of application.

Symbol	Definition	SI units	U.S. customary units
A	Area	m^2	ft^2
A	Factor for determining minimum value of R_1		
A_∞	Free-stream speed of sound		
a	Area	m^2	ft^2
a	Duct or channel width	m	ft
a	Coefficient, general		
B	Height	m	ft
b	Duct or channel height	m	ft
b	Coefficient, general		
C	Coefficient, general		
C	Conductance	m^3/s	ft^3/s
C	Sum of mechanical allowances (thread or groove depth) plus corrosion or erosion allowances	mm	in
C	Cold-spring factor		
C	Constant		
C_a	Capillary number	Dimensionless	Dimensionless
C_1	Estimated self-spring or relaxation factor		
c_p	Constant-pressure specific heat	J/(kg·K)	Btu/(lb·°R)
c_v	Constant-volume specific heat	J/(kg·K)	Btu/(lb·°R)
D	Diameter	m	ft
D, D_0	Outside diameter of pipe	mm	in
d	Diameter	m	ft
E	Modulus of elasticity	N/m^2	lbf/ft^2
E	Quality factor		
E_a	As-installed Young's modulus	MPa	kip/in^2 (ksi)
E_c	Casting quality factor		
E_j	Joint quality factor		
E_m	Minimum value of Young's modulus	MPa	kip/in^2 (ksi)
F	Force	N	lbf
F	Friction loss	(N·m)/kg	(ft·lbf)/lb
F	Correction factor	Dimensionless	Dimensionless
f	Frequency	Hz	l/s
f	Friction factor	Dimensionless	Dimensionless
f	Stress-range reduction factor		
G	Mass velocity	$kg/(s·m^2)$	$lb/(s·ft^2)$
g	Local acceleration due to gravity	m/s^2	ft/s^2
g_c	Dimensional constant	$1.0\ (kg·m)/(N·s^2)$	$32.2\ (lb·ft)/(lbf·s^2)$
H	Depth of liquid	m	ft
H, h	Head of fluid, height	m	ft
H_{ad}	Adiabatic head	N·m/kg	lbf·ft/lbm
h	Flexibility characteristic		
h	Height of truncated cone; depth of head	m	in
i	Specific enthalpy	J/kg	Btu/lb
i	Stress-intensification factor		
i_i	In-plane stress-intensification factor		
i_o	Out-plane stress intensification factor		
I	Electric current	A	A
J	Mechanical equivalent of heat	1.0 (N·m)/J	778 (ft·lbf)/Btu
K	Index, constant or flow parameter		
K	Fluid bulk modulus of elasticity	N/m^2	lbf/ft^2
K_1	Constant in empirical flexibility equation		
k	Ratio of specific heats	Dimensionless	Dimensionless
k	Flexibility factor		
k	Adiabatic exponent c_p/c_v		
L	Length	m	ft
L	Developed length of piping between anchors	m	ft
L	Dish radius	m	in
M	Molecular weight	kg/mol	lb/mol
M_i, m_i	In-plane bending moment	N·mm	in·lbf
M_o	Out-plane bending moment	N·mm	in·lbf
M_t	Torsional moment	N·mm	in·lbf
M_∞	Free stream Mach number		
m	Mass	kg	lb
m	Thickness	m	ft
N	Number of data points or items	Dimensionless	Dimensionless
N	Frictional resistance	Dimensionless	Dimensionless
N	Equivalent full temperature cycles		
N_S	Strouhal number	Dimensionless	Dimensionless
N_{De}	Dean number	Dimensionless	Dimensionless
N_{Fr}	Froude number	Dimensionless	Dimensionless
N_{Re}	Reynolds number	Dimensionless	Dimensionless
N_{We}	Weber number	Dimensionless	Dimensionless
NPSH	Net positive suction head	m	ft
n	Polytropic exponent		
n	Pulsation frequency	Hz	1/s
n	Constant, general		
n	Number of items	Dimensionless	Dimensionless
P	Design gauge pressure	kPa	lbf/in^2
P_{ad}	Adiabatic power	kW	hp
p	Pressure	Pa	lbf/ft^2
p	Power	kW	hp
Q	Heat	J	Btu
Q	Volume	m^3	ft^3
Q	Volume rate of flow (liquids)	m^3/h	gal/min
Q	Volume rate of flow (gases)	m^3/h	ft^3/min (cfm)
q	Volume flow rate	m^3/s	ft^3/s
R	Gas constant	8314 J/(K·mol)	1545 (ft·lbf)/(mol·°R)
R	Radius	m	ft
R	Electrical resistance	Ω	Ω
R	Head reading	m	ft
R	Range of reaction forces or moments in flexibility analysis	N or N·mm	lbf or in·lbf
R	Cylinder radius	m	ft
R	Universal gas constant	J/(kg·K)	(ft·lbf)/(lbm·°R)
R_a	Estimated instantaneous reaction force or moment at installation temperature	N or N·mm	lbf or in·lbf
R_m	Estimated instantaneous maximum reaction force or moment at maximum or minimum metal temperature	N or N·mm	lbf or in·lbf
R_1	Effective radius of miter bend	mm	in

Nomenclature and Units (*Concluded*)

Symbol	Definition	SI units	U.S. customary units
r	Radius	m	ft
r	Pressure ratio	Dimensionless	Dimensionless
r_c	Critical pressure ratio		
r_k	Knuckle radius	m	in
r_2	Mean radius of pipe using nominal wall thickness $\bar{T}$	mm	in
S	Specific surface area	m^2/m^3	ft^2/ft^3
S	Fluid head loss	Dimensionless	Dimensionless
S	Specific energy loss	m/s^2	lbf/lb
S	Speed	m^3/s	ft^3/s
S	Basic allowable stress for metals, excluding factor E, or bolt design stress	MPa	kip/in^2 (ksi)
S_A	Allowable stress range for displacement stress	MPa	kip/in^2 (ksi)
S_E	Computed displacement-stress range	MPa	kip/in^2 (ksi)
S_L	Sum of longitudinal stresses	MPa	kip/in^2 (ksi)
S_T	Allowable stress at test temperature	MPa	kip/in^2 (ksi)
S_b	Resultant bending stress	MPa	kip/in^2 (ksi)
S_c	Basic allowable stress at minimum metal temperature expected	MPa	kip/in^2 (ksi)
S_h	Basic allowable stress at maximum metal temperature expected	MPa	kip/in^2 (ksi)
S_t	Torsional stress	MPa	kip/in^2 (ksi)
s	Specific gravity		
s	Specific entropy	J/(kg·K)	Btu/(lb·°R)
T	Temperature	K (°C)	°R (°F)
T_s	Effective branch-wall thickness	mm	in
$\bar{T}$	Nominal wall thickness of pipe	mm	in
$\bar{T}_b$	Nominal branch-pipe wall thickness	mm	in
$\bar{T}_h$	Nominal header-pipe wall thickness	mm	in
t	Head or shell radius	mm	in
t	Pressure design thickness	mm	in
t	Time	s	s
t_m	Minimum required thickness, including mechanical, corrosion, and erosion allowances	mm	in
t_r	Pad or saddle thickness	mm	in
U	Straight-line distance between anchors	m	ft
u	Specific internal energy	J/kg	Btu/lb
u	Velocity	m/s	ft/s
V	Velocity	m/s	ft/s
V	Volume	m^3	ft^3

Symbol	Definition	SI units	U.S. customary units
v	Specific volume	m^3/kg	ft^3/lb
W	Work	N·m	lbf·ft
W	Weight	kg	lb
w	Weight flow rate	kg/s	lb/s
x	Weight fraction	Dimensionless	Dimensionless
x	Distance or length	m	ft
x	Value of expression $[(p_2/p_1)^{(k-1/k)} - 1]$		
Y	Expansion factor	Dimensionless	Dimensionless
y	Distance or length	m	ft
y	Resultant of total displacement strains	mm	in
Z	Section modulus of pipe	mm^3	in^3
Z	Vertical distance	m	ft
Z_e	Effective section modulus for branch	mm^3	in^3
Z	Gas-compressibility factor	Dimensionless	Dimensionless
z	Vertical distance	m	ft
	Greek symbols		
α	Viscous-resistance coefficient	$1/m^2$	$1/ft^2$
α	Angle	°	°
σ	Half-included angle	°	°
α, β, θ	Angles	°	°
β	Inertial-resistance coefficient	1/m	1/ft
β	Ratio of diameters	Dimensionless	Dimensionless
Γ	Liquid loading	kg/(s·m)	lb/(s·ft)
Γ	Pulsation intensity	Dimensionless	Dimensionless
δ	Thickness	m	ft
ε	Wall roughness	m	ft
ε	Voidage—fractional free volume	Dimensionless	Dimensionless
η	Viscosity, nonnewtonian fluids	Pa·s	lb/(ft·s)
η_{ad}	Adiabatic efficiency		
η_p	Polytropic efficiency		
θ	Angle	°	°
λ	Molecular mean free-path length	m	ft
μ	Viscosity	Pa·s	lb/(ft·s)
ν	Kinematic viscosity	m^2/s	ft^2/s
ρ	Density	kg/m^3	lb/ft^3
σ	Surface tension	N/m	lbf/ft
σ_c	Cavitation number	Dimensionless	Dimensionless
τ	Shear stress	N/m^2	lbf/ft^2
ϕ	Shape factor	Dimensionless	Dimensionless
ϕ	Angle	°	°
ϕ	Flow coefficient		
ψ	Pressure coefficient		
ψ	Sphericity	Dimensionless	Dimensionless

MEASUREMENT OF FLOW

GENERAL REFERENCES: ASME, *Performance Test Code on Compressors and Exhausters*, PTC 10-1997, American Society of Mechanical Engineers (ASME), New York, 1997. Norman A. Anderson, *Instrumentation for Process Measurement and Control*, 3d ed., CRC Press, Boca Raton, Fla., 1997. Roger C. Baker, *Flow Measurement Handbook: Industrial Designs, Operating Principles, Performance, and Applications*, Cambridge University Press, Cambridge, United Kingdom, 2000. Roger C. Baker, *An Introductory Guide to Flow Measurement*, ASME, New York, 2003. Howard S. Bean, ed., *Fluid Meters—Their Theory and Application—Report of the ASME Research Committee on Fluid Meters*, 6th ed., ASME, New York, 1971. Douglas M. Considine, Editor-in-Chief, *Process/Industrial Instruments and Controls Handbook*, 4th ed., McGraw-Hill, New York, 1993. Bela G. Liptak, Editor-in-Chief, *Process Measurement and Analysis*, 4th ed., CRC Press, Boca Raton, Fla., 2003. Richard W. Miller, *Flow Measurement Engineering Handbook*, 3d ed., McGraw-Hill, New York, 1996. Ower and Pankhurst, *The Measurement of Air Flow*, Pergamon, Oxford, United Kingdom, 1966. Brian Price et al., *Engineering Data Book*, 12th ed., Gas Processors Suppliers Association, Tulsa, Okla., 2004. David W. Spitzer, *Flow Measurement*, 2d ed., Instrument Society of America, Research Triangle Park, N.C., 2001. David W. Spitzer, *Industrial Flow Measurement*, 3d ed., Instrument Society of America, Research Triangle Park, N.C., 2005.

INTRODUCTION

The flow rate of fluids is a critical variable in most chemical engineering applications, ranging from flows in the process industries to environmental flows and to flows within the human body. *Flow* is defined as mass flow or volume flow per unit of time at specified temperature and pressure conditions for a given fluid. This subsection deals with the techniques of measuring pressure, temperature, velocities, and flow rates of flowing fluids. For more detailed discussion of these variables, consult Sec. 8. Section 8 introduces methods of measuring flow rate, temperature, and pressure. This subsection builds on the coverage in Sec. 8 with emphasis on measurement of the flow of fluids.

PROPERTIES AND BEHAVIOR OF FLUIDS

Transportation and the storage of fluids (gases and liquids) involves the understanding of the properties and behavior of fluids. The study of fluid dynamics is the study of fluids and their motion in a force field.

Flows can be classified into two major categories: (a) incompressible and (b) compressible flow. Most liquids fall into the incompressible-flow category, while most gases are compressible in nature. A perfect fluid can be defined as a fluid that is nonviscous and nonconducting. Fluid flow, compressible or incompressible, can be classified by the ratio of the inertial forces to the viscous forces. This ratio is represented by the Reynolds number (N_{Re}). At a low Reynolds number, the flow is considered to be laminar, and at high Reynolds numbers, the flow is considered to be turbulent. The limiting types of flow are the inertialess flow, sometimes called Stokes flow, and the inviscid flow that occurs at an infinitely large Reynolds number. Reynolds numbers (dimensionless) for flow in a pipe is given as:

$$N_{Re} = \frac{\rho VD}{\mu} \tag{10-1}$$

where ρ is the density of the fluid, V the velocity, D the diameter, and μ the viscosity of the fluid. In fluid motion where the frictional forces interact with the inertia forces, it is important to consider the ratio of the viscosity μ to the density ρ. This ratio is known as the kinematic viscosity (ν). Tables 10-1 and 10-2 give the kinematic viscosity for several fluids. A flow is considered to be *adiabatic* when there is no transfer of heat between the fluid and its surroundings. An isentropic flow is one in which the entropy of each fluid element remains constant.

To fully understand the mechanics of flow, the following definitions explain the behavior of various types of fluids in both their static and flowing states.

A perfect fluid is a nonviscous, nonconducting fluid. An example of this type of fluid would be a fluid that has a very small viscosity and conductivity and is at a high Reynolds number. An ideal gas is one that obeys the equation of state:

$$\frac{P}{\rho} = RT \tag{10-2}$$

where P = pressure, ρ = density, R is the gas constant per unit mass, and T = temperature.

A flowing fluid is acted upon by many forces that result in changes in pressure, temperature, stress, and strain. A fluid is said to be isotropic when the relations between the components of stress and those of the rate of strain are the same in all directions. The fluid is said to be Newtonian when this relationship is linear. These pressures and temperatures must be fully understood so that the entire flow picture can be described.

The *static pressure* in a fluid has the same value in all directions and can be considered as a scalar point function. It is the pressure of a flowing fluid. It is normal to the surface on which it acts and at any given point has the same magnitude irrespective of the orientation of the surface. The static pressure arises because of the random motion in the fluid of the molecules that make up the fluid. In a diffuser or nozzle, there is an increase or decrease in the static pressure due to the change in velocity of the moving fluid.

Total pressure is the pressure that would occur if the fluid were brought to rest in a reversible adiabatic process. Many texts and engineers use the words *total* and *stagnation* to describe the flow characteristics interchangeably. To be accurate, the stagnation pressure

TABLE 10-1 Density, Viscosity, and Kinematic Viscosity of Water and Air in Terms of Temperature

Temperature		Water			Air at a pressure of 760 mm Hg (14.696 lbf/in^2)		
(°C)	(°F)	Density ρ (lbf sec^2/ft^4)	Viscosity $\mu \times 10^6$ (lbf sec/ft^2)	Kinematic viscosity $\nu \times 10^6$ (ft^2/sec)	Density ρ (lbf sec^2/ft^4)	Viscosity $\mu \times 10^6$ (lbf sec/ft^2)	Kinematic viscosity $\nu \times 10^6$ (ft^2/sec)
−20	−4	—	—	—	0.00270	0.326	122
−10	14	—	—	—	0.00261	0.338	130
0	32	1.939	37.5	19.4	0.00251	0.350	140
10	50	1.939	27.2	14.0	0.00242	0.362	150
20	68	1.935	21.1	10.9	0.00234	0.375	160
40	104	1.924	13.68	7.11	0.00217	0.399	183
60	140	1.907	9.89	5.19	0.00205	0.424	207
80	176	1.886	7.45	3.96	0.00192	0.449	234
100	212	1.861	5.92	3.19	0.00183	0.477	264

Conversion factors: 1 kp sec^2/m^4 = 0.01903 lbf sec^2/ft^4 (= slug/ft^3)
1 lbf sec^2/ft^4 = 32.1719 lb/ft^3 (lb = lb mass; lbf = lb force)
1 kp sec^2/m^4 = 9.80665 kg/m^3 (kg = kg mass; kp = kg force)
1 kg/m^3 = 16.02 lb/ft^3

TABLE 10-2 Kinematic Viscosity

Liquid	Temperature °C	°F	$\nu \times 10^6$ (ft²/s)
Glycerine	20	68	7319
Mercury	0	32	1.35
Mercury	100	212	0.980
Lubricating oil	20	68	4306
Lubricating oil	40	104	1076
Lubricating oil	60	140	323

is the pressure that would occur if the fluid were brought to rest adiabatically or diabatically.

Total pressure will only change in a fluid if shaft work or work of extraneous forces are introduced. Therefore, total pressure would increase in the impeller of a compressor or pump; it would remain constant in the diffuser. Similarly, total pressure would decrease in the turbine impeller but would remain constant in the nozzles.

Static temperature is the temperature of the flowing fluid. Like static pressure, it arises because of the random motion of the fluid molecules. Static temperature is in most practical installations impossible to measure since it can be measured only by a thermometer or thermocouple at rest relative to the flowing fluid that is moving with the fluid. Static temperature will increase in a diffuser and decrease in a nozzle.

Total temperature is the temperature that would occur when the fluid is brought to rest in a reversible adiabatic manner. Just like its counterpart *total pressure, total* and *stagnation temperatures* are used interchangeably by many test engineers.

Dynamic temperature and pressure are the difference between the total and static conditions.

$$P_d = P_T - P_s \tag{10-3}$$

$$T_d = T_T - T_s \tag{10-4}$$

where subscript d refers to dynamic, T to total, and s to static.

Another helpful formula is:

$$P_K = \frac{1}{2}\rho V^2 \tag{10-5}$$

For incompressible fluids, $P_K = P_d$.

TOTAL TEMPERATURE

For most points requiring temperature monitoring, either thermocouples or resistive thermal detectors (RTDs) can be used. Each type of temperature transducer has its own advantages and disadvantages, and both should be considered when temperature is to be measured. Since there is considerable confusion in this area, a short discussion of the two types of transducers is necessary.

Thermocouples The various types of thermocouples provide transducers suitable for measuring temperatures from –330 to 5000°F (–201 to 2760°C). Thermocouples function by producing a voltage proportional to the temperature differences between two junctions of dissimilar metals. By measuring this voltage, the temperature difference can be determined. It is assumed that the temperature is known at one of the junctions; therefore, the temperature at the other junction can be determined. Since the thermocouples produce a voltage, no external power supply is required to the test junction; however, for accurate measurement, a reference junction is required. For a temperature monitoring system, reference junctions must be placed at each thermocouple or similar thermocouple wire installed from the thermocouple to the monitor where there is a reference junction. Properly designed thermocouple systems can be accurate to approximately ±2°F (±1°C).

Resistive Thermal Detectors (RTDs) RTDs determine temperature by measuring the change in resistance of an element due to temperature. Platinum is generally utilized in RTDs because it remains mechanically and electrically stable, resists contaminations, and can be highly refined. The useful range of platinum RTDs is –454–1832°F (–270–1000°C). Since the temperature is determined by the resistance in the element, any type of electrical conductor can be utilized to connect the RTD to the indicator; however, an electrical current must be provided to the RTD. A properly designed temperature monitoring system utilizing RTDs can be accurate ±0.02°F (±0.01°C).

STATIC TEMPERATURE

Since this temperature requires the thermometer or thermocouple to be at rest relative to the flowing fluid, it is impractical to measure. It can be, however, calculated from the measurement of total temperature and total and static pressure.

$$T_S = \frac{T_O}{\left(\frac{P_O}{P_S}\right)^{(k-1)/k}} \tag{10-6}$$

DRY- AND WET-BULB TEMPERATURES

The moisture content or humidity of air has an important effect on the properties of the gaseous mixture. Steam in air at any relative humidity less than 100 percent must exist in a superheated condition. The saturation temperature corresponding to the actual partial pressure of the steam in air is called the dew point. This term arose from the fact that when air at less than 100 percent relative humidity is cooled to the temperature at which it becomes saturated, the air has reached the minimum temperature to which it can be cooled without precipitation of the moisture (dew). Dew point can also be defined as that temperature at which the weight of steam associated with a certain weight of dry air is adequate to saturate that weight of air.

The dry-bulb temperature of air is the temperature that is indicated by an ordinary thermometer. When an air temperature is stated without any modifying term, it is always taken to be the dry-bulb temperature. In contrast to dry-bulb, or air, temperature, the term *wet-bulb temperature of the air*, or simply *wet-bulb temperature*, is employed. When a thermometer, with its bulb covered by a wick wetted with water, is moved through air unsaturated with water vapor, the water evaporates in proportion to the capacity of the air to absorb the evaporated moisture, and the temperature indicated by the thermometer drops below the dry-bulb, or air, temperature. The equilibrium temperature finally reached by the thermometer is known as the wet-bulb temperature. The purpose in measuring both the dry-bulb and wet-bulb temperature of the air is to find the exact humidity characteristics of the air from the readings obtained, either by calculation or by use of a psychrometric chart. Instruments for measuring wet-bulb and dry-bulb temperatures are known as psychrometers. A sling psychrometer consists of two thermometers mounted side by side on a holder, with provision for whirling the whole device through the air. The dry-bulb thermometer is bare, and the wet bulb is covered by a wick which is kept wetted with clean water. After being whirled a sufficient amount of time, the wet-bulb thermometer reaches its equilibrium point, and both the wet-bulb and dry-bulb thermometers are then quickly read. Rapid relative movement of the air past the wet-bulb thermometer is necessary to get dependable readings.

For other methods of measuring the moisture content of gases, see Sec. 8.

PRESSURE MEASUREMENTS

Pressure is defined as the force per unit area. Pressure devices measure with respect to the ambient atmospheric pressure: The absolute pressure P_a is the pressure of the fluid (gauge pressure) plus the atmospheric pressure.

Process pressure-measuring devices may be divided into three groups:

1. Those that are based on the height of a liquid column (manometers)
2. Those that are based on the measurement of the distortion of an elastic pressure chamber (mechanical pressure gauges such as Bourdon-tube gauges and diaphragm gauges)

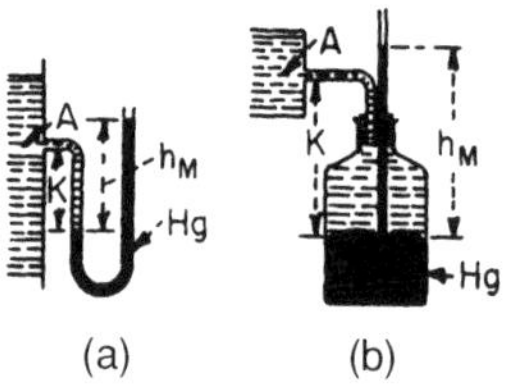

FIG. 10-1 Open manometers.

3. Electric sensing devices (strain gauges, piezoresistive transducers, and piezoelectric transducers)

This subsection contains an expanded discussion of manometric methods. See Sec. 8 for other methods.

Liquid-Column Manometers The **height,** or **head,** $p_n = \rho hg/g_c$ to which a fluid rises in an open vertical tube attached to an apparatus containing a liquid is a direct measure of the pressure at the point of attachment and is frequently used to show the level of liquids in tanks and vessels. This same principle can be applied with U-tube gauges (Fig. 10-1*a*) and equivalent devices (such as that shown in Fig. 10-1*b*) to measure pressure in terms of the head of a fluid other than the one under test. Most of these gauges may be used either as **open** or as **differential manometers.** The manometric fluid that constitutes the measured liquid column of these gauges may be any liquid immiscible with the fluid under pressure. For high vacuums or for high pressures and large pressure differences, the gauge liquid is a high-density liquid, generally mercury; for low pressures and small pressure differences, a low-density liquid (e.g., alcohol, water, or carbon tetrachloride) is used.

The **open U tube** (Fig. 10-1*a*) and the **open gauge** (Fig. 10-1*b*) each show a reading h_M m (ft) of manometric fluid. If the interface of the manometric fluid and the fluid of which the pressure is wanted is K m (ft) below the point of attachment, A, ρ_A is the density of the latter fluid at A, and ρ_M is that of the manometric fluid, then gauge pressure p_A (lbf/ft^2) at A is

$$p_A = (h_M\rho_M - K\rho_A)(g/g_c) \qquad (10\text{-}7)^*$$

where g = local acceleration due to gravity and g_c = dimensional constant. The head H_A at A as meters (feet) of the fluid at that point is

$$h_A = h_M(\rho_M/\rho_A) - K \qquad (10\text{-}8)^*$$

When a gas pressure is measured, unless it is very high, ρ_A is so much smaller than ρ_M that the terms involving K in these formulas are negligible.

The **differential U tube** (Fig. 10-2) shows the pressure difference between taps A and B to be

$$p_A - p_B = [h_M(\rho_M - \rho_A) + K_A\rho_A - K_B\rho_B](g/g_c) \qquad (10\text{-}9)^*$$

where h_M is the difference in height of the manometric fluid in the U tube; K_A and K_B are the vertical distances of the upper surface of the manometric fluid above A and B, respectively; ρ_A and ρ_B are the densities of the fluids at A and B, respectively; and ρ_M is the density of the manometric fluid. If either pressure tap is above the higher level of manometric fluid, the corresponding K is taken to be negative. Valve D, which is kept closed when the gauge is in use, is used to vent off gas which may accumulate at these high points.

*The line leading from the pressure tap to the gauge is assumed to be filled with fluid of the same density as that in the apparatus at the location of the pressure tap; if this is not the case, ρ_A is the density of the fluid actually filling the gauge line, and the value given for h_A must be multiplied by ρ_A/ρ where ρ is the density of the fluid whose head is being measured.

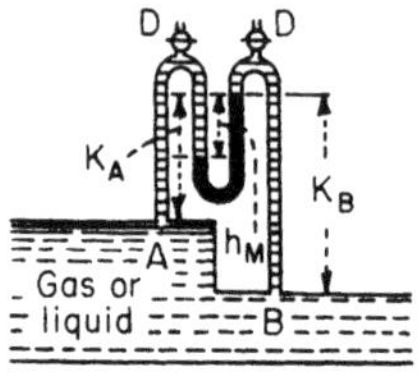

FIG. 10-2 Differential U tube.

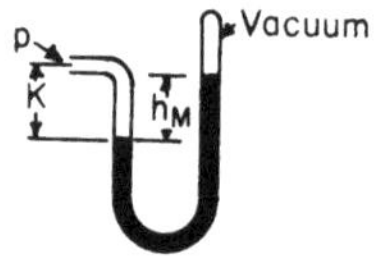

FIG. 10-3 Closed U tube.

The **inverted differential U tube,** in which the manometric fluid may be a gas or a light liquid, can be used to measure liquid pressure differentials, especially for the flow of slurries where solids tend to settle out.

Closed U tubes (Fig. 10-3) using mercury as the manometric fluid serve to measure directly the absolute pressure p of a fluid, provided that the space between the closed end and the mercury is substantially a perfect vacuum.

The **mercury barometer** (Fig. 10-4) indicates directly the absolute pressure of the atmosphere in terms of height of the mercury column. Normal (standard) barometric pressure is 101.325 kPa by definition. Equivalents of this pressure in other units are 760 mm mercury (at 0°C), 29.921 inHg (at 0°C), 14.696 lbf/in^2, and 1 atm. For cases in which barometer readings, when expressed by the height of a mercury column, must be corrected to standard temperature (usually 0°C), appropriate temperature correction factors are given in ASME PTC, op. cit., pp. 23–26; and Weast, *Handbook of Chemistry and Physics,* 62d ed., Chemical Rubber, Cleveland, 1984, pp. E36–E37.

Tube Size for Manometers To avoid capillary error, tube diameter should be sufficiently large and the manometric fluids of such densities that the effect of capillarity is negligible in comparison with the gauge reading. The effect of capillarity is practically negligible for tubes with inside diameters 12.7 mm (½ in) or larger (see ASME PTC, op. cit., p. 15). Small diameters are generally permissible for U tubes because the capillary displacement in one leg tends to cancel that in the other.

The capillary rise in a small vertical open tube of circular cross section dipping into a pool of liquid is given by

$$h = \frac{4\sigma g_c \cos\theta}{gD(\rho_1 - \rho_2)} \qquad (10\text{-}10)$$

Here σ = surface tension, D = inside diameter, ρ_1 and ρ_2 are the densities of the liquid and gas (or light liquid) respectively, g = local acceleration due to gravity, g_c = dimensional constant, and θ is the contact angle subtended by the heavier fluid. For most organic liquids and water, the contact angle θ is zero against glass, provided the glass is wet with a film of the liquid; for mercury against glass, θ = 140° (*International Critical Tables,* vol. IV, McGraw-Hill, New York, 1928, pp. 434–435). For further discussion of capillarity, see Schwartz, *Ind. Eng. Chem.*, **61**(1), 10–21 (1969).

Multiplying Gauges To attain the requisite precision in measurement of small pressure differences by liquid-column manometers,

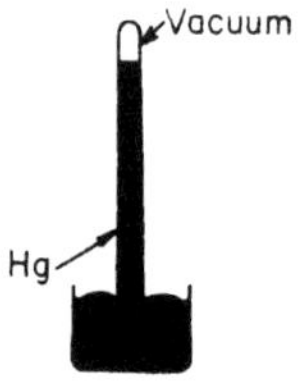

FIG. 10-4 Mercury barometer.

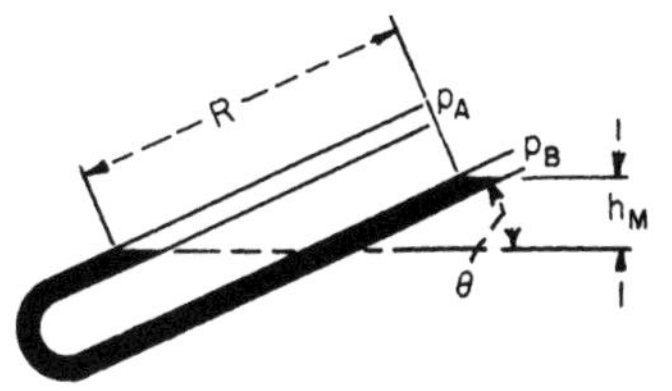

FIG. 10-5 Inclined U tube.

means must often be devised to magnify the readings. Of the schemes that follow, the second and third may give tenfold multiplication; the fourth, as much as thirtyfold. In general, the greater the multiplication, the more elaborate must be the precautions in the use of the gauge if the gain in precision is not to be illusory.

1. *Change of manometric fluid.* In open manometers, choose a fluid of lower density. In differential manometers, choose a fluid such that the difference between its density and that of the fluid being measured is as small as possible.

2. *Inclined U tube* (*Fig. 10-5*). If the reading R m (ft) is taken as shown and R_0 m (ft) is the zero reading, by making the substitution $h_M = (R - R_0) \sin \theta$, the formulas of preceding paragraphs give $(p_A - p_B)$ when the corresponding upright U tube is replaced by one inclined. For precise work, the gauge should be calibrated because of possible variations in tube diameter and slope.

3. *The draft gauge* (*Fig. 10-6*). Commonly used for low gas heads, this gauge has for one leg of the U a reservoir of much larger bore than the tubing that forms the inclined leg. Hence variations of level in the inclined tube produce little change in level in the reservoir. Although h_M may be readily computed in terms of reading R and the dimensions of the tube, calibration of the gauge is preferable; often the changes of level in the reservoir are not negligible, and also variations in tube diameter may introduce serious error into the computation. Commercial gauges are often provided with a scale giving h_M directly in height of water column, provided a particular liquid (often not water) fills the tube; failure to appreciate that the scale is incorrect unless the gauge is filled with the specified liquid is a frequent source of error. If the scale reads correctly when the density of the gauge liquid is ρ_0, then the reading must be multiplied by ρ/ρ_0 if the density of the fluid actually in use is ρ.

4. *Two-fluid U tube* (*Fig. 10-7*). This is a highly sensitive device for measuring small gas heads. Let A be the cross-sectional area of each of the reservoirs and a that of the tube forming the U; let ρ_1 be the density of the lighter fluid and ρ_2 that of the heavier fluid; and if R is the reading and R_0 its value with zero pressure difference, then the pressure difference is

$$p_A - p_B = (R - R_0)\left(\rho_2 - \rho_1 + \frac{a}{A}\rho_1\right)\frac{g}{g_c} \tag{10-11}$$

where g = local acceleration due to gravity and g_c = dimensional constant.

When A/a is sufficiently large, the term $(a/A)\ \rho_1$ in Eq. (10-11) becomes negligible in comparison with the difference $(\rho_2 - \rho_1)$. However, this term should not be omitted without due consideration. In applying Eq. (10-11), the densities of the gauge liquids may not be taken from tables without the possibility of introducing serious error, for each liquid may dissolve appreciable quantities of the other. Before the gauge is filled, the liquids should be shaken together, and the actual densities of the two layers should be measured for the temperature at which the gauge is to be used. When high magnification is being sought, the U tube may have to be enclosed in a constant-temperature bath so that $(\rho_2 - \rho_1)$ may be accurately known. In general, if highest accuracy is desired, the gauge should be calibrated.

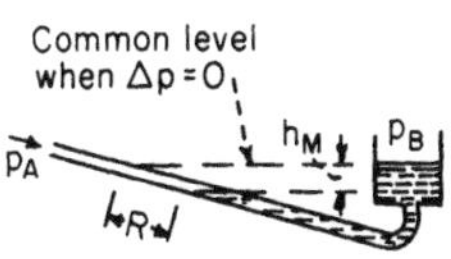

FIG. 10-6 Draft gauge.

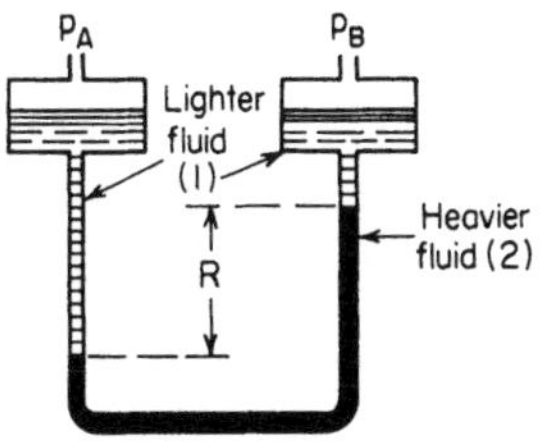

FIG. 10-7 Two-fluid U tube.

Several **micromanometers,** based on the liquid-column principle and possessing extreme precision and sensitivity, have been developed for measuring minute gas-pressure differences and for calibrating low-range gauges. Some of these micromanometers are available commercially. These micromanometers are free from errors due to capillarity and, aside from checking the micrometer scale, require no calibration.

Mechanical Pressure Gauges The **Bourdon-tube gauge** indicates pressure by the amount of flection under internal pressure of an oval tube bent in an arc of a circle and closed at one end. These gauges are commercially available for all pressures below atmospheric and for pressures up to 700 MPa (about 100,000 lbf/in^2) above atmospheric. Details on Bourdon-type gauges are given by Harland [*Mach. Des.*, **40**(22), 69–74 (Sept. 19, 1968)].

A **diaphragm gauge** depends for its indication on the deflection of a diaphragm, usually metallic, when subjected to a difference of pressure between the two faces. These gauges are available for the same general purposes as Bourdon gauges but are not usually employed for high pressures. The aneroid barometer is a type of diaphragm gauge.

Small **pressure transducers with flush-mounted diaphragms** are commercially available for the measurement of either steady or fluctuating pressures up to 100 MPa (about 15,000 lbf/in^2). The metallic diaphragms are as small as 4.8 mm (3⁄16 in) in diameter. The transducer is mounted on the apparatus containing the fluid whose pressure is to be measured so that the diaphragm is flush with the inner surface of the apparatus. Deflection of the diaphragm is measured by unbonded strain gauges and recorded electrically.

With nonnewtonian fluids the pressure measured at the wall with non-flush-mounted pressure gauges may be in error (see subsection "Static Pressure").

Bourdon and diaphragm gauges that show both pressure and vacuum indications on the same dial are called **compound gauges.**

Conditions of Use Bourdon tubes should not be exposed to temperatures over about 65°C (about 150°F) unless the tubes are specifically designed for such operation. When the pressure of a hotter fluid is to be measured, some type of liquid seal should be used to keep the hot fluid from the tube. In using either a Bourdon or a diaphragm gauge to measure gas pressure, if the gauge is below the pressure tap of the apparatus so that liquid can collect in the lead, the gauge reading will be too high by an amount equal to the hydrostatic head of the accumulated liquid.

For measuring pressures of corrosive fluids, slurries, and similar process fluids which may foul Bourdon tubes, a **chemical gauge,** consisting of a Bourdon gauge equipped with an appropriate flexible diaphragm to seal off the process fluid, may be used. The combined volume of the tube and the connection between the diaphragm and the tube is filled with an inert liquid. These gauges are available commercially.

Further details on pressure-measuring devices are found in Sec. 8.

Calibration of Gauges Simple **liquid-column manometers** do not require calibration if they are so constructed as to minimize errors due to capillarity (see subsection "Liquid-Column Manometers"). If the scales used to measure the readings have been checked against a standard, the accuracy of the gauges depends solely upon the precision of determining the position of the liquid surfaces. Hence liquid-column manometers are primary standards used to calibrate other gauges.

For **high pressures** and, with commercial mechanical gauges, even for quite moderate pressures, a deadweight gauge (see ASME PTC, op. cit., pp. 36–41) is commonly used as the primary standard because it is safer and more convenient than use of manometers. When manometers are used as high-pressure standards, an extremely high mercury column may be avoided by connecting a number of the usual U tubes in series. Multiplying gauges are standardized by comparing

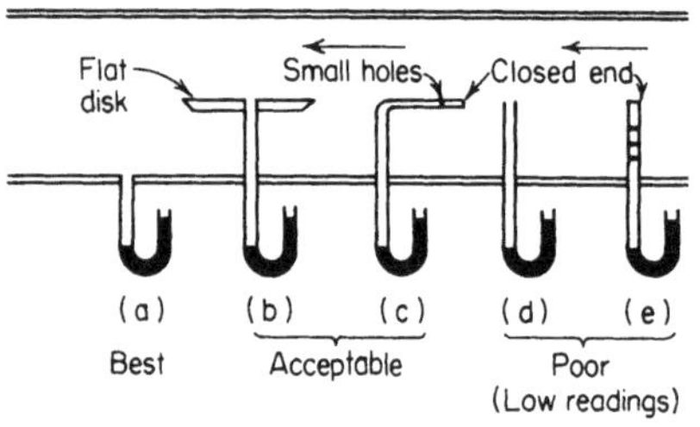

FIG. 10-8 Measurement of static pressure.

them with a micromanometer. Procedure in the calibration of a gauge consists merely of connecting it, in parallel with a standard gauge, to a reservoir wherein constant pressure may be maintained. Readings of the unknown gauge are then made for various reservoir pressures as determined by the standard.

Calibration of **high-vacuum gauges** is described by Sellenger [*Vacuum,* **18**(12), 645–650 (1968)].

STATIC PRESSURE

Local Static Pressure In a moving fluid, the local static pressure is equal to the pressure on a surface which moves with the fluid or to the normal pressure (for newtonian fluids) on a stationary surface which parallels the flow. The pressure on such a surface is measured by making a small hole perpendicular to the surface and connecting the opening to a pressure-sensing element (Fig. 10-8*a*). The hole is known as a piezometer opening or pressure tap.

Measurement of local static pressure is frequently difficult or impractical. If the channel is so small that introduction of any solid object disturbs the flow pattern and increases the velocity, there will be a reduction and redistribution of the static pressure. If the flow is in straight parallel lines, aside from the fluctuations of normal turbulence, the flat disk (Fig. 10-8*b*) and the bent tube (Fig. 10-8*c*) give satisfactory results when properly aligned with the stream. Slight misalignments can cause serious errors. Diameter of the disk should be 20 times its thickness and 40 times the static opening; the face must be flat and smooth, with the knife edges made by beveling the underside. The piezometer tube, such as that in Fig. 10-8*c,* should have openings with size and spacing as specified for a pitot-static tube (Fig. 10-12).

Readings given by open straight tubes (Fig. 10-8*d* and 10-8*e* are too low due to flow separation. Readings of closed tubes oriented perpendicularly to the axis of the stream and provided with side openings (Fig. 10-8*e*) may be low by as much as two velocity heads.

Average Static Pressure In most cases, the object of a static-pressure measurement is to obtain a suitable average value for substitution in Bernoulli's theorem or in an equivalent flow formula. This can be done simply only when the flow is in straight lines parallel to the confining walls, such as in straight ducts at sufficient distance downstream from bends (2 diameters) or other disturbances. For such streams, the sum of static head and gravitational potential head is the same at all points in a cross section taken perpendicularly to the axis of flow. Thus the exact location of a piezometer opening about the periphery of such a cross section is immaterial provided its elevation is known. However, in stating the static pressure, the custom is to give the value at the elevation corresponding to the centerline of the stream.

With flow in curved passages or with swirling flow, determination of a true average static pressure is, in general, impractical. In metering, straightening vanes are often placed upstream of the pressure tap to eliminate swirl. Figure 10-9 shows various flow equalizers and straighteners.

Specifications for Piezometer Taps The size of a static opening should be small compared with the diameter of the pipe and yet large compared with the scale of surface irregularities. For reliable results, it is essential that (1) the surface in which the hole is made be substantially smooth and parallel to the flow for some distance on either side of the opening, and (2) the opening be flush with the surface and possess no "burr" or other irregularity around its edge. Rounding of the edge is often employed to ensure absence of a burr. Pressure readings will be high if the tap is inclined upstream, is rounded excessively on the upstream side, has a burr on the downstream side, or has an excessive countersink or recess. Pressure readings will be low if the tap is inclined downstream, is rounded excessively on the downstream side, has a burr on the upstream side, or protrudes into the flow stream. Errors resulting from these faults can be large.

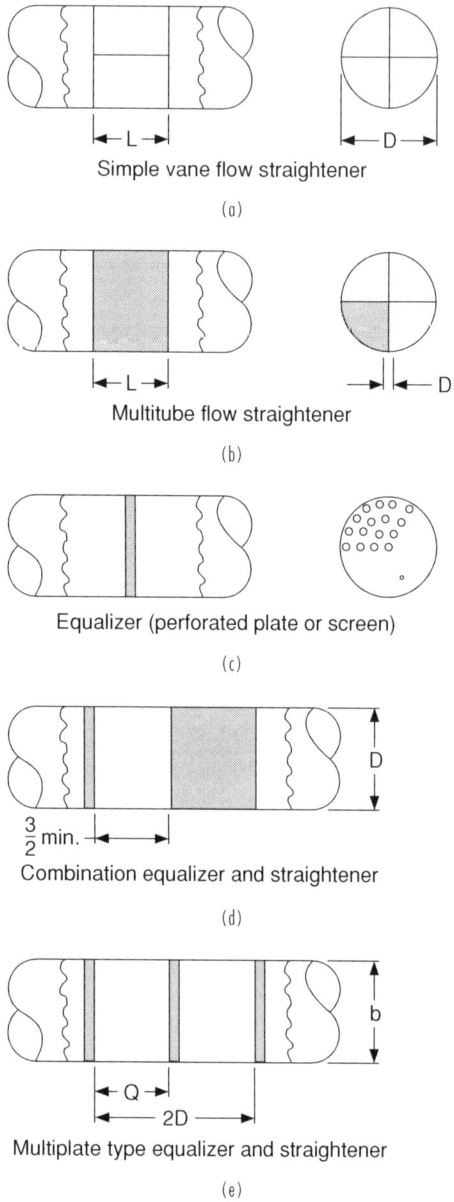

FIG. 10-9 Flow equalizers and straighteners [*Power Test Code 10, Compressors and Exhausters, Amer. Soc. of Mechanical Engineers, 1997*].

Recommendations for **pressure-tap dimensions** are summarized in Table 10-3. Data from several references were used in arriving at these composite values. The length of a pressure-tap opening prior to any enlargement in the tap channel should be at least two tap diameters, preferably three or more.

TABLE 10-3 Pressure-Tap Holes

Nominal inside pipe diameter, in	Maximum diameter of pressure tap, mm (in)	Radius of hole-edge rounding, mm (in)
1	3.18 (1/8)	<0.40 (1/64)
2	6.35 (1/4)	0.40 (1/64)
3	9.53 (3/8)	0.40–0.79 (1/64–1/32)
4	12.7 (1/2)	0.79 (1/32)
8	12.7 (1/2)	0.79–1.59 (1/32–1/16)
16	19.1 (3/4)	0.79–1.59 (1/32–1/16)

A **piezometer ring** is a toroidal manifold into which are connected several sidewall static taps located around the perimeter of a common cross section. Its intent is to give an average pressure if differences in pressure other than those due to static head exist around the perimeter. However, there is generally no assurance that a true average is provided thereby. The principal advantage of the ring is that use of several holes in place of a single hole reduces the possibility of completely plugging the static openings.

For information on prediction of static-hole error, see Shaw, *J. Fluid Mech.,* **7,** 550–564 (1960); Livesey, Jackson, and Southern, *Aircr. Eng.,* **34,** 43–47 (February 1962).

For nonnewtonian fluids, pressure readings with taps may also be low because of fluid-elasticity effects. This error can be largely eliminated by using flush-mounted diaphragms.

For information on the pressure-hole error for nonnewtonian fluids, see Han and Kim, *Trans. Soc. Rheol.,* **17,** 151–174 (1973); Novotny and Eckert, *Trans. Soc. Rheol.,* **17,** 227–241 (1973); and Higashitani and Lodge, *Trans. Soc. Rheol.,* **19,** 307–336 (1975).

VELOCITY MEASUREMENTS

Measurement of flow can be based on the measurement of velocity in ducts or pipes by using devices such as pitot tubes and hot wire anemometers. The local velocity is measured at various sections of a conduit and then averaged for the area under consideration.

$$\frac{w}{\rho} = A \times V = Q \tag{10-12}$$

where w = mass flow rate, lb_m/s, kg/s
ρ = density, lb_m/ft^3, kg/m^3
A = area, ft/s, m/s
V = velocity, ft/s, m/s
Q = volumetric flow rate, ft^3/s, m^3/s

Equation (10-12) shows that the fluid density directly affects the relationship between mass flow rate and both velocity and volumetric flow rate. Liquid temperature affects liquid density and hence volumetric flow rate at a constant mass flow rate. Liquid density is relatively insensitive to pressure. Both temperature and pressure affect gas density and thus volumetric flow rate.

Variables Affecting Measurement Flow measurement methods may sense local fluid velocity, volumetric flow rate, total or cumulative volumetric flow (the integral of volumetric flow rate with respect to elapsed time), mass flow rate, and total mass flow.

Velocity Profile Effects Many variables can influence the accuracy of specific flow measurement methods. For example, the velocity profile in a closed conduit affects many types of flow-measuring devices. The velocity of a fluid varies from zero at the wall and at other stationary solid objects in the flow channel to a maximum at a distance from the wall. In the entry region of a conduit, the velocity field may approach plug flow and a constant velocity across the conduit, dropping to zero only at the wall. As a newtonian fluid progresses down a pipe, a velocity profile develops that is parabolic for laminar flow [Eq. (6-41)] and that approaches plug flow for highly turbulent flow. Once a steady flow profile has developed, the flow is said to be fully developed; the length of conduit necessary to achieve fully developed flow is called the entrance region. For long cylindrical, horizontal pipe ($L < 40D$, where D is the inside diameter of the pipe and L is the upstream length of pipe), the velocity profile becomes fully developed. Velocity profiles in flowing fluids are discussed in greater detail in Sec. 6 (p. 6-11).

For steady-state, isothermal, single-phase, uniform, fully developed newtonian flow in straight pipes, the velocity is greatest at the center of the channel and symmetric about the axis of the pipe. Of those flowmeters that are dependent on the velocity profile, they are usually calibrated for this type of flow. Thus any disturbances in flow conditions can affect flowmeter readings.

Upstream and downstream disturbances in the flow field are caused by valves, elbows, and other types of fittings. Two upstream elbows in two perpendicular planes will impart swirl in the fluid downstream. Swirl, similar to atypical velocity profiles, can lead to erroneous flow measurements. Although the effect is not as great as in upstream flow disturbances, downstream flow disturbances can also lead to erroneous flow measurements.

Other Flow Disturbances Other examples of deviations from fully developed, single-phase newtonian flow include nonnewtonian flow, pulsating flow, cavitation, multiphase flow, boundary layer flows, and nonisothermal flows. See Sec. 6.

Pitot Tubes The combination of pitot tubes in conjunction with sidewall static taps measures local or point velocities by measuring the difference between the total pressure and the static pressure. The pitot tube shown in Fig. 10-10 consists of an impact tube whose opening faces directly into the stream to measure impact pressure, plus one or more sidewall taps to measure local static pressure.

Dynamic pressure may be measured by use of a pitot tube that is a simple impact tube. These tubes measure the pressure at a point where the velocity of the fluid is brought to zero. Pitot tubes must be parallel to the flow. The pitot tube is sensitive to yaw or angle attack. In general angles of attack over 10° should be avoided. In cases where the flow direction is unknown, it is recommended to use a Kiel probe. Figure 10-11 shows a Kiel probe. This probe will read accurately to an angle of about 22° with the flow.

The combined pitot-static tube shown in Fig. 10-12 consists of a jacketed impact tube with one or more rows of holes, 0.51 to 1.02 mm (0.02 to 0.04 in) in diameter, in the jacket to measure the static pressure. Velocity V_0 m/s (ft/s) at the point where the tip is located is given by

$$V_0 = C\sqrt{2g_c\Delta h} = C\sqrt{2g_c(P_T - P_S)/\rho_0} \tag{10-13}$$

where C = coefficient, dimensionless; g_c = dimensional constant; Δh = dynamic pressure ($\Delta h_s g/g_c$), expressed in (N·m)/kg[(ft·lbf)/lb or ft of fluid flowing]; Δh_s = differential height of static liquid column corresponding to Δh; g = local acceleration due to gravity; g_c = dimensional constant; p_i = impact pressure; p_0 = local static pressure; and ρ_0 = fluid density measured at pressure p_0 and the local temperature. With gases at velocities above 60 m/s (about 200 ft/s), compressibility becomes important, and the following equation should be used:

$$V_0 = C\sqrt{\frac{2g_c k}{k-1}\left(\frac{p_0}{\rho_0}\right)\left[\left(\frac{p_i}{p_0}\right)^{(k-1)/k} - 1\right]} \tag{10-14}$$

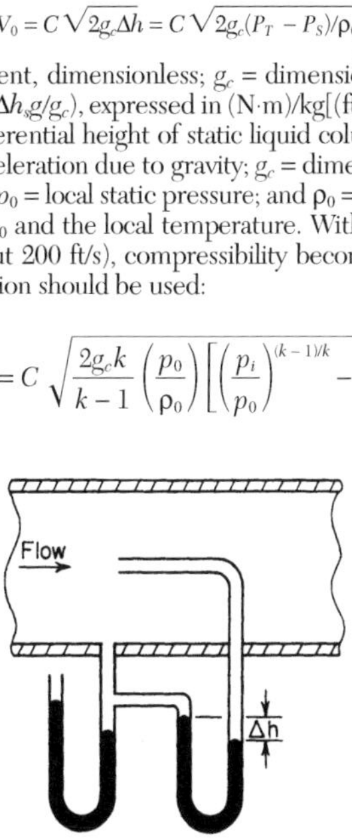

FIG. 10-10 Pitot tube with sidewall static tap.

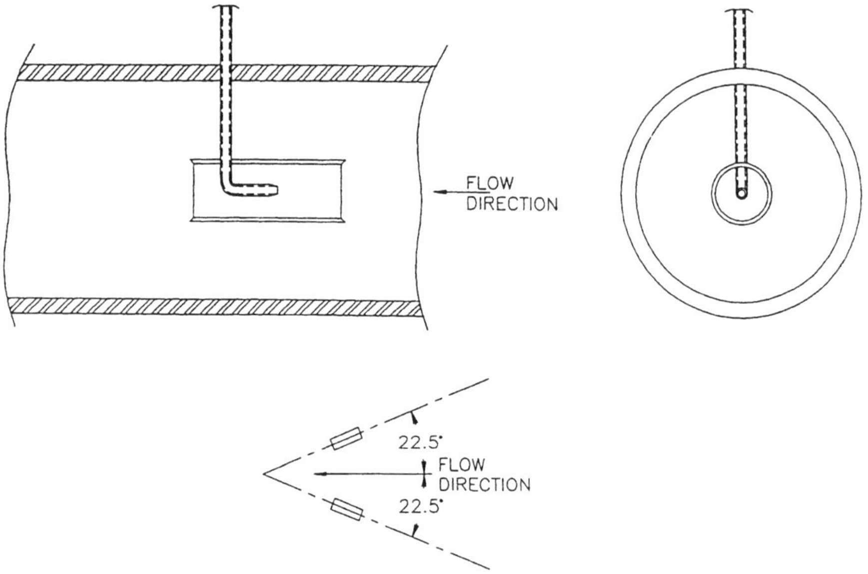

FIG. 10-11 Kiel probe. Accurate measurements can be made at angles up to 22.5° with the flow stream.

where k is the ratio of specific heat at constant pressure to that at constant volume. (See ASME Research Committee on Fluid Meters Report, op. cit., p. 105.) Coefficient C is usually close to 1.00 (±0.01) for simple pitot tubes (Fig. 10-10) and generally ranges between 0.98 and 1.00 for pitot-static tubes (Fig. 10-12).

There are certain limitations on the range of usefulness of pitot tubes. With gases, the differential is very small at low velocities; e.g., at 4.6 m/s (15.1 ft/s) the differential is only about 1.30 mm (0.051 in) of water (20°C) for air at 1 atm (20°C), which represents a lower limit for 1 percent error even when one uses a micromanometer with a precision of 0.0254 mm (0.001 in) of water. Equation does not apply for Mach numbers greater than 0.7 because of the interference of shock waves. For supersonic flow, local Mach numbers can be calculated from a knowledge of the dynamic and true static pressures. The free stream Mach number (M_∞) is defined as the ratio of the speed of the stream (V_∞) to the speed of sound in the free stream:

$$A_\infty = \sqrt{\left(\frac{\partial P}{\partial \rho}\right)_{s=c}} \qquad (10\text{-}15)$$

$$M_\infty = \frac{V_\infty}{\sqrt{\left(\frac{\partial P}{\partial \rho}\right)_{s=c}}} \qquad (10\text{-}16)$$

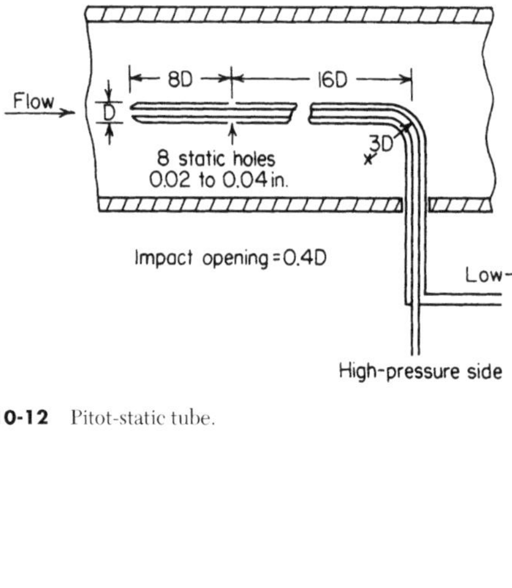

FIG. 10-12 Pitot-static tube.

where S is the entropy. For isentropic flow, this relationship and pressure can be written as:

$$M_\infty = \frac{V_\infty}{\sqrt{kRT_s}} \qquad (10\text{-}17)$$

The relationships between total and static temperature and pressure are given by the following relationship:

$$\frac{T_T}{T_S} = 1 + \frac{k-1}{2} M^2 \qquad (10\text{-}18)$$

$$\frac{P_T}{P_S} = \left(1 + \frac{k-1}{2} M^2\right)^{(k-1)/k} \qquad (10\text{-}19)$$

With **liquids** at low velocities, the effect of the Reynolds number upon the coefficient is important. The coefficients are appreciably less than unity for Reynolds numbers less than 500 for pitot tubes and for Reynolds numbers less than 2300 for pitot-static tubes [see Folsom, *Trans. Am. Soc. Mech. Eng.*, **78**, 1447–1460 (1956)]. Reynolds numbers here are based on the probe outside diameter. Operation at low Reynolds numbers requires prior calibration of the probe.

The pitot-static tube is also sensitive to **yaw** or **angle of attack** than is the simple pitot tube because of the sensitivity of the static taps to orientation. The error involved is strongly dependent upon the exact probe dimensions. In general, angles greater than 10° should be avoided if the velocity error is to be 1 percent or less.

Disturbances upstream of the probe can cause large errors, in part because of the turbulence generated and its effect on the static-pressure measurement. A calming section of at least 50 pipe diameters is desirable. If this is not possible, the use of straightening vanes or a honeycomb is advisable.

The effect of **pulsating flow** on pitot-tube accuracy is treated by Ower et al., op. cit., pp. 310–312. For sinusoidal velocity fluctuations, the ratio of indicated velocity to actual mean velocity is given by the factor $\sqrt{1 + \lambda^2/2}$, where λ is the velocity excursion as a fraction of the mean velocity. Thus, the indicated velocity would be about 6 percent high for velocity fluctuations of ±50 percent, and pulsations greater than ±20 percent should be damped to avoid errors greater than 1 percent. The error increases as the frequency of flow oscillations approaches the natural frequency of the pitot tube and the density of

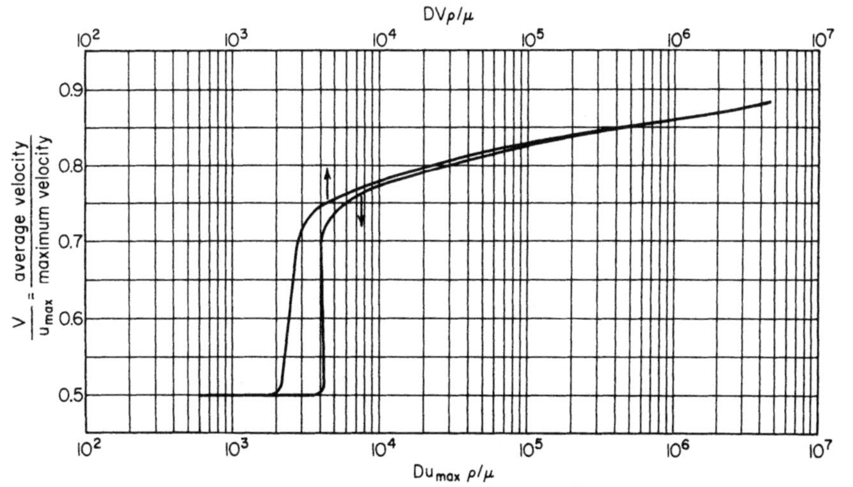

FIG. 10-13 Velocity ration versus Reynolds number for smooth circular pipes. [*Based on data from Rothfus, Archer, Klimas, and Sikchi,* Am. Inst. Chem. Eng. J., *3, 208 (1957).*]

the measuring fluid approaches the density of the process fluid [see Horlock and Daneshyar, *J. Mech. Eng. Sci.,* **15,** 144–152 (1973)].

Pressures substantially lower than true impact pressures are obtained with pitot tubes in turbulent flow of dilute polymer solutions [see Halliwell and Lewkowicz, *Phys. Fluids,* **18,** 1617–1625 (1975)].

Special Tubes A variety of special forms of the pitot tube have been evolved. Folsom (loc. cit.) gives a description of many of these special types together with a comprehensive bibliography. Included are the impact tube for **boundary-layer** measurements and **shielded total-pressure tubes.** The latter are insensitive to angle of attack up to 40°.

Chue [*Prog. Aerosp. Sci.,* **16,** 147–223 (1975)] reviews the use of the pitot tube and allied pressure probes for impact pressure, static pressure, dynamic pressure, flow direction and local velocity, skin friction, and flow measurements.

A reversed pitot tube, also known as a **pitometer,** has one pressure opening facing upstream and the other facing downstream. Coefficient C for this type is on the order of 0.85. This gives about a 40 percent increase in pressure differential as compared with standard pitot tubes and is an advantage at low velocities. There are commercially available very compact types of pitometers which require relatively small openings for their insertion into a duct.

The **pitot-venturi** flow element is capable of developing a pressure differential 5 to 10 times that of a standard pitot tube. This is accomplished by employing a pair of concentric venturi elements in place of the pitot probe. The low-pressure tap is connected to the throat of the inner venturi, which in turn discharges into the throat of the outer venturi. For a discussion of performance and application of this flow element, see Stoll, *Trans. Am. Soc. Mech. Eng.,* **73,** 963–969 (1951).

Traversing for Mean Velocity Mean velocity in a duct can be obtained by dividing the cross section into a number of equal areas, finding the local velocity at a representative point in each, and averaging the results. In the case of **rectangular passages,** the cross section is usually divided into small squares or rectangles and the velocity is found at the center of each. In circular pipes, the cross section is divided into several equal annular areas as shown in Fig. 10-13. Readings of velocity are made at the intersections of a diameter and the set of circles which bisect the annuli and the central circle.

For an N-point traverse on a circular cross section, make readings on each side of the cross section at

$$100 \times \sqrt{(2n - 1)/N} \text{ percent} \quad (n = 1, 2, 3 \text{ to } N/2)$$

of the pipe radius from the center. Traversing several diameters spaced at equal angles about the pipe is required if the velocity distribution is unsymmetrical. With a normal velocity distribution in a circular pipe, a 10-point traverse theoretically gives a mean velocity 0.3 percent high; a 20-point traverse, 0.1 percent high.

For normal velocity distribution in straight circular pipes at locations preceded by runs of at least 50 diameters without pipe fittings or other obstructions, the graph in Fig. 10-13 shows the ratio of mean velocity V to velocity at the center u_{max} plotted against the Reynolds number, where D = inside pipe diameter, ρ = fluid density, and μ = fluid viscosity, all in consistent units. Mean velocity is readily determined from this graph and a pitot reading at the center of the pipe if the quantity $Du_{max}\rho/\mu$ is less than 2000 or greater than 5000. The method is unreliable at intermediate values of the Reynolds number.

Methods for determining mean flow rate from probe measurements under nonideal conditions are described by Mandersloot, Hicks, and Langejan [*Chem. Eng. (London*), no. 232, CE370-CE380 (1969)].

The **hot-wire anemometer** consists essentially of an electrically heated fine wire (generally platinum) exposed to the gas stream whose velocity is being measured. An increase in fluid velocity, other things being equal, increases the rate of heat flow from the wire to the gas, thereby tending to cool the wire and alter its electrical resistance. In a constant-current anemometer, gas velocity is determined by measuring the resulting wire resistance; in the constant-resistance type, gas velocity is determined from the current required to maintain the wire temperature, and thus the resistance, constant. The difference in the two types is primarily in the electric circuits and instruments employed.

The hot-wire anemometer can, with suitable calibration, accurately measure velocities from about 0.15 m/s (0.5 ft/s) to supersonic velocities and detect velocity fluctuations with frequencies up to 200,000 Hz. Fairly rugged, inexpensive units can be built for the measurement of mean velocities in the range of 0.15 to 30 m/s (about 0.5 to 100 ft/s). More elaborate, compensated units are commercially available for use in unsteady flow and turbulence measurements. In calibrating a hot-wire anemometer, it is preferable to use the same gas, temperature, and pressure as will be encountered in the intended application. In this case the quantity $I^2R_w/\Delta t$ can be plotted against $\sqrt{V}$, where I = hot-wire current, R_w = hot-wire resistance, Δt = difference between the wire temperature and the gas bulk temperature, and V = mean local velocity. A procedure is given by Wasan and Baid [*Am. Inst. Chem. Eng. J.,* **17,** 729–731 (1971)] for use when it is impractical to calibrate with the same gas composition or conditions of temperature and pressure. Andrews, Bradley, and Hundy [*Int. J. Heat Mass Transfer,* **15,** 1765–1786 (1972)] give a calibration correlation for measurement of small gas velocities. The hot-wire anemometer is treated in considerable detail in Dean, op. cit., chap. VI; in Ladenburg et al., op. cit., art. F-2; by Grant and Kronauer, *Symposium on Measurement in Unsteady Flow*, American Society of Mechanical Engineers, New York, 1962, pp. 44–53; ASME Research Committee on Fluid Meters Report, op. cit., pp. 105–107; and by Compte-Bellot, *Ann. Rev. Fluid Mech.*, **8**, pp. 209–231 (1976).

The hot-wire anemometer can be modified for liquid measurements, although difficulties are encountered because of bubbles and dirt adhering to the wire. See Stevens, Borden, and Strausser, David Taylor Model Basin Rep. 953, December 1956; Middlebrook and Piret, *Ind. Eng. Chem.*, **42**, 1511–1513 (1950); and Piret et al., *Ind. Eng. Chem.*, **39**, 1098–1103 (1947).

The **hot-film anemometer** has been developed for applications in which use of the hot-wire anemometer presents problems. It consists of a platinum-film sensing element deposited on a glass substrate. Various geometries can be used. The most common involves a wedge with a 30° included angle at the end of a tapered rod. The wedge is commonly 1 mm (0.039 in) long and 0.2 mm (0.0079 in) wide on each face. Compared with the hot wire, it is less susceptible to fouling by bubbles or dirt when used in liquids, has greater mechanical strength when used with gases at high velocities and high temperatures, and can give a higher signal-to-noise ratio. For additional information see Ling and Hubbard, *J. Aeronaut. Sci.*, **23**, 890–891 (1956); and Ling, *J. Basic Eng.*, **82**, 629–634 (1960).

The **heated-thermocouple anemometer** measures gas velocity from the cooling effect of the gas stream flowing across the hot junctions of a thermopile supplied with constant electrical power input. Alternate junctions are maintained at ambient temperature, thus compensating for the effect of ambient temperature. For details see Bunker, *Proc. Instrum. Soc. Am.*, **9**, pap. 54-43-2 (1954).

A glass-coated bead **thermistor anemometer** can be used for the measurement of low fluid velocities, down to 0.001 m/s (0.003 ft/s) in air and 0.0002 m/s (0.0007 ft/s) in water [see Murphy and Sparks, *Ind. Eng. Chem. Fundam.*, **7**, 642–645 (1968)].

The **laser-Doppler anemometer** measures local fluid velocity from the change in frequency of radiation, between a stationary source and a receiver, due to scattering by particles along the wave path. A laser is commonly used as the source of incident illumination. The measurements are essentially independent of local temperature and pressure. This technique can be used in many different flow systems with transparent fluids containing particles whose velocity is actually measured. For a brief review of the laser-Doppler technique see Goldstein, *Appl. Mech. Rev.*, **27**, 753–760 (1974). For additional details see Durst, Melling, and Whitelaw, *Principles and Practice of Laser-Doppler Anemometry*, Academic, New York, 1976.

FLOWMETERS

In the process industries, flow measurement devices are the largest market in the process instrumentation field. Two web sites for process equipment and instrumentation, www.globalspec.com, and www.thomasnet.com, both list more than 800 companies that offer flow measurement products. There are more than one hundred types of flowmeters commercially available. The aforementioned web sites not only facilitate selection and specification of commercial flowmeters, but also provide electronic access to manufacturers' technical literature.

Devices that measure flow can be categorized in two areas as follows:

1. All types of measuring devices in which the material passes without being divided into isolated quantities. Movement of the material is usually sensed by a primary measuring element which activates a secondary device. The flow rate is then inferred from the response of the secondary device by means of known physical laws or from empirical relationships.

2. A positive-displacement meter, which applies to a device in which the flow is divided into isolated measured volumes. The number of fillings of these known volumes are measured with respect to time.

The most common application of flow measurement in process plants is flow in pipes, ducts, and tubing. Table 10-4 lists widely used flowmeters for these closed conduits as well as the two major classes of open-channel flowmeters. Table 10-4 also lists many other types of flowmeters that are discussed later in this subsection.

This subsection summarizes selection and installation of flowmeters, including the measurement of pressure and velocities of fluids when the flow measurement technique requires it.

INDUSTRY GUIDELINES AND STANDARDS

Because flow measurement is important, many engineering societies and trade organizations have developed flow-related guidelines, standards, and other publications (Table 10-5). The reader should consult the appropriate standards when specifying, installing, and calibrating flow measurement systems.

There are also numerous articles in scholarly journals, trade magazines, and manufacturers' literature related to flow measurement.

Different types of flowmeters differ markedly in their degrees of sensitivity to flow disturbances. In the most extreme cases, obtaining highly accurate flow measurements with certain types of flowmeters may require 60*D* upstream straight pipe and 20*D* downstream. Valves can be particularly problematic because their effects on a flowmeter vary with valve position. Numerous types of flow straighteners or conditioners, as shown in Fig. 10-9, can significantly reduce the required run of straight pipe upstream of a given flowmeter.

CLASSIFICATION OF FLOWMETERS

Table 10-4 lists the major classes of flowmeters, along with common examples of each. Brief descriptions are provided in this subsection, followed by more details in subsequent subsections.

Differential Pressure Meters Differential pressure meters or head meters measure the change in pressure across a special flow element. The differential pressure increases with increasing flow rate. The pitot tubes described previously work on this principle. Other examples include orifices [see also Eqs. (6-111) and (8-102), and Fig. 10-14], nozzles (Fig. 10-19), targets, venturis (see also Sec. 8 and Fig. 10-17), and elbow meters. Averaging pitot tubes produce a pressure differential that is based on multiple measuring points across the flow path.

Differential pressure meters are widely used. Temperature, pressure, and density affect gas density and readings of differential pressure meters. For that reason, many commercial flowmeters that are based on measurement of differential pressure often have integral temperature and absolute pressure measurements in addition to differential pressure. They also frequently have automatic temperature and pressure compensation.

Velocity Meters Velocity meters measure fluid velocity. Examples include electromagnetic, propeller, turbine, ultrasonic Doppler, ultrasonic transit time, and vortex meters. Section 8 describes the principles of operation of electromagnetic, turbine, ultrasonic, and vortex flowmeters.

Mass Meters Mass flowmeters measure the rate of mass flow through a conduit. Examples include Coriolis flowmeters and thermal mass flowmeters. Coriolis flowmeters can measure fluid density simultaneously with mass flow rate. This permits calculation of volumetric flow rate as well. Section 8 includes brief descriptions of Coriolis and thermal mass flowmeters.

Volumetric Meters Volumetric meters (also called positive-displacement flowmeters) are devices that mechanically divide a fluid stream into discrete, known volumes and count the number of volumes that pass through the device. See Spitzer (2005, op. cit.).

Variable-Area Meters Variable-area meters, which are also called rotameters, offer popular and inexpensive flow measurement devices. These meters employ a float inside a tube that has an internal cross-sectional area that increases with distance upward in the flow path through the tube. As the flow rate increases, the float rises in the tube to provide a larger area for the flowing fluid to pass.

Open-Channel Flow Measurement Open-channel flow measurements are usually based on measurement of liquid level in a flow channel constructed of a specified geometry. The two most common flow channels used are weirs and flumes. See Spitzer (2005, op. cit.).

TABLE 10-4 Comparison of Flowmeter Technologies

Flowmeter technology	Accuracy* (+/–)	Turndown	Fluids†	Pipe sizes,‡ in	Maximum pressure,‡ psig	Temperature range,‡ °F	Pipe run	Relative pressure loss
				Differential Pressure Meters				
Pitot		8 : 1	L, G	1 to 96	2.76E-4 to 200	–200 to 750		L
Averaging pitot	1% R	8 : 1	L, G, S	0.25 to 72	8800	–20 to 2370		L
Orifice							Long	
Square-edged	0.5 to 1.5% R	4 : 1	L, G, S	0.5 to 40	8800	–4 to 2300		M
Eccentric	2% R	4 : 1	L, G, S	0.5 to 40	8800	–4 to 2300		M
Segmental	2% R	4 : 1	L, G, S, SL	0.5 to 40	8800	–4 to 2300		M
Orifice and multi-variable flow transmitter	0.5 to 1% R	10 : 1	L, G, S	> 0.5	4000	1000		M
Venturi	0.5 to 1.5% R	10 : 1	L, G, S, SL	1 to 120	8800	–4 to 2300		L
Flow nozzle	0.5 to 2% R	8 : 1	L, G, S	2 to 80	>1000	<1000		M
V cone	0.5% F	10 : 1	L, G, S, SL	0.25 to 120	6000	To 400		M
Wedge	0.5 to 5% R	10 : 1	L, G, S, SL	0.5 to 24	>600	‡		M
				Velocity Meters				
Correlation	0.5% R	10 : 1	L, G, SL	1 to 60	Piping limits	To 600	Long	L
Electromagnetic	0.2 to 2% R	10 : 1	L	0.15 to 60	5000	–40 to 350	Short	L
Propeller	2% R	15 : 1	L	2 to 12	230	0 to 300		M
Turbine	0.15 to 1% R	10 : 1	L, G	0.5 to 30	6000	–450 to 600	Short	M
Ultrasonic Doppler	1 to 30% R	50 : 1	L, G, SL	0.5 to 200	6000	–40 to 250	Long	L
Ultrasonic transit time	0.5 to 5% R	Down to zero flow	L, G	1 to 540	6000	–40 to 650	Long	L
Vortex	0.5 to 2% R	20 : 1	L, G, S	0.5 to 16	1500	–330 to 800	Short	M
				Mass Meters				
Coriolis	0.1 to 0.3% R	10 : 1 to 80 : 1	L, G	0.06 to 12	5700	–400 to 800	None	L, M
Thermal (for gases)	1% F	50 : 1	G	0.125 to 8	4500	32 to 572	Short	L
Thermal (for liquids)	0.5% F	50 : 1	L	0.06 to 0.25	4500	40 to 165	Short	L
Volumetric	0.15 to 2% R	10 : 1	L	0.25 to 16	2000	–40 to 600	None	M to H
Variable area	1 to 5% F	10 : 1	L, G	0.125 to 6	6000	<1000	None	M
				Open-Channel Flowmeters				
Weirs	2 to 5% R	25 : 1	L	Wide range	NA	NA		L
Flumes	3 to 10% R	40 : 1	L	Wide range	NA	NA		L

*F = full scale, R = rate. †L = liquid, G = gas, S = steam, SL = slurry.
‡Dependent on the material selection and application. Readers should consult manufacturers for current capabilities.
Adapted from J. Pomroy, *Chemical Engineering*, pp. 94–102, May 1996; J. W. Dolenc, *Chemical Engineering Progress*, pp. 22–32, Jan. 1996; R. C. Baker, *Introductory Guide to Flow Measurement*, American Society of Mechanical Engineers, New York, 2003; R. W. Miller, *Flow Measurement Engineering Handbook*, 3d ed., McGraw-Hill, New York, 1996; D. W. Spitzer, *Industrial Flow Measurement*, 3d ed., The Instrumentation, Systems, and Automation Society, Research Triangle Park, N.C., 2005; and manufacturers' literature at www.globalspec.com.

DIFFERENTIAL PRESSURE FLOWMETERS

General Principles If a constriction is placed in a closed channel carrying a stream of fluid, there will be an increase in velocity, and hence an increase in kinetic energy, at the point of constriction. From an energy balance, as given by Bernoulli's theorem [see Sec. 6, subsection "Energy Balance," Eq. (6-16)], there must be a corresponding reduction in pressure. Rate of discharge from the constriction can be calculated by knowing this pressure reduction, the area available for flow at the constriction, the density of the fluid, and the coefficient of discharge C. The last-named is defined as the ratio of actual flow to the theoretical flow and makes allowance for stream contraction and frictional effects. The metering characteristics of commonly used differential pressure meters are reviewed and grouped by Halmi [*J. Fluids Eng.*, **95**, 127–141 (1973)].

The term **static head** generally denotes the pressure in a fluid due to the head of fluid above the point in question. Its magnitude is given by the application of Newton's law (force = mass × acceleration). In the case of **liquids** (constant density), the static head p_h Pa (lbf/ft^2) is given by

$$p_h = h\rho g/g_c \tag{10-20}$$

where h = head of liquid above the point, m (ft); ρ = liquid density; g = local acceleration due to gravity; and g_c = dimensional constant.

The head developed in a compressor or pump is the energy force per unit mass. In the measuring systems it is often misnamed as (ft) while the units are really ft-lb/lbm or kilojoules.

For a compressor or turbine, it is represented by the following relationship:

$$E = U_1V_{\theta 1} - U_2V_{\theta 2} \tag{10-21}$$

TABLE 10-5 Guidelines, Standards, and Other Publications Related to Flow Measurement

Technical society	Number of guidelines and standards*
American Gas Association (AGA)	2
American Petroleum Institute (API)	11
American Society of Heating, Refrigeration, and Air Conditioning Engineers (ASHRAE)	5
American Society of Mechanical Engineers (ASME)	18
ASTM International (ASTM)	17
British Standards Institution (BSI)	100
Deutsches Institut fur Normung E. V. (DIN)	48
International Electrotechnical Commission (IEC)	6
Instrumentation, Systems, and Automation Society (ISA)	3
International Organization for Standardization (ISO)	212
SAE International (SAE)	6

*Number of documents identified by searching for *flow measurement* on http://global.ihs.com, the web site of a clearinghouse of industry guidelines, codes, and standards.

where U is the blade speed and V_θ is the tangential velocity component of absolute velocity. This equation is known as the Euler equation.

Orifice Meters A **square-edged** or **sharp-edged** orifice, as shown in Fig. 10-14, is a clean-cut square-edged hole with straight walls perpendicular to the flat upstream face of a thin plate placed crosswise of the channel. The stream issuing from such an orifice attains its minimum cross section (vena contracta) at a distance downstream of the orifice which varies with the ratio β of orifice to pipe diameter (see Fig. 10-15).

For a centered circular orifice in a pipe, the pressure differential is customarily measured between one of the following pressure-tap pairs. Except in the case of flange taps, all measurements of distance from the orifice are made from the upstream face of the plate.

1. *Corner taps.* Static holes drilled one in the upstream and one in the downstream flange, with the openings as close as possible to the orifice plate.
2. *Radius taps.* Static holes located one pipe diameter upstream and one-half pipe diameter downstream from the plate.
3. *Pipe taps.* Static holes located 2½ pipe diameters upstream and eight pipe diameters downstream from the plate.

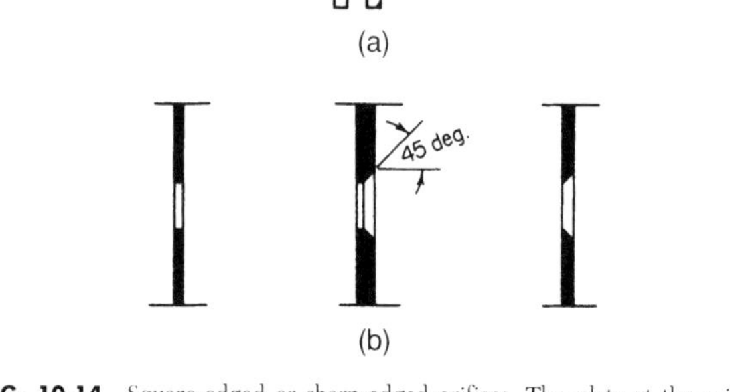

FIG. 10-14 Square-edged or sharp-edged orifices. The plate at the orifice opening must not be thicker than one-thirtieth of the pipe diameter, one-eighth of the orifice diameter, or one-fourth of the distance from the pipe wall to the edge of the opening. (*a*) Pipe-line orifice. (*b*) Types of plates.

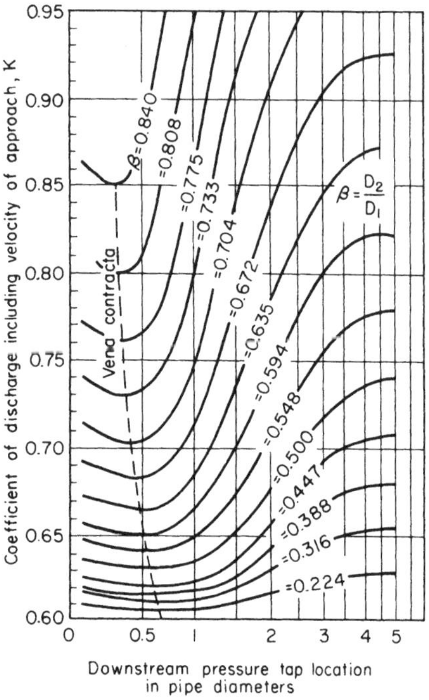

FIG. 10-15 Coefficient of discharge for square-edged circular orifices for $N_{Re} > 30{,}000$ with the upstream tap located between one and two pipe diameters from the orifice plate. [*Spitzglass*, Trans. Am. Soc. Mech. Eng., ***44,*** *919 (1922).*]

4. *Flange taps.* Static holes located 25.4 mm (1 in) upstream and 25.4 mm (1 in) downstream from the plate.
5. *Vena-contracta taps.* The upstream static hole is one-half to two pipe diameters from the plate. The downstream tap is located at the position of minimum pressure (see Fig. 10-15).

Radius taps are best from a practical standpoint; the downstream pressure tap is located at about the mean position of the vena contracta, and the upstream tap is sufficiently far upstream to be unaffected by distortion of the flow in the immediate vicinity of the orifice (in practice, the upstream tap can be as much as two pipe diameters from the plate without affecting the results). Vena-contracta taps give the largest differential head for a given rate of flow but are inconvenient if the orifice size is changed from time to time. Corner taps offer the sometimes great advantage that the pressure taps can be built into the plate carrying the orifice. Thus the entire apparatus can be quickly inserted in a pipe line at any convenient flanged joint without having to drill holes in the pipe. Flange taps are similarly convenient, since by merely replacing standard flanges with special orifice flanges, suitable pressure taps are made available. Pipe taps give the lowest differential pressure, the value obtained being close to the permanent pressure loss.

The practical working equation for weight rate of discharge, adopted by the ASME Research Committee on Fluid Meters for use with either gases or liquids, is

$$w = q_1\rho_1 = CYA_2\sqrt{\frac{2g_c(p_1 - p_2)\rho_1}{1 - \beta^4}}$$

$$= KYA_2\sqrt{2g_c(p_1 - p_2)\rho_1} \qquad (10\text{-}22)$$

where A_2 = cross-sectional area of throat; C = coefficient of discharge, dimensionless; g_c = dimensional constant; $K = C/\sqrt{1 - \beta^4}$, dimensionless; p_1, p_2 = pressure at upstream and downstream static pressure taps respectively; q_1 = volumetric rate of discharge measured at upstream pressure and temperature; w = weight rate of discharge; Y = expansion factor, dimensionless; β = ratio of throat diameter to pipe diameter, dimensionless; and ρ_1 = density at upstream pressure and temperature.

For the case of subsonic flow of a gas ($r_c < r < 1.0$), the expansion factor Y for orifices is approximated by

$$Y = 1 - [(1 - r)/k](0.41 + 0.35\beta^4) \quad (10\text{-}23)$$

where r = ratio of downstream to upstream static pressure (p_2/p_1), k = ratio of specific heats (c_p/c_v), and β = diameter ratio. (See also Fig. 10-18.) Values of Y for supercritical flow of a gas ($r < r_c$) through orifices are given by Benedict [*J. Basic Eng.*, **93,** 121–137 (1971)]. For the case of **liquids,** expansion factor Y is unity, and Eq. (10-27) should be used, since it allows for any difference in elevation between the upstream and downstream taps.

Coefficient of discharge C for a given orifice type is a function of the Reynolds number N_{Re} (based on orifice diameter and velocity) and diameter ratio β. At Reynolds numbers greater than about 30,000, the coefficients are substantially constant. For square-edged or sharp-edged concentric circular orifices, the value will fall between 0.595 and 0.620 for vena-contracta or radius taps for β up to 0.8 and for flange taps for β up to 0.5. Figure 10-15 gives the coefficient of discharge K, including the velocity-of-approach factor ($1/\sqrt{1 - \beta^4}$), as a function of β and the location of the downstream tap. Precise values of K are given in *ASME PTC,* op. cit., pp. 20–39, for flange taps, radius taps, vena-contracta taps, and corner taps. Precise values of C are given in the ASME Research Committee on Fluid Meters Report, op. cit., pp. 202–207, for the first three types of taps.

The discharge coefficient of sharp-edged orifices was shown by Benedict, Wyler, and Brandt [*J. Eng. Power,* **97,** 576–582 (1975)] to increase with edge roundness. Typical as-purchased orifice plates may exhibit deviations on the order of 1 to 2 percent from ASME values of the discharge coefficient.

In the transition region (N_{Re} between 50 and 30,000), the coefficients are generally higher than the above values. Although calibration is generally advisable in this region, the curves given in Fig. 10-16 for corner and vena-contracta taps can be used as a guide. In the laminar-flow region ($N_{Re} < 50$), the coefficient C is proportional to $\sqrt{N_{Re}}$. For $1 < N_{Re} < 100$, Johansen [*Proc. R. Soc. (London),* **A121,** 231–245 (1930)] presents discharge-coefficient data for sharp-edged orifices with corner taps. For $N_{Re} < 1$, Miller and Nemecek [ASME Paper 58-A-106 (1958)] present correlations giving coefficients for sharp-edged orifices and short-pipe orifices (L/D from 2 to 10). For short-pipe orifices (L/D from 1 to 4), Dickerson and Rice [*J. Basic Eng.,* **91,** 546–548 (1969)] give coefficients for the intermediate range ($27 < N_{Re} < 7000$). See also subsection "Contraction and Entrance Losses."

Permanent pressure loss across a concentric circular orifice with radius or vena-contracta taps can be approximated for turbulent flow by

$$(p_1 - p_4)/(p_1 - p_2) = 1 - \beta^2 \quad (10\text{-}24)$$

where p_1, p_2 = upstream and downstream pressure-tap readings respectively, p_4 = fully recovered pressure (four to eight pipe diameters downstream of the orifice), and β = diameter ratio. See *ASME PTC,* op. cit., Fig. 5.

See Benedict, *J. Fluids Eng.,* **99,** 245–248 (1977), for a general equation for pressure loss for orifices installed in pipes or with plenum inlets. Orifices show higher loss than nozzles or venturis. Permanent pressure loss for laminar flow depends on the Reynolds number in addition to β. See Alvi, Sridharan, and Lakshmana Rao, loc. cit., for details.

For the case of **critical flow** through a square- or sharp-edged concentric circular orifice (where $r \le r_c$, as discussed earlier in this subsection), use Eqs. (10-31), (10-32), and (10-33) as given for critical-flow nozzles. However, unlike nozzles, the flow through a sharp-edged orifice continues to increase as the downstream pressure drops below that

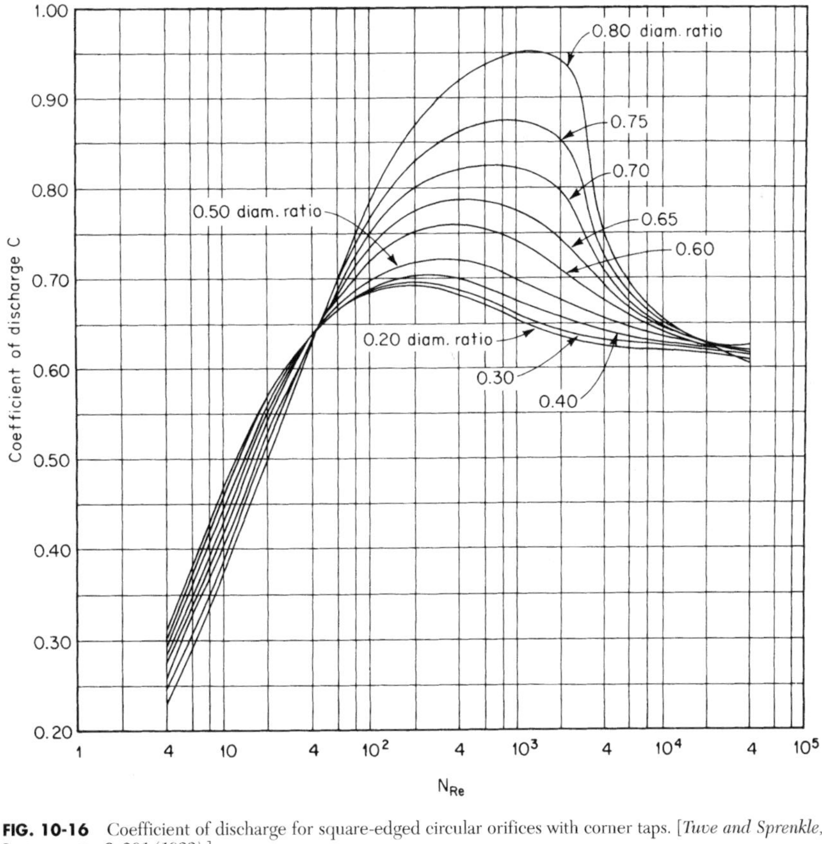

FIG. 10-16 Coefficient of discharge for square-edged circular orifices with corner taps. [*Tuve and Sprenkle,* Instruments, *6, 201 (1933).*]

corresponding to the critical pressure ratio r_c. This is due to an increase in the cross section of the vena contracta as the downstream pressure is reduced, giving a corresponding increase in the coefficient of discharge. At $r = r_c$, C is about 0.75, while at $r \cong 0$, C has increased to about 0.84. See Grace and Lapple, loc. cit.; and Benedict, *J. Basic Eng.*, **93,** 99–120 (1971).

Measurements by Harris and Magnall [*Trans. Inst. Chem. Eng. (London)*, **50,** 61–68 (1972)] with a venturi ($\beta = 0.62$) and orifices with radius taps ($\beta = 0.60 - 0.75$) indicate that the discharge coefficient for **nonnewtonian fluids,** in the range N'_{Re} (generalized Reynolds number) 3500 to 100,000, is approximately the same as for newtonian fluids at the same Reynolds number.

Quadrant-edge orifices have holes with rounded edges on the upstream side of the plate. The quadrant-edge radius is equal to the thickness of the plate at the orifice location. The advantages claimed for this type versus the square- or sharp-edged orifice are constant-discharge coefficients extending to lower Reynolds numbers and less possibility of significant changes in coefficient because of erosion or other damage to the inlet shape.

Values of discharge coefficient C and Reynolds numbers limit for constant C are presented in Table 10-6, based on Ramamoorthy and Seetharamiah [*J. Basic Eng.*, **88,** 9–13 (1966)] and Bogema and Monkmeyer (*J. Basic Eng.*, **82,** 729–734 (1960)]. At Reynolds numbers above those listed for the upper limits, the coefficients rise abruptly. As Reynolds numbers decrease below those listed for the lower limits, the coefficients pass through a hump and then drop off. According to Bogema, Spring, and Ramamoorthy [*J. Basic Eng.*, **84,** 415–418 (1962)], the hump can be eliminated by placing a fine-mesh screen about three pipe diameters upstream of the orifice. This reduces the lower N_{Re} limit to about 500.

Permanent pressure loss across quadrant-edge orifices for turbulent flow is somewhat lower than given by Eq. (10-24). See Alvi, Sridharan, and Lakshmana Rao, loc. cit., for values of discharge coefficient and permanent pressure loss in laminar flow.

Slotted orifices offer significant advantages over a standard square-edged orifice with an identical open area for homogeneous gases or liquids [G. L. Morrison and K. R. Hall, *Hydrocarbon Processing* **79,** 12, 65–72 (2000)]. The slotted orifice flowmeter only requires compact header configurations with very short upstream pipe lengths and maintains accuracy in the range of 0.25 percent with no flow conditioner. Permanent head loss is less than or equal to that of a standard orifice that has the same β ratio. Discharge coefficients for the slotted orifice are much less sensitive to swirl or to axial velocity profiles. A slotted orifice plate can be a "drop in" replacement for a standard orifice plate.

Segmental and **eccentric orifices** are frequently used for gas metering when there is a possibility that entrained liquids or solids would otherwise accumulate in front of a concentric circular orifice. This can be avoided if the opening is placed on the lower side of the pipe. For liquid flow with entrained gas, the opening is placed on the upper side. The pressure taps should be located on the opposite side of the pipe from the opening.

Coefficient C for a square-edged eccentric circular orifice (with opening tangent to pipe wall) varies from about 0.61 to 0.63 for β's from 0.3 to 0.5, respectively, and pipe Reynolds numbers > 10,000 for either vena-contracta or flange taps (where β = diameter ratio). For square-edged segmental orifices, the coefficient C falls generally between 0.63 and 0.64 for $0.3 \le \beta \le 0.5$ and pipe Reynolds numbers > 10,000, for vena-contracta or flange taps, where β = diameter ratio for an equivalent circular orifice = $\sqrt{\alpha}$ (α = ratio of orifice to pipe cross-sectional areas). Values of expansion factor Y are slightly higher than for concentric circular orifices, and the location of the vena contracta is moved farther downstream as compared with concentric circular orifices. For further details, see ASME Research Committee on Fluid Meters Report, op. cit., pp. 210–213.

TABLE 10-6 Discharge Coefficients for Quadrant-Edge Orifices

β	C‡	K‡	Limiting N^*_{Re} for constant coefficient† Lower	Upper
0.225	0.770	0.771	5,000	60,000
0.400	0.780	0.790	5,000	150,000
0.500	0.824	0.851	4,000	200,000
0.600	0.856	0.918	3,000	120,000
0.630	0.885	0.964	3,000	105,000

*Based on pipe diameter and velocity.
†For a precision of about ±0.5 percent.
‡Can be used with corner taps, flange taps, or radius taps.

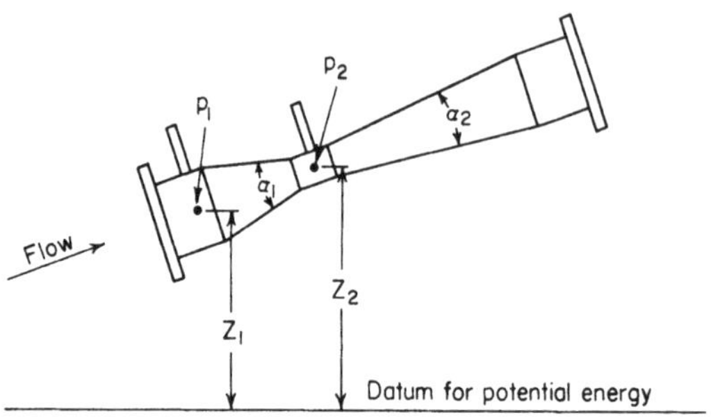

FIG. 10-17 Herschel-type venturi tube.

For permanent pressure loss with segmental and eccentric orifices with laminar pipe flow see Lakshmana Rao and Sridharan, *Proc. Am. Soc. Civ. Eng., J. Hydraul. Div.*, **98** (HY 11), 2015–2034 (1972).

Annular orifices can also be used to advantage for gas metering when there is a possibility of entrained liquids or solids and for liquid metering with entrained gas present in small concentrations. Coefficient K was found by Bell and Bergelin [*Trans. Am. Soc. Mech. Eng.*, **79,** 593–601 (1957)] to range from about 0.63 to 0.67 for annulus Reynolds numbers in the range of 100 to 20,000 respectively for values of $2L/(D - d)$ less than 1 where L = thickness of orifice at outer edge, D = inside pipe diameter, and d = diameter of orifice disk. The annulus Reynolds number is defined as

$$N_{Re} = (D - d)(G/\mu) \tag{10-25}$$

where G = mass velocity ρV through orifice opening and μ = fluid viscosity. The above coefficients were determined for β's (= d/D) in the range of 0.95 to 0.996 and with pressure taps located 19 mm (¾ in) upstream of the disk and 230 mm (9 in) downstream in a 5.25-in-diameter pipe.

Venturi Meters The standard Herschel-type venturi meter consists of a short length of straight tubing connected at either end to the pipe line by conical sections (see Fig. 10-17). Recommended proportions (*ASME PTC*, op. cit., p. 17) are entrance cone angle $\alpha_1 = 21 \pm 2°$, exit cone angle $\alpha_2 = 5$ to 15°, throat length = one throat diameter, and upstream tap located 0.25 to 0.5 pipe diameter upstream of the entrance cone. The straight and conical sections should be joined by smooth curved surfaces for best results. **Rate of discharge** of either gases or liquids through a venturi meter is given by Eq. (10-22).

For the flow of **gases,** expansion factor Y, which allows for the change in gas density as it expands adiabatically from p_1 to p_2, is given by

$$Y = \sqrt{r^{2/k}\left(\frac{k}{k-1}\right)\left(\frac{1-r^{(k-1)/k}}{1-r}\right)\left(\frac{1-\beta^4}{1-\beta^4 r^{2/k}}\right)} \tag{10-26}$$

for venturi meters and flow nozzles, where $r = p_2/p_1$ and k = specific heat ratio c_p/c_v. Values of Y computed from Eq. (10-26) are given in Fig. 10-18 as a function of r, k, and β.

For the flow of **liquids,** expansion factor Y is unity. The change in potential energy in the case of an inclined or vertical venturi meter must be allowed for. Equation (10-22) is accordingly modified to give

$$w = q_1\rho = CA_2\sqrt{\frac{[2g_c(p_1 - p_2) + 2g\rho(Z_1 - Z_2)]\rho}{1-\beta^4}} \tag{10-27}$$

where g = local acceleration due to gravity and Z_1, Z_2 = vertical heights above an arbitrary datum plane corresponding to the centerline pressure-reading locations for p_1 and p_2 respectively.

Value of the **discharge coefficient** C for a **Herschel-type venturi meter** depends upon the Reynolds number and to a minor extent upon the size of the venturi, increasing with diameter. A plot of C versus pipe Reynolds number is given in *ASME PTC*, op. cit., p. 19. A value of 0.984 can be used for pipe Reynolds numbers larger than 200,000.

Permanent pressure loss for a Herschel-type venturi tube depends upon diameter ratio β and discharge cone angle α_2. It ranges from 10 to 15 percent of the pressure differential $(p_1 - p_2)$ for small angles (5 to 7°) and from 10 to 30 percent for large angles (15°), with the larger losses occurring at low values of β (see *ASME PTC*, op. cit., p. 12). See Benedict, *J. Fluids Eng.*, **99,** 245–248 (1977), for a general equation for pressure loss for venturis installed in pipes or with plenum inlets.

For flow measurement of **steam and water mixtures** with a Herschel-type venturi in 2½-in- and 3-in-diameter pipes, see Collins and Gacesa, *J. Basic Eng.*, **93,** 11–21 (1971).

A variety of **short-tube** venturi meters are available commercially. They require less space for installation and are generally (although not always) characterized by a greater pressure loss than the corresponding Herschel-type venturi meter. Discharge coefficients vary widely for different types, and individual calibration is recommended if the manufacturer's calibration is not available. Results of tests on the Dall flow tube are given by Miner [*Trans. Am. Soc. Mech. Eng.*, **78,** 475–479 (1956)] and Dowdell [*Instrum. Control Syst.*, **33,** 1006–1009 (1960)]; and on the Gentile flow tube (also called Beth flow tube or Foster flow tube) by Hooper [*Trans. Am. Soc. Mech. Eng.*, **72,** 1099–1110 (1950)].

The use of a **multiventuri system** (in which an inner venturi discharges into the throat of an outer venturi) to increase both the differential pressure for a given flow rate and the signal-to-loss ratio is described by Klomp and Sovran [*J. Basic Eng.*, **94,** 39–45 (1972)].

Flow Nozzles A simple form of flow nozzle is shown in Fig. 10-19. It consists essentially of a short cylinder with a flared approach section. The approach cross section is preferably elliptical in shape but may be conical. Recommended contours for long-radius flow nozzles are given in *ASME PTC*, op. cit., p. 13. In general, the length of the straight portion of the throat is about one-half throat diameter, the upstream pressure tap is located about one pipe diameter from the nozzle inlet face, and the downstream pressure tap about one-half pipe diameter from the inlet face. For subsonic flow, the pressures at points 2 and 3 will be practically identical. If a conical inlet is preferred, the inlet and throat geometry specified for a Herschel-type venturi meter can be used, omitting the expansion section.

Rate of discharge through a flow nozzle for subcritical flow can be determined by the equations given for venturi meters, Eq. (10-22) for gases and Eq. (10-27) for liquids. The expansion factor Y for nozzles is the same as that for venturi meters [Eq. (10-26), Fig. 10-18]. The value of the discharge coefficient C depends primarily upon the pipe Reynolds number and to a lesser extent upon the diameter ratio β. Curves of recommended coefficients for long-radius flow nozzles with pressure taps located one pipe diameter upstream and one-half pipe diameter downstream of the inlet face of the nozzle are given in *ASME PTC*, op. cit., p. 15. In general, coefficients range from 0.95 at a pipe Reynolds number of 10,000 to 0.99 at 1,000,000.

The performance characteristics of pipe-wall-tap nozzles (Fig. 10-19) and throat-tap nozzles are reviewed by Wyler and Benedict [*J. Eng. Power*, **97,** 569–575 (1975)].

Permanent pressure loss across a subsonic flow nozzle is approximated by

$$p_1 - p_4 = \frac{1 - \beta^2}{1 + \beta^2}(p_1 - p_2) \tag{10-28}$$

where p_1, p_2, p_4 = static pressures measured at the locations shown in Fig. 10-19; and β = ratio of nozzle throat diameter to pipe diameter, dimensionless. Equation (10-28) is based on a momentum balance assuming constant fluid density (see Lapple et al., *Fluid and Particle Mechanics*, University of Delaware, Newark, 1951, p. 13).

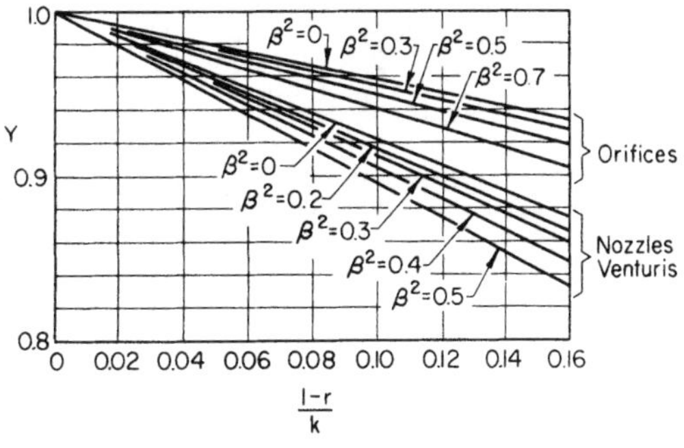

FIG. 10-18 Values of expansion factor Y for orifices, nozzles, and venturis.

See Benedict, loc. cit., for a general equation for pressure loss for nozzles installed in pipes or with plenum inlets. Nozzles show higher loss than venturis. Permanent pressure loss for laminar flow depends on the Reynolds number in addition to β. For details, see Alvi, Sridharan, and Lakshamana Rao, *J. Fluids Eng.*, **100,** 299–307 (1978).

Critical Flow Nozzle For a given set of upstream conditions, the rate of discharge of a gas from a nozzle will increase for a decrease in the absolute pressure ratio p_2/p_1 until the linear velocity in the throat reaches that of sound in the gas at that location. The value of p_2/p_1 for which the acoustic velocity is just attained is called the critical pressure ratio r_c. The actual pressure in the throat will not fall below $p_1 r_c$ even if a much lower pressure exists downstream.

The **critical pressure ratio** r_c can be obtained from the following theoretical equation, which assumes a perfect gas and a frictionless nozzle:

$$r_c^{(1-k)/k} + \left(\frac{k-1}{2}\right)\beta^4 r_c^{2/k} = \frac{k+1}{2} \tag{10-29}$$

This reduces, for $\beta \le 0.2$, to

$$r_c = \left(\frac{2}{k+1}\right)^{k/(k-1)} \tag{10-30}$$

where k = ratio of specific heats c_p/c_v and β = diameter ratio. A table of values of r_c as a function of k and β is given in the ASME Research Committee on Fluid Meters Report, op. cit., p. 68. For small values of β, r_c = 0.487 for k = 1.667, 0.528 for k = 1.40, 0.546 for k = 1.30, and 0.574 for k = 1.15.

Under **critical flow conditions,** only the upstream conditions p_1, v_1, and T_1 need be known to determine flow rate, which, for $\beta \le 0.2$, is given by

$$w_{max} = CA_2\sqrt{g_c k\left(\frac{p_1}{v_1}\right)\left(\frac{2}{k+1}\right)^{(k+1)/(k-1)}} \tag{10-31}$$

For a **perfect gas,** this corresponds to

$$w_{max} = CA_2 p_1\sqrt{g_c k\left(\frac{M}{RT_1}\right)\left(\frac{2}{k+1}\right)^{(k+1)/(k-1)}} \tag{10-32}$$

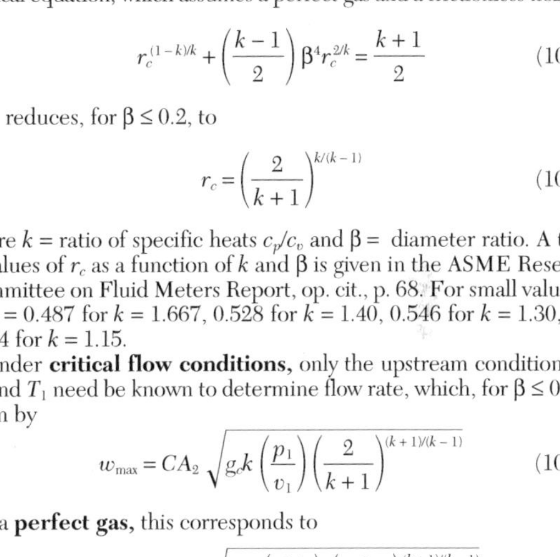

FIG. 10-19 Flow-nozzle assembly.

For air, Eq. (10-31) reduces to

$$w_{max} = C_1 C A_2 p_1/\sqrt{T_1} \qquad (10\text{-}33)$$

where A_2 = cross-sectional area of throat; C = coefficient of discharge, dimensionless; g_c = dimensional constant; k = ratio of specific heats, c_p/c_v; M = molecular weight; p_1 = pressure on upstream side of nozzle; R = gas constant; T_1 = absolute temperature on upstream side of nozzle; v_1 = specific volume on upstream side of nozzle; C_1 = dimensional constant, 0.0405 SI units (0.533 U.S. customary units); and w_{max} = maximum-weight flow rate.

Discharge coefficients for critical flow nozzles are, in general, the same as those for subsonic nozzles. See Grace and Lapple, *Trans. Am. Soc. Mech. Eng.*, **73,** 639–647 (1951); and Szaniszlo, *J. Eng. Power,* **97,** 521–526 (1975). Arnberg, Britton, and Seidl [*J. Fluids Eng.*, **96,** 111–123 (1974)] present discharge-coefficient correlations for circular-arc venturi meters at critical flow. For the calculation of the flow of natural gas through nozzles under critical-flow conditions, see Johnson, *J. Basic Eng.*, **92,** 580–589 (1970).

Elbow Meters A pipe elbow can be used as a flowmeter for liquids if the differential centrifugal head generated between the inner and outer radii of the bend is measured by means of pressure taps located midway around the bend. Equation (10-27) can be used, except that the pressure-difference term $(p_1 - p_2)$ is now taken to be the differential centrifugal pressure and β is taken as zero if one assumes no change in cross section between the pipe and the bend. The discharge coefficient should preferably be determined by calibration, but as a guide it can be estimated within ±6 percent for circular pipe for Reynolds numbers greater than 10^5 from $C = 0.98\sqrt{R_c/2D}$, where R_c = radius of curvature of the centerline and D = inside pipe diameter in consistent units. See Murdock, Foltz, and Gregory, *J. Basic Eng.*, **86,** 498–506 (1964); or the ASME Research Committee on Fluid Meters Report, op. cit., pp. 75–77.

Accuracy Square-edged orifices and venturi tubes have been so extensively studied and standardized that reproducibilities within 1 to 2 percent can be expected between standard meters when new and clean. This is therefore the order of reliability to be had, if one assumes (1) accurate measurement of meter differential, (2) selection of the coefficient of discharge from recommended published literature, (3) accurate knowledge of fluid density, (4) accurate measurement of critical meter dimensions, (5) smooth upstream face of orifice, and (6) proper location of the meter with respect to other flow-disturbing elements in the system. Care must also be taken to avoid even slight corrosion or fouling during use.

Presence of **swirling flow** or an **abnormal velocity distribution** upstream of the metering element can cause serious metering error unless calibration in place is employed or sufficient straight pipe is inserted between the meter and the source of disturbance. Table 10-7 gives the minimum lengths of straight pipe required to avoid appreciable error due to the presence of certain fittings and valves either upstream or downstream of an orifice or nozzle. These values were extracted from plots presented by Sprenkle [*Trans. Am. Soc. Mech. Eng.*, **67,** 345–360 (1945)]. Table 10-7 also shows the reduction in spacing made possible by the use of straightening vanes between the fittings and the meter. Entirely adequate straightening vanes can be provided by fitting a bundle of thin-wall tubes within the pipe. The center-to-center distance between tubes should not exceed one-fourth of the pipe diameter, and the bundle length should be at least 8 times this distance.

The distances specified in Table 10-7 will be conservative if applied to venturi meters. For specific information on requirements for venturi meters, see a discussion by Pardoe appended to Sprenkle (op. cit.). Extensive data on the effect of installation on the coefficients of venturi meters are given elsewhere by Pardoe [*Trans. Am. Soc. Mech. Eng.*, **65,** 337–349 (1943)].

In the presence of **flow pulsations,** the indications of head meters such as orifices, nozzles, and venturis will often be undependable for several reasons. First, the measured pressure differential will tend to be high, since the pressure differential is proportional to the square of flow rate for a head meter, and the square root of the mean differential pressure is always greater than the mean of the square roots of the differential pressures. Second, there is a phase shift as the wave passes through the metering restriction which can affect the differential. Third, pulsations can be set up in the manometer leads themselves. Frequency of the pulsation also plays a part. At low frequencies, the meter reading can generally faithfully follow the flow pulsations, but at high frequencies it cannot. This is due to inertia of the fluid in the manometer leads or of the manometric fluid, whereupon the meter would give a reading intermediate between the maximum and minimum flows but having no readily predictable relation to the mean flow. Pressure transducers with flush-mounted diaphragms can be used together with high-speed recording equipment to provide accurate records of the pressure profiles at the upstream and downstream pressure taps, which can then be analyzed and translated into a mean flow rate.

The rather general practice of producing a steady differential reading by placing restrictions in the manometer leads can result in a reading which, under a fixed set of conditions, may be useful in control of an operation but which has no readily predictable relation to the actual average flow. If calibration is employed to compensate for the presence of pulsations, complete reproduction of operating conditions, including source of pulsations and waveform, is necessary to ensure reasonable accuracy.

According to Head [*Trans. Am. Soc. Mech. Eng.*, **78,** 1471–1479 (1956)], a pulsation-intensity limit of $\Gamma = 0.1$ is recommended as a practical pulsation threshold below which the performance of all types of flowmeters will differ negligibly from steady-flow performance (an error of less than 1 percent in flow due to pulsation). Γ is the peak-to-trough flow variation expressed as a fraction of the average flow rate. According to the ASME Research Committee on Fluid Meters Report (op. cit., pp. 34–35), the fractional metering error E for **liquid flow** through a head meter is given by

$$(1 + E)^2 = 1 + \Gamma^2/8 \qquad (10\text{-}34)$$

When the pulsation amplitude is such as to result in a greater-than-permissible metering error, consideration should be given to installation of a pulsation damper between the source of pulsations and the

TABLE 10-7 Locations of Orifices and Nozzles Relative to Pipe Fittings

Distances in pipe diameters, D_1

Type of fitting upstream	$\frac{D_2}{D_1}$	Distance, upstream fitting to orifice: Without straightening vanes	Distance, upstream fitting to orifice: With straightening vanes	Distance, vanes to orifice	Distance, nearest downstream fitting from orifice
Single 90° ell, tee, or cross used as ell	0.2	6			2
	0.4	6			
	0.6	9	9		
	0.8	20	12	8	4
2 short-radius 90° ells in form of S	0.2	7			2
	0.4	8	8		
	0.6	13	10	6	
	0.8	25	15	11	4
2 long- or short-radius 90° ells in perpendicular planes	0.2	15	9	5	2
	0.4	18	10	6	
	0.6	25	11	7	
	0.8	40	13	9	4
Contraction or enlargement	0.2	8	Vanes have no advantage		2
	0.4	9			
	0.6	10			
	0.8	15			4
Globe valve or stop check	0.2	9	9	5	2
	0.4	10	10	6	
	0.6	13	10	6	
	0.8	21	13	9	4
Gate valve, wide open, or plug cocks	0.2	6	Same as globe valve		2
	0.4	6			
	0.6	8			
	0.8	14			4

flowmeter. References to methods of pulsation-damper design are given in the subsection "Unsteady-State Behavior."

Pulsations are most likely to be encountered in discharge lines from reciprocating pumps or compressors and in lines supplying steam to reciprocating machinery. For **gas flow,** a combination involving a surge chamber and a constriction in the line can be used to damp out the pulsations to an acceptable level. The surge chamber is generally located as close to the pulsation source as possible, with the constriction between the surge chamber and the metering element. This arrangement can be used for either a suction or a discharge line. For such an arrangement, the metering error has been found to be a function of the Hodgson number N_H, which is defined as

$$N_H = Qn\ \Delta p_s/qp_s \tag{10-35}$$

where Q = volume of surge chamber and pipe between metering element and pulsation source; n = pulsation frequency; Δp_s = permanent pressure drop between metering element and surge chamber; q = average volume flow rate, based on gas density in the surge chamber; and p_s = pressure in surge chamber.

Herning and Schmid [*Z. Ver. Dtsch. Ing.,* **82,** 1107–1114 (1938)] presented charts for a simplex double-acting compressor for the prediction of metering error as a function of the Hodgson number and s, the ratio of piston discharge time to total time per stroke. Table 10-8*a* gives the minimum Hodgson numbers required to reduce the metering error to 1 percent as given by the charts (for specific heat ratios between 1.28 and 1.37). Schmid [*Z. Ver. Dtsch. Ing.,* **84,** 596–598 (1940)] presented similar charts for a duplex double-acting compressor and a triplex double-acting compressor for a specific heat ratio of 1.37. Table 10-8*b* gives the minimum Hodgson numbers corresponding to a 1 percent metering error for these cases. The value of $Q\ \Delta p_s$ can be calculated from the appropriate Hodgson number, and appropriate values of Q and Δp_s selected so as to satisfy this minimum requirement.

VELOCITY METERS

Anemometers An anemometer may be any instrument for measurement of gas velocity, e.g., a pitot tube, but usually the term refers to one of the following types.

The vane **anemometer** is a delicate revolution counter with jeweled bearings, actuated by a small windmill, usually 75 to 100 mm (about 3 to 4 in) in diameter, constructed of flat or slightly curved radially disposed vanes. Gas velocity is determined by using a stopwatch to find the time interval required to pass a given number of meters (feet) of gas as indicated by the counter. The velocity so obtained is inversely proportional to gas density. If the original calibration was carried out in a gas of density ρ_0 and the density of the gas stream being metered is ρ_1, the true gas velocity can be found as follows: From the calibration curve for the instrument, find $V_{t,0}$ corresponding to the quantity $V_m\sqrt{\rho_1/\rho_0}$, where V_m = measured velocity. Then the actual velocity $V_{t,1}$ is equal to $V_{t,0}\sqrt{\rho_0/\rho_1}$. In general, when working with air, the effects of

TABLE 10-8*a* Minimum Hodgson Numbers

Simplex double-acting compressor

s	N_H	s	N_H
0.167	1.31	0.667	0.60
0.333	1.00	0.833	0.43
0.50	0.80	1.00	0.34

TABLE 10-8*b* Minimum Hodgson Numbers

Duplex double-acting compressor		Triplex double-acting compressor	
s	N_H	s	N_H
0.167	1.00	0.167	0.85
0.333	0.70	0.333	0.30
0.50	0.30	0.50	0.15
0.667	0.10	0.667	0.06
0.833	0.05	0.833	0.00
1.00	0.00	1.00	0.00

atmospheric-density changes can be neglected for all velocities above 1.5 m/s (about 5 ft/s). In all cases, care must be taken to hold the anemometer well away from one's body or from any object not normally present in the stream.

Vane anemometers can be used for gas-velocity measurements in the range of 0.3 to 45 m/s (about 1 to 150 ft/s), although a given instrument generally has about a twentyfold velocity range. Bearing friction has to be minimized in instruments designed for accuracy at the low end of the range, while ample rotor and vane rigidity must be provided for measurements at the higher velocities. Vane anemometers are sensitive to shock and cannot be used in corrosive atmospheres. Therefore, accuracy is questionable unless a recent calibration has been made and the history of the instrument subsequent to calibration is known. For additional information, see Ower et al., op. cit., chap. VIII.

Turbine Flowmeters They consist of a straight flow tube containing a turbine which is free to rotate on a shaft supported by one or more bearings and located on the centerline of the tube. Means are provided for magnetic detection of the rotational speed, which is proportional to the volumetric flow rate. Its use is generally restricted to clean, noncorrosive fluids. Additional information on construction, operation, range, and accuracy can be obtained from Baker, pp. 215–252, 2000; Miller, op. cit.; and Spitzer, pp. 303–317, 2005.

The **current meter** is generally used for measuring velocities in open channels such as rivers and irrigation channels. There are two types, the cup meter and the propeller meter. The former is more widely used. It consists of six conical cups mounted on a vertical axis pivoted at the ends and free to rotate between the rigid arms of a U-shaped clevis to which a vaned tailpiece is attached. The wheel rotates because of the difference in drag for the two sides of the cup, and a signal proportional to the revolutions of the wheel is generated. The velocity is determined from the count over a period of time. The current meter is generally useful in the range of 0.15 to 4.5 m/s (about 0.5 to 15 ft/s) with an accuracy of ±2 percent. For additional information see Creager and Justin, *Hydroelectric Handbook,* 2d ed., Wiley, New York, 1950, pp. 42–46.

Other important classes of velocity meters include electromagnetic flowmeters and ultrasonic flowmeters. Both are described in Sec. 8.

MASS FLOWMETERS

General Principles There are two main types of mass flowmeters: (1) the so-called true mass flowmeter, which responds directly to mass flow rate, and (2) the inferential mass flowmeter, which commonly measures volume flow rate and fluid density separately. A variety of types of true mass flowmeters have been developed, including the following: (*a*) the Magnus-effect mass flowmeter, (*b*) the axial-flow, transverse-momentum mass flowmeter, (*c*) the radial-flow, transverse-momentum mass flowmeter, (*d*) the gyroscopic transverse-momentum mass flowmeter, and (*e*) the thermal mass flowmeter. Type *b* is the basis for several commercial mass flowmeters, one version of which is briefly described here.

Axial-Flow Transverse-Momentum Mass Flowmeter This type is also referred to as an angular-momentum mass flowmeter. One embodiment of its principle involves the use of axial flow through a driven impeller and a turbine in series. The impeller imparts angular momentum to the fluid, which in turn causes a torque to be imparted to the turbine, which is restrained from rotating by a spring. The torque, which can be measured, is proportional to the rotational speed of the impeller and the mass flow rate.

Inferential Mass Flowmeter There are several types in this category, including the following:

1. *Head meters with density compensation.* Head meters such as orifices, venturis, or nozzles can be used with one of a variety of densitometers [e.g., based on (*a*) buoyant force on a float, (*b*) hydraulic coupling, (*c*) voltage output from a piezoelectric crystal, or (*d*) radiation absorption]. The signal from the head meter, which is proportional to ρV^2 (where ρ = fluid density and V = fluid velocity), is multiplied by ρ given by the densitometer. The square root of the product is proportional to the mass flow rate.

2. *Head meters with velocity compensation.* The signal from the head meter, which is proportional to ρV^2, is divided by the signal from a velocity meter to give a signal proportional to the mass flow rate.

3. *Velocity meters with density compensation.* The signal from the velocity meter (e.g., turbine meter, electromagnetic meter, or sonic velocity meter) is multiplied by the signal from a densitometer to give a signal proportional to the mass flow rate.

Coriolis Mass Flowmeter This type, described in Sec. 8, offers simultaneous direct measurement of both mass flow rate and fluid density. The coriolis flowmeter is insensitive to upstream and downstream flow disturbances, but its performance is adversely affected by the presence of even a few percent of a gas when measuring a liquid flow.

VARIABLE-AREA METERS

General Principles The underlying principle of an ideal area meter is the same as that of a head meter of the orifice type (see subsection "Orifice Meters"). The stream to be measured is throttled by a constriction, but instead of observing the variation with flow of the differential head across an orifice of fixed size, the constriction of an area meter is so arranged that its size is varied to accommodate the flow while the differential head is held constant.

A simple example of an area meter is a gate valve of the rising-stem type provided with static-pressure taps before and after the gate and a means for measuring the stem position. In most common types of area meters, the variation of the opening is automatically brought about by the motion of a weighted piston or float supported by the fluid. Two different cylinder- and piston-type area meters are described in the ASME Research Committee on Fluid Meters Report, op. cit., pp. 82–83.

Rotameters The rotameter, an example of which is shown in Fig. 10-20, has become one of the most popular flowmeters in the chemical-process industries. It consists essentially of a plummet, or "float," which is free to move up or down in a vertical, slightly tapered tube having its small end down. The fluid enters the lower end of the tube and causes the float to rise until the annular area between the float and the wall of the tube is such that the pressure drop across this constriction is just sufficient to support the float. Typically, the tapered tube is of glass and carries etched upon it a nearly linear scale on which the position of the float may be visually noted as an indication of the flow.

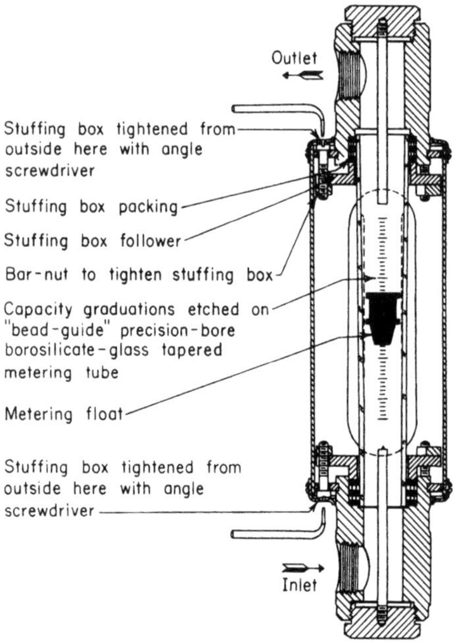

FIG. 10-20 Rotameter.

Interchangeable precision-bore glass tubes and metal metering tubes are available. Rotameters have proved satisfactory both for gases and for liquids at high and at low pressures. A single instrument can readily cover a tenfold range of flow, and by providing floats of different densities a two-hundredfold range is practicable. Rotameters are available with pneumatic, electric, and electronic transmitters for actuating remote recorders, integrators, and automatic flow controllers (see Considine, op. cit., pp. 4-35–4-36, and Sec. 8 of this *Handbook*).

Rotameters require no straight runs of pipe before or after the point of installation. Pressure losses are substantially constant over the whole flow range. In experimental work, for greatest precision, a rotameter should be calibrated with the fluid which is to be metered. However, most modern rotameters are precision-made so that their performance closely corresponds to a master calibration plot for the type in question. Such a plot is supplied with the meter upon purchase.

According to Head [*Trans. Am. Soc. Mech. Eng.*, **76,** 851–862 (1954)], flow rate through a rotameter can be obtained from

$$w = q\rho = KD_f\sqrt{\frac{W_f(\rho_f - \rho)\rho}{\rho_f}} \tag{10-36}$$

and

$$K = \phi\left[\frac{D_t}{D_f}, \frac{\mu}{\sqrt{\dfrac{W_f(\rho_f - \rho)\rho}{\rho_f}}}\right] \tag{10-37}$$

where w = weight flow rate; q = volume flow rate; ρ = fluid density; K = flow parameter, $m^{1/2}/s$ ($ft^{1/2}/s$); D_f = float diameter at constriction; W_f = float weight; ρ_f = float density; D_t = tube diameter at point of constriction; and μ = fluid viscosity. The appropriate value of K is obtained from a composite correlation of K versus the parameters shown in Eq. (10-37) corresponding to the float shape being used. The relation of D_t to the rotameter reading is also required for the tube taper and size being used.

The ratio of flow rates for two different fluids A and B at the same rotameter reading is given by

$$\frac{w_A}{w_B} = \frac{K_A}{K_B}\sqrt{\frac{(\rho_f - \rho_A)\rho_A}{(\rho_f - \rho_B)\rho_B}} \tag{10-38}$$

A measure of self-compensation, with respect to weight rate of flow, for fluid-density changes can be introduced through the use of a float with a density twice that of the fluid being metered, in which case an increase of 10 percent in ρ will produce a decrease of only 0.5 percent in w for the same reading. The extent of immunity to changes in fluid viscosity depends upon the shape of the float.

According to Baird and Cheema [*Can. J. Chem. Eng.*, **47,** 226–232 (1969)], the presence of square-wave pulsations can cause a rotameter to overread by as much as 100 percent. The higher the pulsation frequency, the less the float oscillation, although the error can still be appreciable even when the frequency is high enough so that the float is virtually stationary. Use of a damping chamber between the pulsation source and the rotameter will reduce the error.

Additional information on rotameter theory is presented by Fischer [*Chem. Eng.*, **59**(6), 180–184 (1952)], Coleman [*Trans. Inst. Chem. Eng.*, **34,** 339–350 (1956)], and McCabe, Smith, and Harriott (*Unit Operations of Chemical Engineering,* 4th ed., McGraw-Hill, New York, 1985, pp. 202–205).

TWO-PHASE SYSTEMS

It is generally preferable to meter each of the individual components of a two-phase mixture separately prior to mixing, since it is difficult to meter such mixtures accurately. Problems arise because of fluctuations in composition with time and variations in composition over the cross section of the channel. Information on metering of such mixtures can be obtained from the following sources.

Gas-Solid Mixtures Carlson, Frazier, and Engdahl [*Trans. Am. Soc. Mech. Eng.*, **70,** 65–79 (1948)] describe the use of a **flow nozzle** and a **square-edged orifice** in series for the measurement of both the gas rate and the solids rate in the flow of a finely divided solid-in-gas mixture. The nozzle differential is sensitive to the flow of both phases, whereas the orifice differential is not influenced by the solids flow.

Farbar [*Trans. Am. Soc. Mech. Eng.*, **75,** 943–951 (1953)] describes how a **venturi meter** can be used to measure solids flow rate in a gas-solids mixture when the gas rate is held constant. Separate calibration curves (solids flow versus differential) are required for each gas rate of interest.

Cheng, Tung, and Soo [*J. Eng. Power,* **92,** 135–149 (1970)] describe the use of an **electrostatic probe** for measurement of solids flow in a gas-solids mixture.

Goldberg and Boothroyd [*Br. Chem. Eng.*, **14,** 1705–1708 (1969)] describe several types of solids-in-gas flowmeters and give an extensive bibliography.

Gas-Liquid Mixtures An empirical equation was developed by Murdock [*J. Basic Eng.*, **84,** 419–433 (1962)] for the measurement of gas-liquid mixtures using **sharp-edged orifice** plates with either radius, flange, or pipe taps.

An equation for use with **venturi meters** was given by Chisholm [*Br. Chem. Eng.*, **12,** 454–457 (1967)]. A procedure for determining steam quality via pressure-drop measurement with upflow through either venturi meters or sharp-edged orifice plates was given by Collins and Gacesa [*J. Basic Eng.*, **93,** 11–21 (1971)].

Liquid-Solid Mixtures Liptak [*Chem. Eng.*, **74**(4), 151–158 (1967)] discusses a variety of techniques that can be used for the measurement of solids-in-liquid suspensions or slurries. These include metering pumps, weigh tanks, magnetic flowmeter, ultrasonic flowmeter, gyroscope flowmeter, etc.

Shirato, Gotoh, Osasa, and Usami [*J. Chem. Eng. Japan*, **1,** 164–167 (January 1968)] present a method for determining the mass flow rate of suspended solids in a liquid stream wherein the liquid velocity is measured by an electromagnetic flowmeter and the flow of solids is calculated from the pressure drops across each of two vertical sections of pipe of different diameter through which the suspension flows in series.

FLOWMETER SELECTION

Web sites for process equipment and instrumentation, such as www.globalspec.com and www.thomasnet.com, are valuable tools when selecting a flowmeter. These search engines can scan the flowmeters manufactured by more than 800 companies for specific products that meet the user's specifications. Table 10-4 was based in part on information from these web sites. Note that the accuracies claimed are achieved only under ideal conditions when the flowmeters are properly installed and calibrated for the application.

The purpose of this subsection is to summarize the preferred applications as well as the advantages and disadvantages of some of the common flowmeter technologies.

Table 10-9 divides flowmeters into four classes. Flowmeters in class I depend on wetted moving parts that can wear, plug, or break. The potential for catastrophic failure is a disadvantage. However, in clean fluids, class I flowmeters have often proved reliable and stable when properly installed, calibrated, and maintained.

Class II flowmeters have no wetted moving parts to break and are thus not subject to catastrophic failure. However, the flow surfaces such as orifice plates may wear, eventually biasing flow measurements. Other disadvantages of some flowmeters in this class include high pressure drop and susceptibility to plugging. Very dirty and abrasive fluids should be avoided.

Because class III flowmeters have neither moving parts nor obstructions to flow, they are suitable for dirty and abrasive fluids provided that appropriate materials of construction are available.

Class IV flowmeters have sensors mounted external to the pipe, and would thus seem to be ideal, but problems of accuracy and sensitivity have been encountered in early devices. These comparatively new technologies are under development, and these problems may be overcome in the future.

Section 8 outlines the following criteria for selection of measurement devices: measurement span, performance, reliability, materials of construction, prior use, potential for releasing process materials to the environment, electrical classification, physical access, invasive or noninvasive, and life-cycle cost.

Spitzer, op. cit., 2005, cites four intended end uses of the flowmeter: rate indication, control, totalization, and alarm. Thus high accuracy may be important for rate indication, while control may just need good repeatability. Volumetric flow or mass flow indication is another choice.

Baker, op. cit., 2003, identifies the type of fluid (liquid or gas, slurry, multiphase), special fluid constraints (clean or dirty, hygienic, corrosive, abrasive, high flammability, low lubricity, fluids causing scaling). He lists the following flowmeter constraints: accuracy or measurement uncertainty, diameter range, temperature range, pressure range, viscosity range, flow range, pressure loss caused by the flowmeter, sensitivity to installation, sensitivity to pipework supports, sensitivity to pulsation, whether the flowmeter has a clear bore, whether a clamp-on version is available, response time, and ambient conditions. Finally, Baker identifies these environmental considerations: ambient temperature, humidity, exposure to weather, level of electromagnetic radiation, vibration, tamperproof for domestic use, and classification of area requiring explosionproof, intrinsic safety, etc.

Note that the accuracies cited in Table 10-4 can be achieved by those flowmeters only under ideal conditions of application, installation, and calibration. This subsection has only given an introduction to issues to consider in the choice of a flowmeter for a given application. See Baker, op. cit., 2003; Miller, op. cit., 1996; and Spitzer, op. cit., 2005, for further guidance and to obtain application-specific data from flowmeter vendors.

TABLE 10-9 Flowmeter Classes

Class I: Flowmeters with wetted moving parts	Class II: Flowmeters with no wetted moving parts
Positive displacement Turbine Variable-area	Differential pressure Vortex Target Thermal
Class III: Obstructionless flowmeters	**Class IV: Flowmeters with sensors mounted external to the pipe**
Coriolis mass Electromagnetic Ultrasonic	Clamp-on ultrasonic Correlation

Adapted from Spitzer, op. cit., 2005.

WEIRS

Liquid flow in an open channel may be metered by means of a weir, which consists of a dam over which, or through a notch in which, the liquid flows. The terms "rectangular weir," "triangular weir," etc., generally refer to the shape of the notch in a notched weir. All weirs considered here have flat upstream faces that are perpendicular to the bed and walls of the channel.

Sharp-edged weirs have edges like those of square or sharp-edged orifices (see subsection "Orifice Meters"). Notched weirs are ordinarily sharp-edged. Weirs not in the sharp-edged class are, for the most part, those described as **broad-crested weirs.**

The head h_0 on a weir is the liquid-level height above the crest or base of the notch. The head must be measured sufficiently far upstream to avoid the drop in level occasioned by the overfall which begins at a distance about $2h_0$ upstream from the weir. Surface-level measurements should be made a distance of $3h_0$ or more upstream, preferably by using a stilling box equipped with a high-precision level gauge, e.g., a hook gauge or float gauge.

With sharp-edged weirs, the sheet of discharging liquid, called the "nappe," contracts as it leaves the opening and free discharge occurs. Rounding the upstream edge will reduce the contraction and increase the flow rate for a given head. A clinging nappe may result

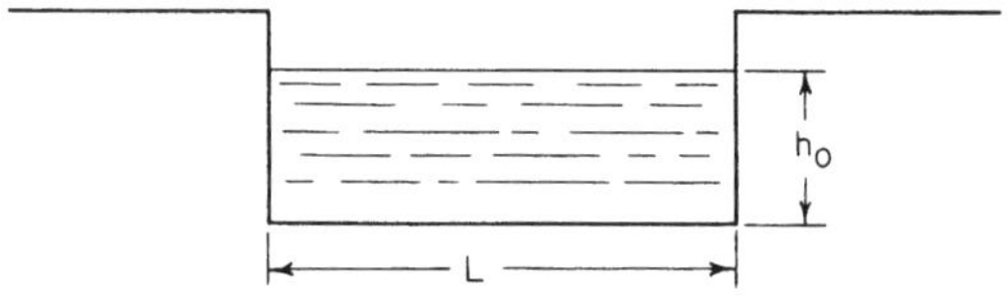

FIG. 10-21 Rectangular weir.

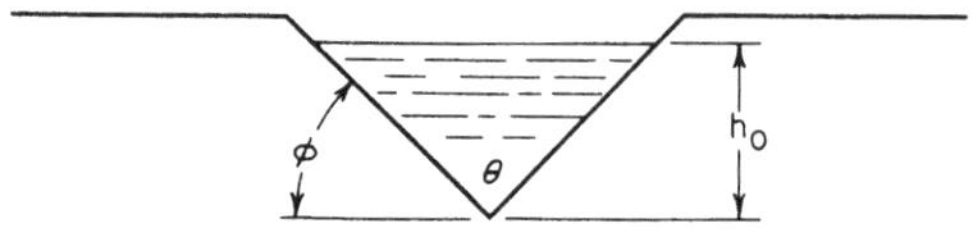

FIG. 10-22 Triangular weir.

if the head is very small, if the edge is well rounded, or if air cannot flow in beneath the nappe. This, in turn, results in an increase in the discharge rate for a given head as compared with that for a free nappe. For further information on the effect of the nappe, see Gibson, *Hydraulics and Its Applications,* 5th ed., Constable, London, 1952; and Chow, *Open-Channel Hydraulics,* McGraw-Hill, New York, 1959.

Flow through a **rectangular weir** (Fig. 10-21) is given by

$$q = 0.415(L - 0.2h_0)h_0^{1.5}\sqrt{2g} \quad (10\text{-}39)$$

where q = volume flow rate, L = crest length, h_0 = weir head, and g = local acceleration due to gravity. This is known as the modified Francis formula for a rectangular sharp-edged weir with two end corrections; it applies when the velocity-of-approach correction is small. The Francis formula agrees with experiments within 3 percent if (1) L is greater than $2h_0$, (2) velocity of approach is 0.6 m/s (2 ft/s) or less, (3) height of crest above bottom of channel is at least $3h_0$, and (4) h_0 is not less than 0.09 m (0.3 ft).

Narrow rectangular notches ($h_0 > L$) have been found to give about 93 percent of the discharge predicted by the Francis formula. Thus

$$q = 0.386Lh_0^{1.5}\sqrt{2g} \quad (10\text{-}40)$$

In this case, no end corrections are applied even though the formula applies only for sharp-edged weirs. See Schoder and Dawson, *Hydraulics,* McGraw-Hill, New York, 1934, p. 175, for further details.

The **triangular-notch weir** has the advantage that a single notch can accommodate a wide range of flow rates, although this in turn reduces its accuracy. The discharge for sharp- or square-edged weirs is given by

$$q = (0.31h_0^{2.5}\sqrt{2g})/\tan\phi \quad (10\text{-}41)$$

See Eq. (10-39) for nomenclature. Angle ϕ is illustrated in Fig. 10-22. Equations (10-39), (10-40), and (10-41) are applicable only to the flow of water. However, for the case of triangular-notch weirs Lenz [*Trans. Am. Soc. Civ. Eng.,* **108,** 759–802 (1943)] has presented correlations predicting the effect of viscosity over the range of 0.001 to 0.15 Pa·s (1 to 150 cP) and surface tension over the range of 0.03 to 0.07 N/m (30 to 70 dyn/cm). His equation predicts about an 8 percent increase in flow for a liquid of 0.1-Pa·s (100-cP) viscosity compared with water at 0.001 Pa·s (1 cP) and about a 1 percent increase for a liquid with one-half of the surface tension of water. For fluids of moderate viscosity, Ranga Raju and Asawa [*Proc. Am. Soc. Civ. Eng., J. Hydraul. Div.,* **103** (HY 10), 1227–1231 (1977)] find that the effect of viscosity and surface tension on the discharge flow rate for rectangular and triangular-notch (ϕ = 45°) weirs can be neglected when

$$(N_{Re})^{0.2}(N_{We})^{0.6} > 900 \quad (10\text{-}42)$$

where N_{Re} (Reynolds number) = $\sqrt{gh_0^3}/v$, g = local acceleration due to gravity, h_0 = weir head, ν = kinematic viscosity; N_{We} (Weber number) = $\rho g h_0^2/g_c\sigma$, ρ = density, g_c = dimensional constant, and σ = surface tension.

For the flow of high-viscosity liquids over rectangular weirs, see Slocum, *Can. J. Chem. Eng.,* **42,** 196–200 (1964). His correlation is based on data for liquids with viscosities in the range of 2.5 to 500 Pa·s (25 to 5000 cP), in which range the discharge decreases markedly for a given head as viscosity is increased.

Information on other types of weirs can be obtained from Addison, op. cit.; Gibson, *Hydraulics and Its Applications,* 5th ed., Constable, London, 1952; Henderson, *Open Channel Flow,* Macmillan, New York, 1966; Linford, *Flow Measurement and Meters,* Spon, London, 1949; Lakshmana Rao, "Theory of Weirs," in *Advances in Hydroscience,* vol. 10, Academic, New York, 1975; and Merritt, *Standard Handbook for Civil Engineers,* 2d ed., McGraw-Hill, New York, 1976.

PUMPS AND COMPRESSORS

General References: Meherwan P. Boyce, P.E., *Centrifugal Compressors: A Basic Guide,* Pennwell Books, Tulsa, Okla., 2002; Royce N. Brown, *Compressors: Selection and Sizing,* 3d ed., Gulf Professional Publishing, Houston, Tex., 2005; James Corley, "The Vibration Analysis of Pumps: A Tutorial," Fourth International Pump Symposium, Texas A & M University, Houston, Tex., May 1987; John W. Dufor and William E. Nelson, *Centrifugal Pump Sourcebook,* McGraw-Hill, New York, 1992; *Engineering Data Book,* 12th ed., vol. I, Secs. 12 and 13, Gas Processors Suppliers Association, Tulsa, Okla., 2004; Paul N. Garay, P.E., *Pump Application Desk Book,* Fairmont Press, 1993; *Process Pumps,* ITT Fluid Technology Corporation, 1992; Igor J. Karassik et al., *Pump Handbook,* 3d ed., McGraw-Hill, New York, 2001; Val S. Lobanoff and Robert R. Ross, *Centrifugal Pumps: Design and Application,* 2d ed., Gulf Professional Publishing, Houston, Tex., 1992; A. J. Stephanoff, *Centrifugal and Axial Flow Pumps: Theory, Design, and Application,* 2d ed., Krieger Publishing, Melbourne, Fla., 1992.

INTRODUCTION

The following subsections deal with pumps and compressors. A pump or compressor is a physical contrivance that is used to deliver fluids from one location to another through conduits. The term *pump* is used when the fluid is a liquid, while the term *compressor* is used when the fluid is a gas. The basic requirements to define the application are suction and delivery pressures, pressure loss in transmission, and flow rate. Special requirements may exist in food, pharmaceutical, nuclear, and other industries that impose material selection requirements of the pump.

The primary means of transfer of energy to the fluid that causes flow are gravity, displacement, centrifugal force, electromagnetic force, transfer of momentum, mechanical impulse, and a combination of these energy-transfer mechanisms. Displacement and centrifugal force are the most common energy-transfer mechanisms in use.

Pumps and compressors are designed per technical specifications and standards developed over years of operating and maintenance experience. Table 10-10 lists some of these standards for pumps and compressors and for related equipment such as lubrication systems and gearboxes which, if not properly specified, could lead to many operational and maintenance problems with the pumps and compressors. These standards specify design, construction, maintenance, and testing details such as terminology, material selection, shop inspection and tests, drawings, clearances, construction procedures, and so on.

There are four (4) major types of pumps: (1) positive displacement, (2) dynamic (kinetic), (3) lift, and (4) electromagnetic. Piston pumps are positive displacement pumps. The most common centrifugal

TABLE 10-10 Standards Governing Pumps and Compressors

ASME Standards, American Society of Mechanical Engineers, New York
- B73.1-2001, *Specification for Horizontal End Suction Centrifugal Pumps for Chemical Process*
- B73.2-2003, *Specification for Vertical In-Line Centrifugal Pumps for Chemical Process*
- PTC 10, 1997 *Test Code on Compressors and Exhausters*
- PTC 11, 1984 *Fans*
- B19.3-1991, *Safety Standard for Compressors for Process Industries*

API Standards, American Petroleum Institute, Washington
- API Standard 610, *Centrifugal Pumps for Petroleum, Petrochemical, and Natural Gas Industries*, Adoption of ISO 13709, October 2004
- API Standard 613, *Special Purpose Gear Units for Petroleum, Chemical and Gas Industry Services*, February 2003
- API Standard 614, *Lubrication, Shaft-Sealing, and Control-Oil Systems and Auxiliaries for Petroleum, Chemical and Gas Industry Services*, April 1999
- API Standard 616, *Gas Turbines for the Petroleum, Chemical, and Gas Industry Services*, August 1998
- API Standard 617, *Axial and Centrifugal Compressors and Expanders—Compressors for Petroleum, Chemical, and Gas Industry Services*, June 2003
- API Standard 618, *Reciprocating Compressors for Petroleum, Chemical, and Gas Industry Services*, June 1995
- API Standard 619, *Rotary-Type Positive Displacement Compressors for Petroleum, Petrochemical, and Natural Gas Industries*, December 2004
- API Standard 670, *Machinery Protection Systems*, November 2003
- API Standard 671, *Special Purpose Couplings for Petroleum, Chemical, and Gas Industry Services*, October 1998
- API Standard 672, *Packaged, Integrally Geared, Centrifugal Air Compressors for Petroleum, Chemical, and Gas Industry Services*, March 2004
- API Standard 673, *Centrifugal Fans for Petroleum, Chemical, and Gas Industry Services*, October 2002
- API Standard 674, *Positive Displacement Pumps—Reciprocating*, June 1995
- API Standard 675, *Positive Displacement Pumps—Controlled Volume*, March 2000
- API Standard 677, *General Purpose Gear Units for Petroleum, Chemical, and Gas Industry Services*, April 2006
- API Standard 680, *Packaged Reciprocating Plant and Instrument Air Compressors for General Refinery Services*, October 1987
- API Standard 681, *Liquid Ring Vacuum Pumps and Compressors for Petroleum, Chemical, and Gas Industry Services*, June 2002
- API Standard 682, *Pumps—Shaft Sealing Systems for Centrifugal and Rotary Pumps*, September 2004
- API Standard 685, *Sealless Centrifugal Pumps for Petroleum, Heavy Duty Chemical, and Gas Industry Services*, October 2000

Hydraulic Institute, Parsippany, N.J. (www.pumps.org)
- ANSI/HI Pump Standards, 2005 (covers centrifugal, vertical, rotary, and reciprocating pumps)

National Fire Protection Association, Quincy, Mass. (www.nfpa.org)
- Standards for pumps used in fire protection systems

pumps are of dynamic type; ancient bucket-type pumps are lift pumps; and electromagnetic pumps use electromagnetic force and are common in modern reactors. Canned pumps are also becoming popular in the petrochemical industry because of the drive to minimize fugitive emissions. Figure 10-23 shows pump classification:

TERMINOLOGY

Displacement Discharge of a fluid from a vessel by partially or completely displacing its internal volume with a second fluid or by mechanical means is the principle upon which a great many fluid-transport devices operate. Included in this group are reciprocating-piston and diaphragm machines, rotary-vane and gear types, fluid piston compressors, acid eggs, and air lifts.

The large variety of displacement-type fluid-transport devices makes it difficult to list characteristics common to each. However, for most types it is correct to state that (1) they are adaptable to high-pressure operation, (2) the flow rate through the pump is variable (auxiliary damping systems may be employed to reduce the magnitude of pressure pulsation and flow variation), (3) mechanical considerations limit maximum throughputs, and (4) the devices are capable of efficient performance at extremely low-volume throughput rates.

Centrifugal Force Centrifugal force is applied by means of the centrifugal pump or compressor. Though the physical appearance of the many types of centrifugal pumps and compressors varies greatly, the basic function of each is the same, i.e., to produce kinetic energy by the action of centrifugal force and then to convert this energy into pressure by efficiently reducing the velocity of the flowing fluid.

In general, centrifugal fluid-transport devices have these characteristics: (1) discharge is relatively free of pulsation; (2) mechanical design lends itself to high throughputs, capacity limitations are rarely a problem; (3) the devices are capable of efficient performance over a wide range of pressures and capacities even at constant-speed operation; (4) discharge pressure is a function of fluid density; and (5) these are relatively small high-speed devices and less costly.

A device which combines the use of centrifugal force with mechanical impulse to produce an increase in pressure is the axial-flow compressor or pump. In this device the fluid travels roughly parallel to the shaft through a series of alternately rotating and stationary radial blades having airfoil cross sections. The fluid is accelerated in the axial direction by mechanical impulses from the rotating blades; concurrently, a positive-pressure gradient in the radial direction is established in each stage by centrifugal force. The net pressure rise per stage results from both effects.

Electromagnetic Force When the fluid is an electrical conductor, as is the case with molten metals, it is possible to impress an electromagnetic field around the fluid conduit in such a way that a driving force that will cause flow is created. Such pumps have been developed for the handling of heat-transfer liquids, especially for nuclear reactors.

Transfer of Momentum Deceleration of one fluid (motivating fluid) in order to transfer its momentum to a second fluid (pumped fluid) is a principle commonly used in the handling of corrosive materials, in pumping from inaccessible depths, or for evacuation. Jets and eductors are in this category.

Absence of moving parts and simplicity of construction have frequently justified the use of jets and eductors. However, they are relatively inefficient devices. When air or steam is the motivating fluid, operating costs may be several times the cost of alternative types of fluid-transport equipment. In addition, environmental considerations in today's chemical plants often inhibit their use.

Mechanical Impulse The principle of mechanical impulse when applied to fluids is usually combined with one of the other means of imparting motion. As mentioned earlier, this is the case in axial-flow compressors and pumps. The turbine or regenerative-type pump is another device which functions partially by mechanical impulse.

Measurement of Performance The amount of useful work that any fluid-transport device performs is the product of (1) the mass rate of fluid flow through it and (2) the total pressure differential measured immediately before and after the device, usually expressed in the height of column of fluid equivalent under adiabatic conditions. The first of these quantities is normally referred to as **capacity,** and the second is known as **head.**

Capacity This quantity is expressed in the following units. In SI units capacity is expressed in cubic meters per hour (m^3/h) for both liquids and gases. In U.S. customary units it is expressed in U.S. gallons per minute (gal/min) for liquids and in cubic feet per minute (ft^3/min) for gases. Since all these are volume units, the density or specific gravity must be used for conversion to mass rate of flow. When gases are being handled, capacity must be related to a pressure and a temperature, usually the conditions prevailing at the machine inlet. It is important to note that all heads and other terms in the following equations are expressed in height of column of liquid.

Total Dynamic Head The total dynamic head H of a pump is the total discharge head h_d minus the total suction head h_s.

Total Suction Head This is the reading h_{gs} of a gauge at the suction flange of a pump (corrected to the pump centerline*), plus the barometer reading and the velocity head h_{vs} at the point of gauge attachment:

$$h_s = h_{gs} + \text{atm} + h_{vs} \qquad (10\text{-}43)$$

If the gauge pressure at the suction flange is less than atmospheric, requiring use of a vacuum gauge, this reading is used for h_{gs} in Eq. (10-43) with a negative sign.

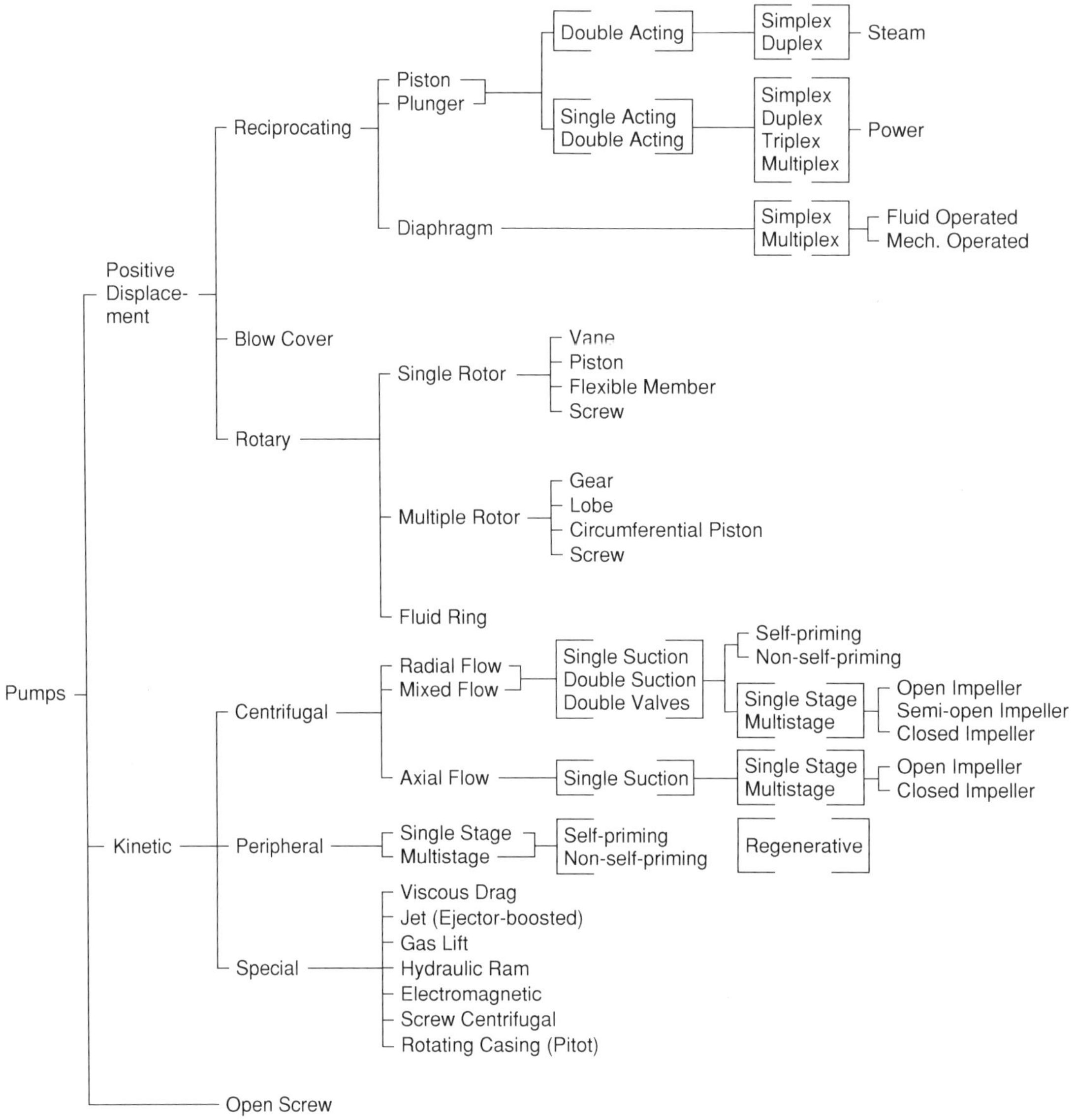

FIG. 10-23 Classification of pumps. (*Courtesty of Hydraulic Institute.*)

Before installation it is possible to estimate the total suction head as follows:

$$h_s = h_{ss} - h_{fs} \tag{10-44}$$

where h_{ss} = static suction head and h_{fs} = suction friction head.

Static Suction Head The static suction head h_{ss} is the vertical distance measured from the free surface of the liquid source to the pump centerline plus the absolute pressure at the liquid surface.

Total Discharge Head The total discharge head h_d is the reading h_{gd} of a gauge at the discharge flange of a pump (corrected to the pump centerline*), plus the barometer reading and the velocity head h_{vd} at the point of gauge attachment:

$$h_d = h_{gd} + \text{atm} + h_{vd} \tag{10-45}$$

Again, if the discharge gauge pressure is below atmospheric, the vacuum-gauge reading is used for h_{gd} in Eq. (10-45) with a negative sign.

Before installation it is possible to estimate the total discharge head from the static discharge head h_{sd} and the discharge friction head h_{fd} as follows:

$$h_d = h_{sd} + h_{fd} \tag{10-46}$$

Static Discharge Head The static discharge head h_{sd} is the vertical distance measured from the free surface of the liquid in the receiver to the pump centerline,* plus the absolute pressure at the liquid surface. **Total static head** h_{ts} is the difference between discharge and suction static heads.

*On vertical pumps, the correction should be made to the eye of the suction impeller.

Velocity Since most liquids are practically incompressible, the relation between the quantity flowing past a given point in a given time and the velocity of flow is expressed as follows:

$$Q = Av \tag{10-47}$$

This relationship in SI units is as follows:

$$v \text{ (for circular conduits)} = 3.54\ Q/d^2 \tag{10-48}$$

where v = average velocity of flow, m/s; Q = quantity of flow, m^3/h; and d = inside diameter of conduit, cm.

This same relationship in U.S. customary units is

$$v \text{ (for circular conduits)} = 0.409\ Q/d^2 \tag{10-49}$$

where v = average velocity of flow, ft/s; Q = quantity of flow, gal/min; and d = inside diameter of conduit, in.

Velocity Head This is the vertical distance by which a body must fall to acquire the velocity v.

$$h_v = v^2/2g \tag{10-50}$$

Viscosity (See Sec. 6 for further information.) In flowing liquids the existence of internal friction or the internal resistance to relative motion of the fluid particles must be considered. This resistance is called viscosity. The viscosity of liquids usually decreases with rising temperature. Viscous liquids tend to increase the power required by a pump, to reduce pump efficiency, head, and capacity, and to increase friction in pipe lines.

Friction Head This is the pressure required to overcome the resistance to flow in pipe and fittings. It is dealt with in detail in Sec. 6.

Work Performed in Pumping To cause liquid to flow, work must be expended. A pump may raise the liquid to a higher elevation, force it into a vessel at higher pressure, provide the head to overcome pipe friction, or perform any combination of these. Regardless of the service required of a pump, all energy imparted to the liquid in performing this service must be accounted for; consistent units for all quantities must be employed in arriving at the work or power performed.

When arriving at the performance of a pump, it is customary to calculate its **power output,** which is the product of (1) the total dynamic head and (2) the mass of liquid pumped in a given time. In SI units power is expressed in kilowatts; horsepower is the conventional unit used in the United States.

In SI units,

$$\text{kW} = HQ\rho/3.670 \times 10^5 \tag{10-51}$$

where kW is the pump power output, kW; H = total dynamic head, N·m/kg (column of liquid); Q = capacity, m^3/h; and ρ = liquid density, kg/m^3.

When the total dynamic head H is expressed in pascals, then

$$\text{kW} = HQ/3.599 \times 10^6 \tag{10-52}$$

In U.S. customary units,

$$\text{hp} = HQs/3.960 \times 10^3 \tag{10-53}$$

where hp is the pump-power output, hp; H = total dynamic head, lbf·ft/lbm (column of liquid); Q = capacity, U.S. gal/min; and s = liquid specific gravity.

When the total dynamic head H is expressed in pounds-force per square inch, then

$$\text{hp} = HQ/1.714 \times 10^3 \tag{10-54}$$

The **power input** to a pump is greater than the **power output** because of internal losses resulting from friction, leakage, etc. The efficiency of a pump is therefore defined as

$$\text{Pump efficiency} = (\text{power output})/(\text{power input}) \tag{10-55}$$

PUMP SELECTION

When selecting pumps for any service, it is necessary to know the liquid to be handled, the total dynamic head, the suction and discharge heads, and, in most cases, the temperature, viscosity, vapor pressure, and specific gravity. In the chemical industry, the task of pump selection is frequently further complicated by the presence of solids in the liquid and liquid corrosion characteristics requiring special materials of construction. Solids may accelerate erosion and corrosion, have a tendency to agglomerate, or require delicate handling to prevent undesirable degradation.

Range of Operation Because of the wide variety of pump types and the number of factors which determine the selection of any one type for a specific installation, the designer must first eliminate all but those types of reasonable possibility. Since range of operation is always an important consideration, Fig. 10-24 should be of assistance. The boundaries shown for each pump type are at best approximate. In most cases, following Fig. 10-24 will select the pump that is best suited for a given application. Low-capacity pumps with high discharge head requirements are best served by positive-displacement pumps. Reciprocating pumps and rotary pumps such as gear and roots rotor-type pumps are examples of positive-displacement pumps. Displacement pumps provide high heads at low capacities which are beyond the capability of centrifugal pumps. Displacement pumps achieve high pressure with low velocities and are thus suited for high-viscosity service and slurry.

The centrifugal pump operates over a very wide range of flows and pressures. For low heads but high flows the axial pump is best suited. Both the centrifugal and axial flow pumps impart energy to the fluid by the rotational speed of the impeller and the velocity it imparts to the fluid.

NET POSITIVE SUCTION HEAD

Net positive suction head available $(NPSH)_A$ is the difference between the total absolute suction pressure at the pump suction nozzle when the pump is running and the vapor pressure at the flowing liquid temperature. All pumps require the system to provide adequate $(NPSH)_A$. In a positive-displacement pump the $(NPSH)_A$ should be large enough to open the suction valve, to overcome the friction losses within the pump liquid end, and to overcome the liquid acceleration head.

Suction Limitations of a Pump Whenever the pressure in a liquid drops below the vapor pressure corresponding to its temperature, the liquid will vaporize. When this happens within an operating pump, the vapor bubbles will be carried along to a point of higher pressure, where they suddenly collapse. This phenomenon is known as **cavitation.** Cavitation in a pump should be avoided, as it is accompanied by metal removal, vibration, reduced flow, loss in efficiency, and noise. When the absolute suction pressure is low, cavitation may occur in the pump inlet and damage result in the pump suction and on the impeller vanes near the inlet edges. To avoid this phenomenon, it is necessary to maintain a **required net positive suction head**

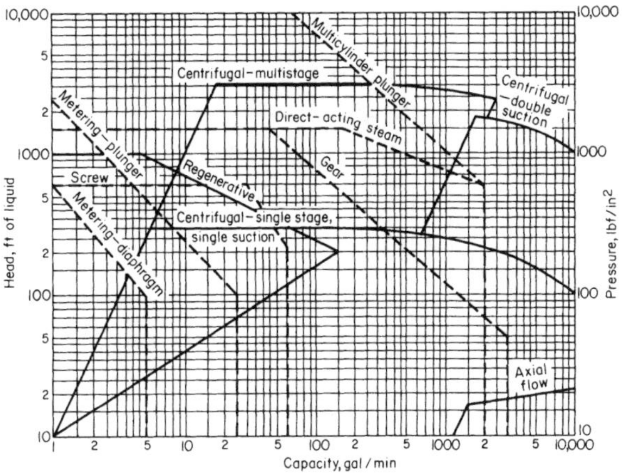

FIG. 10-24 Pump coverage chart based on normal ranges of operation of commercially available types. Solid lines: use left ordinate, head scale. Broken lines: use right ordinate, pressure scale. To convert gallons per minute to cubic meters per hour, multiply by 0.2271; to convert feet to meters, multiply by 0.3048; and to convert pounds-force per square inch to kilopascals, multiply by 6.895.

$(\text{NPSH})_R$, which is the equivalent total head of liquid at the pump centerline less the vapor pressure p. Each pump manufacturer publishes curves relating $(\text{NPSH})_R$ to capacity and speed for each pump.

When a pump installation is being designed, the **available net positive suction head** $(\text{NPSH})_A$ must be equal to or greater than the $(\text{NPSH})_R$ for the desired capacity. The $(\text{NPSH})_A$ can be calculated as follows:

$$(\text{NPSH})_A = h_{ss} - h_{fs} - p \tag{10-56}$$

If $(\text{NPSH})_A$ is to be checked on an existing installation, it can be determined as follows:

$$(\text{NPSH})_A = \text{atm} + h_{gs} - p + h_{vs} \tag{10-57}$$

Practically, the NPSH required for operation without cavitation and vibration in the pump is somewhat greater than the theoretical. The actual $(\text{NPSH})_R$ depends on the characteristics of the liquid, the total head, the pump speed, the capacity, and impeller design. Any suction condition which reduces $(\text{NPSH})_A$ below that required to prevent cavitation at the desired capacity will produce an unsatisfactory installation and can lead to mechanical difficulty.

The following two equations usually provide an adequate design margin between $(\text{NPSH})_A$ and $(\text{NPSH})_R$:

$$(\text{NPSH})_A = (\text{NPSH})_R + 5\text{ ft} \tag{10-58}$$

$$(\text{NPSH})_A = 1.35(\text{NPSH})_R \tag{10-59}$$

Use the larger value of $(\text{NPSH})_A$ calculated with Eqs. (10-58) and (10-59).

NPSH Requirements for Other Liquids NPSH values depend on the fluid being pumped. Since water is considered a standard fluid for pumping, various correction methods have been developed to evaluate NPSH when pumping other fluids. The most recent of these corrective methods has been developed by the Hydraulic Institute and is shown in Fig. 10-25.

The chart shown in Fig. 10-25 is for pure liquids. Extrapolation of data beyond the ranges indicated in the graph may not produce accurate results. Figure 10-25 shows the variation of vapor pressure and NPSH reductions for various hydrocarbons and hot water as a function of temperature. Certain rules apply while using this chart. When using the chart for hot water, if the NPSH reduction is greater than one-half of the NPSH required for cold water, deduct one-half of cold water NPSH to obtain the corrected NPSH required. On the other hand, if the value read on the chart is less than one-half of cold water NPSH, deduct this chart value from the cold water NPSH to obtain the corrected NPSH.

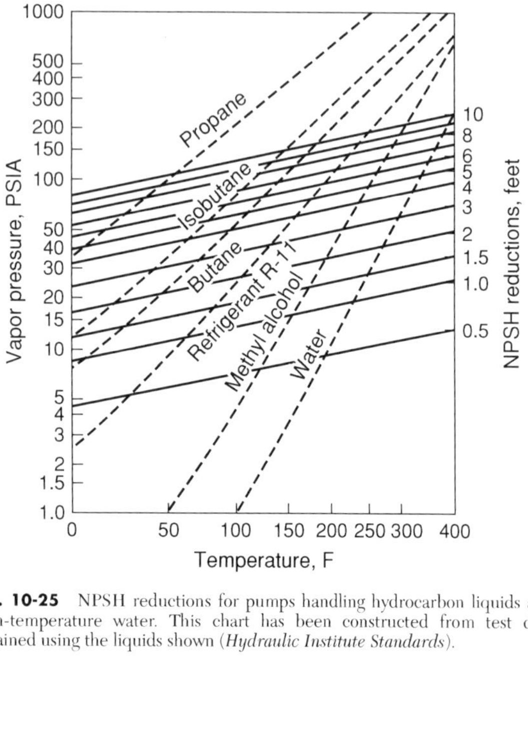

FIG. 10-25 NPSH reductions for pumps handling hydrocarbon liquids and high-temperature water. This chart has been constructed from test data obtained using the liquids shown (*Hydraulic Institute Standards*).

Example 1: NPSH Calculation Suppose a selected pump requires a minimum NPSH of 16 ft (4.9 m) when pumping cold water; What will be the NPSH limitation to pump propane at 55°F (12.8°C) with a vapor pressure of 100 psi? Using the chart in Fig. 10-25, NPSH reduction for propane gives 9.5 ft (2.9 m). This is greater than one-half of cold water NPSH of 16 ft (4.9 m). The corrected NPSH is therefore 8 ft (2.2 m) or one-half of cold water NPSH.

PUMP SPECIFICATIONS

Pump specifications depend upon numerous factors but mostly on application. Typically, the following factors should be considered while preparing a specification.

1. Application, scope, and type
2. Service conditions
3. Operating conditions
4. Construction application-specific details and special considerations
   a. Casing and connections
   b. Impeller details
   c. Shaft
   d. Stuffing box details—lubrications, sealing, etc.
   e. Bearing frame and bearings
   f. Baseplate and couplings
   g. Materials
   h. Special operating conditions and miscellaneous items

Table 10-11 is based on the API and ASME codes and illustrates a typical specification for centrifugal pumps.

POSITIVE-DISPLACEMENT PUMPS

Positive-displacement pumps and those that approach positive displacement will ideally produce whatever head is impressed upon them by the system restrictions to flow. The maximum head attainable is determined by the power available in the drive (slippage neglected) and the strength of the pump parts. A pressure relief valve on the discharge side should be set to open at a safe pressure for the casing and the internal components of the pump such as piston rods, cylinders, crankshafts, and other components which would be pressurized. In the case of a rotary pump, the total dynamic head developed is uniquely determined for any given flow by the speed at which it rotates.

In general, overall efficiencies of positive-displacement pumps are higher than those of centrifugal equipment because internal losses are minimized. On the other hand, the flexibility of each piece of equipment in handling a wide range of capacities is somewhat limited.

Positive-displacement pumps may be of either the **reciprocating** or the **rotary** type. In all positive-displacement pumps, a cavity or cavities are alternately filled and emptied of the pumped fluid by the action of the pump.

Reciprocating Pumps There are three classes of reciprocating pumps: **piston pumps, plunger pumps,** and **diaphragm pumps.** Basically, the action of the liquid-transferring parts of these pumps is the same, a cylindrical piston, plunger, or bucket or a round diaphragm being caused to pass or flex back and forth in a chamber. The device is equipped with valves for the inlet and discharge of the liquid being pumped, and the operation of these valves is related in a definite manner to the motions of the piston. In all modern-design reciprocating pumps, the suction and discharge valves are operated by pressure difference. That is, when the pump is on its suction stroke and the pump cavity is increasing in volume, the pressure is lowered within the pump cavity, permitting the higher suction pressure to open the suction valve and allowing liquid to flow into the pump. At the same time, the higher discharge-line pressure holds the discharge valve closed. Likewise on the discharge stroke, as the pump cavity is decreasing in volume, the higher pressure developed in the pump cavity holds the suction valve closed and opens the discharge valve to expel liquid from the pump into the discharge line.

The *overall efficiency* of these pumps varies from about 50 percent for the small pumps to about 90 percent or more for the larger sizes.

TABLE 10-11 Typical Pump Specification

Specification	Description
1.0	Scope:
	This specification covers horizontal, end suction, vertically split, single-stage centrifugal pumps with top centerline discharge and "back pullout" feature.
2.0	Service Conditions:
	Pump shall be designed to operate satisfactorily with a reasonable service life when operated either intermittently or continuously in typical process applications.
3.0	Operating Conditions:
	Capacity __________ U.S. gallons per minute __________ Head (_____ ft total head) (_____ psig). Speed _____ rpm Suction Pressure (_____ ft head) (positive) (lift) (_____ psig) Liquid to be handled ______________________________ Specific gravity ___________ Viscosity (______________) Temperature of liquid at inlet _________°F Solids content _________% _________ Max. size
4.0	Pump Construction:
	4.1 *Casing.* Casing shall be vertically split with self-venting top centerline discharge, with an integral foot located directly under the casing for added support. All casings shall be of the "back pullout" design with suction and discharge nozzles cast integrally. Casings shall be provided with bosses in suction and discharge nozzles, and in bottom of casing for gauge taps and drain tap. (Threaded taps with plugs shall be provided for these features.)
	4.2 *Casing Connections.* Connections shall be A.N.S.I. flat-faced flanges. [Cast iron (125) (250) psig rated] [Duron metal, steel, alloy steel (150) (300) psig rated]
	4.3 *Casing Joint Gasket.* A confined-type nonasbestos gasket suitable for corrosive service shall be provided at the casing joint.
	4.4 *Impeller.* Fully-open impeller with front edge having contoured vanes curving into the suction for minimum NPSH requirements and maximum efficiency shall be provided. A hex head shall be cast in the eye of the impeller to facilitate removal, and eliminate need for special impeller removing tool. All impellers shall have radial "pump-out" vanes on the back side to reduce stuffing box pressure and aid in eliminating collection of solids at stuffing box throat. Impellers shall be balanced within A.N.S.I. guidelines to ISO tolerances.
	4.4.1 *Impeller Clearance Adjustment.* All pumps shall have provisions for adjustment of axial clearance between the leading edge of the impeller and casing. This adjustment shall be made by a precision microdial adjustment at the outboard bearing housing, which moves the impeller forward toward the suction wall of the casing.
	4.5 *Shafts.* Shafts shall be suitable for hook-type sleeve. Shaft material shall be (SAE 1045 steel on Duron and 316 stainless steel pumps) or (AISI 316 stainless steel on CD-4MCu pumps and #20 stainless steel pumps). Shaft deflection shall not exceed .005 at the vertical centerline of the impeller.
	4.6 *Shaft Sleeve.* Renewable hook-type shaft sleeve that extends through the stuffing box and gland shall be provided. Shaft sleeve shall be (316 stainless steel), (#20 stainless steel) or (XH-800 Ni-chrome-boron coated 316 stainless steel with coated surface hardness of approximately 800 Brinnell).
	4.7 *Stuffing Box.* Stuffing box shall be suitable for packing, single (inside or outside) or double-inside mechanical seal without modifications. Stuffing box shall be accurately centered by machined rabbit fits on case and frame adapter.
	4.7.1 *Packed Stuffing Box.* The standard packed stuffing box shall consist of five rings of graphited nonasbestos packing; a stainless steel packing base ring in the bottom of the box to prevent extrusion of the packing past the throat; a Teflon seal cage, and a two-piece 316 stainless steel packing gland to ensure even pressure on the packing. Ample space shall be provided for repacking the stuffing box.
	4.7.1.1 *Lubrication-Packed Stuffing Box.* A tapped hole shall be provided in the stuffing box directly over the seal cage for lubrication and cooling of the packing. Lubrication liquid shall be supplied (from an external source) (through a by-pass line from the pump discharge nozzle).
	4.7.2 *Stuffing Box with Mechanical Seal.* Mechanical seal shall be of the (single inside) (single outside) (double inside) (cartridge) type and (balanced) (unbalanced). Stuffing box is to be (standard) (oversize) (oversize tapered). Suitable space shall be provided in the standard and oversized stuffing box for supplying a (throttle bushing) (dilution control bushing) with single seals. Throttle bushings and dilution control bushings shall be made of (glass-filled Teflon) (a suitable metal material).
	4.7.2.1 *Lubrication—Stuffing Box with Mechanical Seals.* Suitable tapped connections shall be provided to effectively lubricate, cool, flush, quench, etc., as required by the application or recommendations of the mechanical seal manufacturer.
4.8	Bearing Frame and Bearings:
	4.8.1 *Bearing Frame.* Frames shall be equipped with axial radiating fins extending the length of the frame to aid in heat dissipation. Frame shall be provided with ductile iron outboard bearing housing. Both ends of the frame shall be provided with lip-type oil seals and labyrinth-type deflectors of metallic reinforced synthetic rubber to prevent the entrance of contaminants.
	4.8.2 *Bearings.* Pump bearings shall be heavy-duty, antifriction ball-type on both ends. The single row inboard bearing, nearest the impeller, shall be free to float within the frame and shall carry only radial load. The double row outboard bearing (F4-G1 and F4-I1) or duplex angular contact bearing (F4-H1), coupling end, shall be locked in place to carry radial and axial thrust loads. Bearings shall be designed for a minimum life of 20,000 hours in any normal pump operating range.
4.9	*Bearing Lubrication.* Ball bearings shall be oil-mist—lubricated by means of a slinger. The oil slinger shall be mounted on the shaft between the bearings to provide equal lubrication to both bearings. Bulls-eye oil-sight glasses shall be provided on both sides of the frame to provide a positive means of checking the proper oil level from either side of the pump. A tapped and plugged hole shall also be provided in both sides of the frame to mount bottle-type constant-level oilers where desired. A tapped and plugged hole shall be provided on both sides for optional straight-through oil cooling device.
5.0	Baseplate and Coupling:
	5.1 *Baseplate.* Baseplates shall be rigid and suitable for mounting pump and motor. Baseplates shall be of channel steel construction.
	5.2 *Coupling.* Coupling shall be flexible-spacer type. Coupling shall have at least three–and–one-half–inch spacer length for ease of rotating element removal. Both coupling hubs shall be provided with flats 180° apart to facilitate removal of impeller. Coupling shall not require lubrication.*
6.0	Mechanical Modifications Required for High Temperature:
	6.1 *Modifications Required, Temperature Range 250–350°F.* Pumps for operation in this range shall be provided with a water-jacketed stuffing box.
	6.2 *Modifications Required, Temperature Range 351–550°F (Maximum).* Pumps for operation in this range shall be provided with a water-jacketed stuffing box and a water-cooled bearing frame.
7.0	Materials: Pump materials shall be selected to suit the particular service requirements.
	7.1 *Cast Iron—316 SS Fitted.* 15" only; pump shall have cast iron casing and stuffing box cover. 316 SS metal impeller; shaft shall be 1045 steel with 316 SS sleeve.
	7.2 *All Duron Metal.* All pump materials shall be Duron metal. Shaft shall be 1045 steel, with 316 SS sleeve. 316 SS metal impeller optional.
	7.3 *All AISI 316 Stainless Steel.* All pump materials shall be AISI 316 stainless steel. Shaft should be 1045 steel, with 316 SS sleeve.
	7.4 *All #20 Stainless Steel.* All pump materials shall be #20 SS stainless steel. Shaft shall be 316 SS, with #20 SS sleeve.
	7.5 *All CD-4MCu.* All pump materials shall be CD-4MCu. Shaft shall be 316 SS, with #20 SS sleeve.
8.0	Miscellaneous:
	8.1 *Nameplates.* All nameplates and other data plates shall be stainless steel, suitably secured to the pump.
	8.2 *Hardware.* All machine bolts, stud nuts, and capscrews shall be of the hex-head type.
	8.3 *Rotation.* Pump shall have clockwise rotation viewed from its driven end.
	8.4 *Parts Numbering.* Parts shall be completely identified with a numerical system (no alphabetical letters) to facilitate parts inventory control and stocking. Each part shall be properly identified by a separate number, and those parts that are identical shall have the same number to effect minimum spare parts inventory.

*Omit if not applicable.

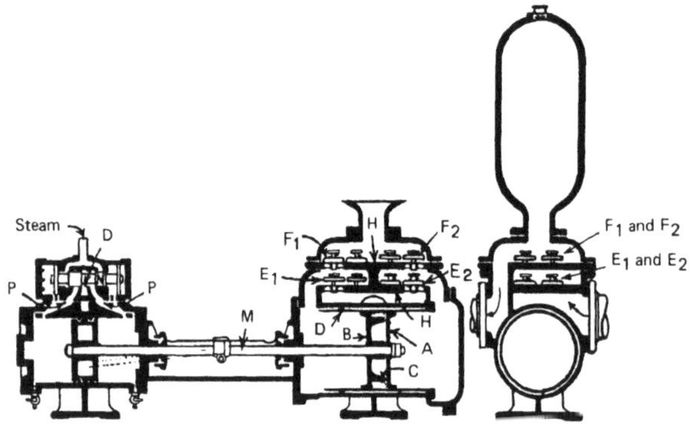

FIG. 10-26 Double-acting steam-driven reciprocating pump.

As shown in Fig. 10-26, reciprocating pumps, except when used for metering service, are frequently provided on the discharge side with gas-charged chambers, the purpose of which is to limit pressure pulsation and to provide a more uniform flow in the discharge line. In many installations, surge chambers are required on the suction side as well. Piping layouts should be studied to determine the most effective size and location. If surge chambers are used, provision should be made to keep the chamber charged with gas. A surge chamber filled with liquid is of no value. A liquid-level gauge is desirable to permit checking the amount of gas in the chamber.

Reciprocating pumps may be of **single-cylinder** or **multicylinder** design. Multicylinder pumps have all cylinders in parallel for increased capacity. Piston-type pumps may be single-acting or double-acting; i.e., pumping may be accomplished from one or both ends of the piston. Plunger pumps are always single-acting. The tabulation in Table 10-12 provides data on the flow variation of reciprocating pumps of various designs.

Piston Pumps There are two ordinary types of piston pumps, simplex double-acting pumps and duplex double-acting pumps.

Simplex Double-Acting Pumps These pumps may be direct-acting (i.e., direct-connected to a steam cylinder) or power-driven (through a crank and flywheel from the crosshead of a steam engine). Figure 10-26 is a direct-acting pump, designed for use at pressures up to 0.690 MPa (100 lbf/in^2). In this figure, the piston consists of disks A and B, with packing rings C between them. A bronze liner for the water cylinder is shown at D. Suction valves are E_1 and E_2. Discharge valves are F_1 and F_2.

Duplex Double-Acting Pumps These pumps differ primarily from those of the simplex type in having two cylinders whose operation is coordinated. They may be direct-acting, steam-driven, or power-driven with crank and flywheel.

A duplex outside-end-packed **plunger pump** with pot valves, of the type used with hydraulic presses and for similar service, is shown in Fig. 10-27. In this drawing, plunger A is direct-connected to rod B, while plunger C is operated from the rod by means of yoke D and tie rods.

Plunger pumps differ from piston pumps in that they have one or more constant-diameter plungers reciprocating through packing glands and displacing liquid from cylinders in which there is considerable radial clearance. They are always single-acting, in the sense that only one end of the plunger is used in pumping the liquid.

TABLE 10-12 Flow Variation of Reciprocating Pumps

Number of cylinders	Single- or double-acting	Flow variation per stroke from mean, percent
Single	Single	+220 to −100
Single	Double	+60 to −100
Duplex	Single	+24.1 to −100
Duplex	Double	+6.1 to −21.5
Triplex	Single and double	+1.8 to −16.9
Quintuplex	Single	+1.8 to −5.2

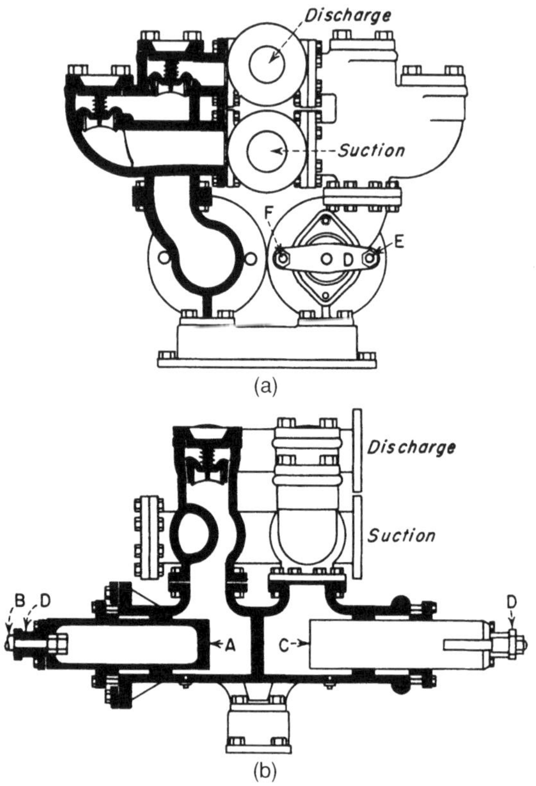

FIG. 10-27 Duplex single-acting plunger pump.

Plunger pumps are available with one, two, three, four, five, or even more cylinders. Simplex and duplex units are often built in a horizontal design. Those with three or more cylinders are usually of vertical design. The driver may be an electric motor, a steam or gas engine, or a steam turbine. This is the common type of **power pump.** An example, arranged for belt drive, is shown in Fig. 10-28 from which the action may be readily traced.

Occasionally plunger pumps are constructed with opposed cylinders and plungers connected by yokes and tie rods; this arrangement, in effect, constitutes a double-acting unit.

Simplex plunger pumps mounted singly or in gangs with a common drive are quite commonly used as **metering** or **proportioning pumps** (Fig. 10-29). Frequently a variable-speed drive or a stroke-adjusting mechanism is provided to vary the flow as desired. These pumps are designed to measure or control the flow of liquid within a deviation of ±2 percent with capacities up to 11.35 m^3/h (50 gal/min) and pressures as high as 68.9 MPa (10,000 lbf/in^2).

Diaphragm Pumps These pumps perform similarly to piston and plunger pumps, but the reciprocating driving member is a flexible diaphragm fabricated of metal, rubber, or plastic. The chief advantage of this arrangement is the elimination of all packing and seals exposed to the liquid being pumped. This, of course, is an important asset for equipment required to handle hazardous or toxic liquids.

A common type of low-capacity diaphragm pump designed for metering service employs a plunger working in oil to actuate a metallic or plastic diaphragm. Built for pressures in excess of 6.895 MPa (1000 lbf/in^2) with flow rates up to about 1.135 m^3/h (5 gal/min) per cylinder, such pumps possess all the characteristics of plunger-type metering pumps with the added advantage that the pumping head can be mounted in a remote (even a submerged) location entirely separate from the drive.

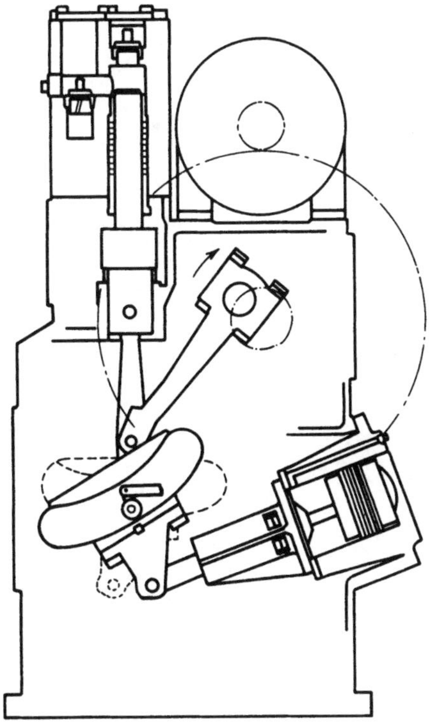

FIG. 10-28 Adrich-Groff variable-stroke power pump. (*Courtesy of Ingersoll-Rand.*)

Figure 10-30 shows a high-capacity 22.7-m^3/h (100-gal/min) pump with actuation provided by a mechanical linkage.

Pneumatically Actuated Diaphragm Pumps (Fig. 10-31) These pumps require no power source other than plant compressed air. They must have a flooded suction, and the pressure is, of course, limited to the available air pressure. Because of their slow speed and large valves, they are well suited to the gentle handling of liquids for which degradation of suspended solids should be avoided.

A major consideration in the application of diaphragm pumps is the realization that diaphragm failure will probably occur eventually. The consequences of such failure should be realistically appraised before selection, and maintenance procedures should be established accordingly.

Rotary Pumps In rotary pumps the liquid is displaced by rotation of one or more members within a stationary housing. Because internal clearances, although minute, are a necessity in all but a few special types, capacity decreases somewhat with increasing pump

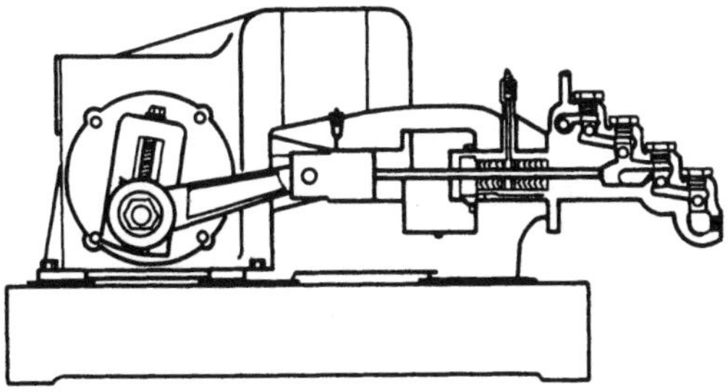

FIG. 10-29 Plunger-type metering pump. (*Courtesy of Milton Roy Co.*)

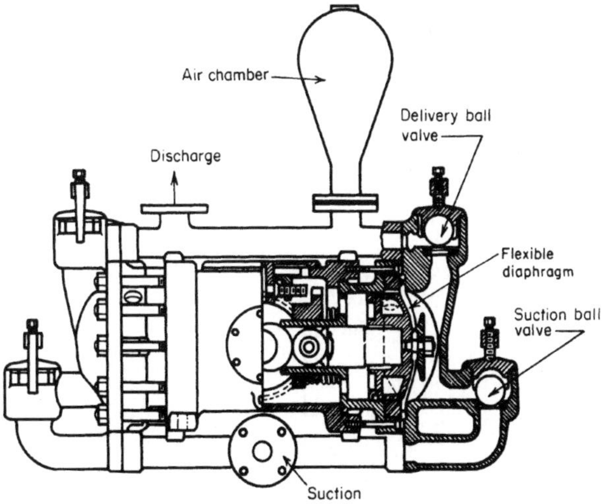

FIG. 10-30 Mechanically actuated diaphragm pump.

differential pressure. Therefore, these pumps are not truly positive-displacement pumps. However, for many other reasons they are considered as such.

The selection of materials of construction for rotary pumps is critical. The materials must be corrosion-resistant, compatible when one part is running against another, and capable of some abrasion resistance.

Gear Pumps When two or more impellers are used in a rotary-pump casing, the impellers will take the form of toothed-gear wheels as in Fig. 10-32, of helical gears, or of lobed cams. In each case, these impellers rotate with extremely small clearance between them and between the surfaces of the impellers and the casing. In Fig. 10-32, the two toothed impellers rotate as indicated by the arrows; the suction connection is at the bottom. The pumped liquid flows into the spaces between the impeller teeth as these cavities pass the suction opening. The liquid is then carried around the casing to the discharge opening, where it is forced out of the impeller teeth mesh. The arrows indicate this flow of liquid.

Rotary pumps are available in two general classes, interior-bearing and exterior-bearing. The **interior-bearing type** is used for handling

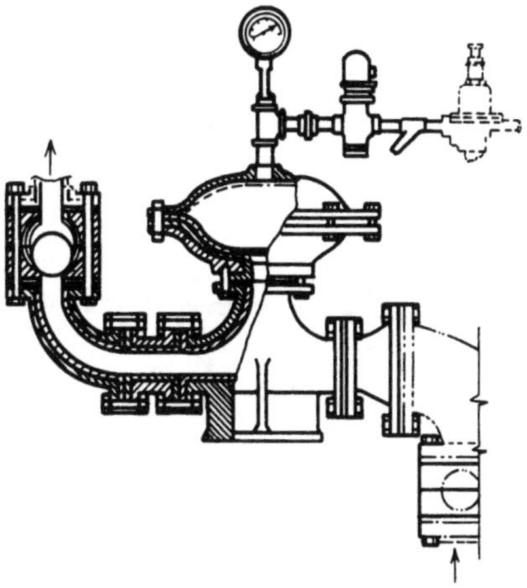

FIG. 10-31 Pneumatically actuated diaphragm pump for slurry service. (*Courtesy of Dorr-Oliver Inc.*)

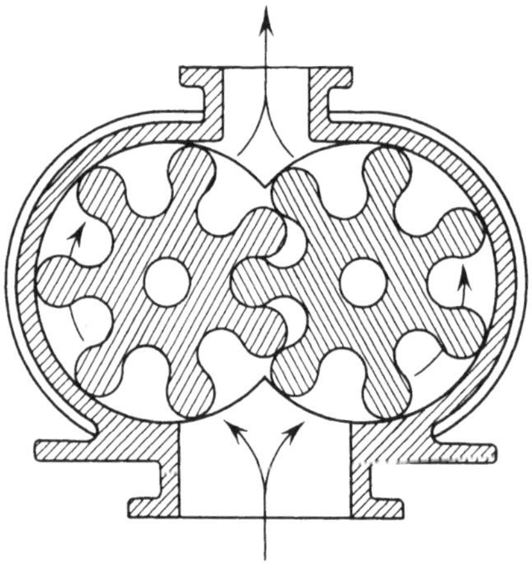

FIG. 10-32 Positive-displacement gear-type rotary pump.

liquids of a lubricating nature, and the **exterior-bearing type** is used with nonlubricating liquids. The interior-bearing pump is lubricated by the liquid being pumped, and the exterior-bearing type is oil-lubricated.

The use of spur gears in gear pumps will produce in the discharge pulsations having a frequency equivalent to the number of teeth on both gears multiplied by the speed of rotation. The amplitude of these disturbances is a function of tooth design. The pulsations can be reduced markedly by the use of rotors with helical teeth. This in turn introduces end thrust, which can be eliminated by the use of double-helical or herringbone teeth.

Screw Pumps A modification of the helical gear pump is the screw pump. Both gear and screw pumps are positive-displacement pumps. Figure 10-33 illustrates a two-rotor version in which the liquid is fed to either the center or the ends, depending upon the direction of rotation, and progresses axially in the cavities formed by the meshing threads or teeth. In three-rotor versions, the center rotor is the driving member while the other two are driven. Figure 10-34 shows still another arrangement, in which a metal rotor of unique design rotates without clearance in an elastomeric stationary sleeve.

Screw pumps, because of multiple dams that reduce slip, are well adapted for producing higher pressure rises, for example, 6.895 MPa (1000 lbf/in^2), especially when handling viscous liquids such as heavy oils. The all-metal pumps are generally subject to the same limitations on handling abrasive solids as conventional gear pumps. In addition, the wide bearing spans usually demand that the liquid have considerable lubricity to prevent metal-to-metal contact.

Among the liquids handled by rotary pumps are mineral oils, vegetable oils, animal oils, greases, glucose, viscose, molasses, paints, varnish, shellac, lacquers, alcohols, catsup, brine, mayonnaise, sizing, soap, tanning liquors, vinegar, and ink. Some screw-type units are specially designed for the gentle handling of large solids suspended in the liquid.

Fluid-Displacement Pumps In addition to pumps that depend on the mechanical action of pistons, plungers, or impellers to move the liquid, other devices for this purpose employ displacement by a secondary fluid. This group includes air lifts and acid eggs.

The air lift is a device for raising liquid by means of compressed air. In the past it was widely used for pumping wells, but it has been less widely used since the development of efficient centrifugal pumps. It operates by introducing compressed air into the liquid near the bottom of the well. The air-and-liquid mixture, being lighter than liquid alone, rises in the well casing. The advantage of this system of pumping lies in the fact that there are no moving parts in the well. The pumping equipment is an air compressor, which can be located on the surface.

A simplified sketch of an air lift for this purpose is shown in Fig. 10-35. Ingersoll-Rand has developed empirical information on air-lift performance which is available upon request.

An important application of the gas-lift principle involves the extraction of oil from wells. There are several references to both practical and theoretical work involving gas lift performance and related problems. Recommended sources are American Petroleum Institute, *Drilling and Production Practices,* 1952, pp. 257–317, and 1939, p. 266; *Trans. Am. Soc. Mining Metall. Eng.,* **92,** 296–313 (1931), **103,** 170–186 (1933), **118,** 56–70 (1936), **192,** 317–326 (1951), **189,** 73–82 (1950), and **198,** 271–278 (1953); *Trans Am. Soc. Mining Metall.,* and *Pet. Eng.,* **213** (1958), and **207,** 17–24 (1956); and *Univ. Wisconsin Bull., Eng. Ser.,* **6,** no. 7 (1911, reprinted 1914).

An **acid egg,** or **blowcase,** consists of an egg-shaped container which can be filled with a charge of liquid that is to be pumped. This container is fitted with an inlet pipe for the charge, an outlet pipe for the discharge, and a pipe for the admission of compressed air or gas, as illustrated in Fig. 10-36. Pressure of air or gas on the surface of the liquid forces it out of the discharge pipe. Such pumps can be hand-operated or arranged for semiautomatic or automatic operation.

CENTRIFUGAL PUMPS

The centrifugal pump is the type most widely used in the chemical industry for transferring liquids of all types—raw materials, materials in manufacture, and finished products—as well as for general services of water supply, boiler feed, condenser circulation, condensate return, etc. These pumps are available through a vast range of sizes, in capacities from 0.5 m^3/h to 2×10^4 m^3/h (2 gal/min to 10^5 gal/min), and for discharge heads (pressures) from a few meters to approximately 48 MPa (7000 lbf/in^2). The size and type best suited to a particular application can be determined only by an engineering study of the problem.

The primary advantages of a centrifugal pump are simplicity, low first cost, uniform (nonpulsating) flow, small floor space, low maintenance expense, quiet operation, and adaptability for use with a motor or a turbine drive.

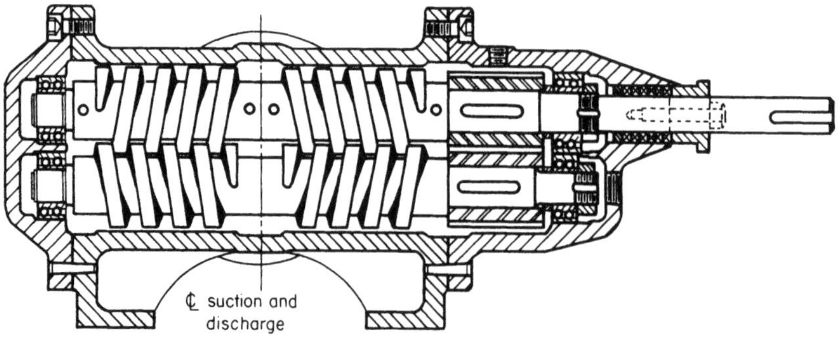

FIG. 10-33 Two-rotor screw pump. (*Courtesy of Warren Quimby Pump Co.*)

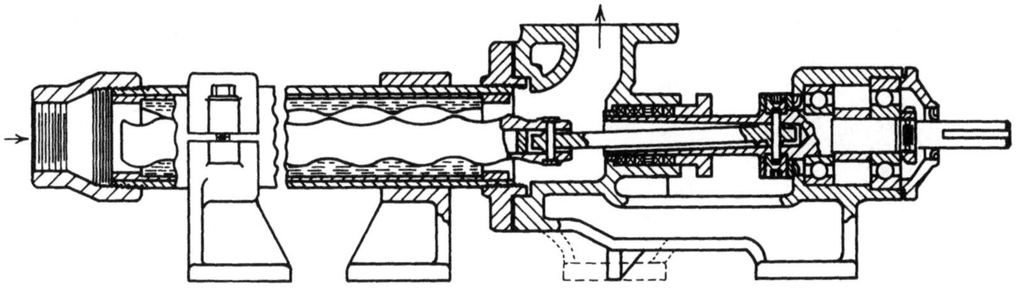

FIG. 10-34 Single-rotor screw pump with an elastomeric lining. (*Courtesy of Moyno Pump Division, Robbins & Myers, Inc.*)

A centrifugal pump, in its simplest form, consists of an impeller rotating within a casing. The **impeller** consists of a number of blades, either open or shrouded, mounted on a shaft that projects outside the casing. Its axis of rotation may be either horizontal or vertical, to suit the work to be done. **Closed-type,** or **shrouded,** impellers are generally the most efficient. **Open-** or **semiopen-type** impellers are used for viscous liquids or for liquids containing solid materials and on many small pumps for general service. Impellers may be of the **single-suction** or the **double-suction** type—single if the liquid enters from one side, double if it enters from both sides.

Casings There are three general types of casings, but each consists of a chamber in which the impeller rotates, provided with inlet and exit for the liquid being pumped. The simplest form is the **circular casing,** consisting of an annular chamber around the impeller; no attempt is made to overcome the losses that will arise from eddies and shock when the liquid leaving the impeller at relatively high velocities enters this chamber. Such casings are seldom used.

Volute casings take the form of a spiral increasing uniformly in cross-sectional area as the outlet is approached. The volute efficiently converts the velocity energy imparted to the liquid by the impeller into pressure energy.

A third type of casing is used in **diffuser-type** or turbine pumps. In this type, **guide vanes** or **diffusers** are interposed between the impeller discharge and the casing chamber. Losses are kept to a minimum in a well-designed pump of this type, and improved efficiency is obtained over a wider range of capacities. This construction is often used in multistage high-head pumps.

Action of a Centrifugal Pump Briefly, the action of a centrifugal pump may be shown by Fig. 10-37. Power from an outside source is applied to shaft *A*, rotating the impeller *B* within the stationary casing *C*. The blades of the impeller in revolving produce a reduction in pressure at the entrance or eye of the impeller. This causes liquid to flow into the impeller from the suction pipe *D*. This liquid is forced outward along the blades at increasing tangential velocity. The velocity head it has acquired when it leaves the blade tips is changed to pressure head as the liquid passes into the volute chamber and thence out the discharge *E*.

Centrifugal Pump Characteristics Figure 10-38 shows a typical characteristic curve of a centrifugal pump. It is important to note that at any fixed speed the pump will operate along this curve and at no other points. For instance, on the curve shown, at 45.5 m^3/h (200 gal/min) the pump will generate 26.5-m (87-ft) head. If the head is increased to 30.48 m (100 ft), 27.25 m^3/h (120 gal/min) will be delivered. It is not possible to reduce the capacity to 27.25 m^3/h (120 gal/min) at 26.5-m (87-ft) head unless the discharge is throttled so that 30.48 m (100 ft) is actually generated within the pump. On pumps with variable-speed drivers such as steam turbines, it is possible to change the characteristic curve, as shown by Fig. 10-39.

As shown in Eq. (10-50), the head depends upon the velocity of the fluid, which in turn depends upon the capability of the impeller

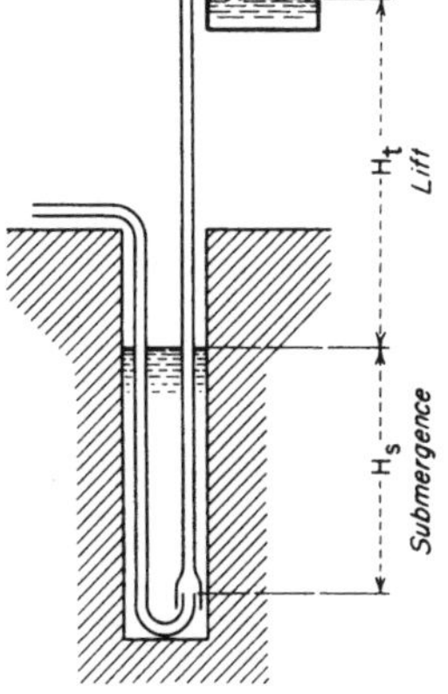

FIG. 10-35 Simplified sketch of an air lift, showing submergence and total head.

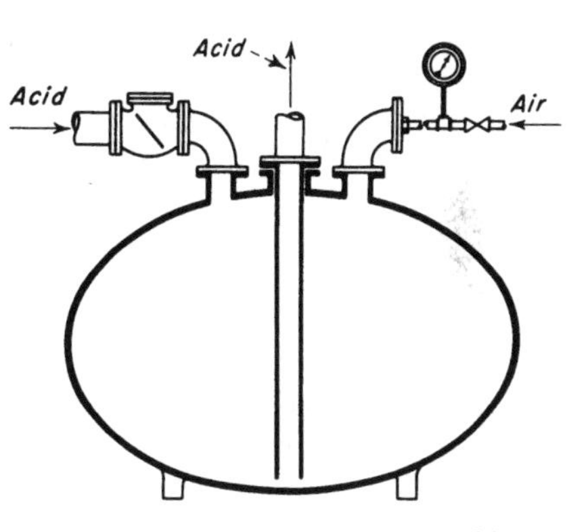

FIG. 10-36 A form of acid egg. External controls required for automatic operation are not shown.

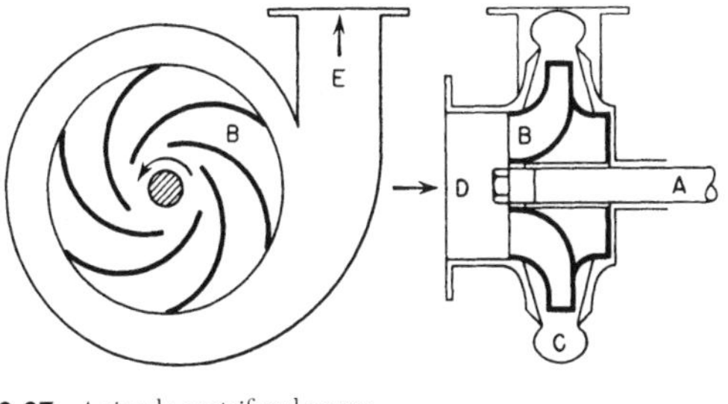

FIG. 10-37 A simple centrifugal pump.

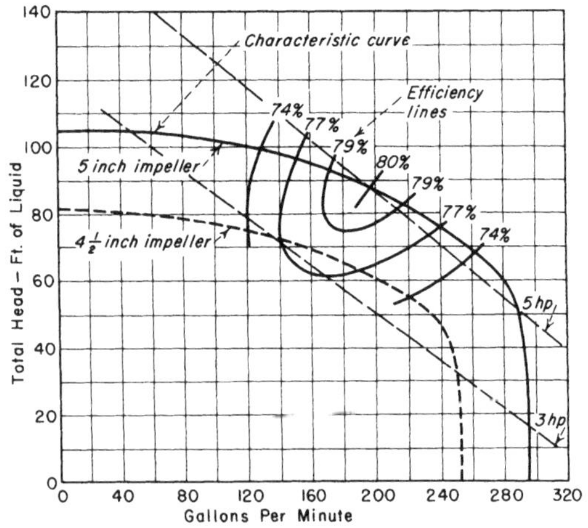

FIG. 10-38 Characteristic curve of a centrifugal pump operating at a constant speed of 3450 r/min. To convert gallons per minute to cubic meters per hour, multiply by 0.2271; to convert feet to meters, multiply by 0.3048; to convert horsepower to kilowatts, multiply by 0.746; and to convert inches to centimeters, multiply by 2.54.

to transfer energy to the fluid. This is a function of the fluid viscosity and the impeller design. It is important to remember that the head produced will be the same for any liquid of the same viscosity. The pressure rise, however, will vary in proportion to the specific gravity.

For quick pump selection, manufacturers often give the most essential performance details for a whole range of pump sizes. Figure 10-40 shows typical performance data for a range of process pumps based on suction and discharge pipes and impeller diameters. The performance data consists of pump flow rate and head. Once a pump meets a required specification, then a more detailed performance data for the particular pump can be easily found based on the curve reference number. Figure 10-41 shows a more detailed pump performance curve that includes, in addition to pump head and flow, the brake horsepower required, NPSH required, number of vanes, and pump efficiency for a range of impeller diameters.

If detailed manufacturer-specified performance curves are not available for a different size of the pump or operating condition, a best estimate of the off-design performance of pumps can be obtained through similarity relationship or the affinity laws. These are:

1. Capacity (Q) is proportional to impeller rotational speed (N).
2. Head (h) varies as square of the impeller rotational speed.
3. Brake horsepower (BHP) varies as the cube of the impeller rotational speed.

These equations can be expressed mathematically and appear in Table 10-13.

System Curves In addition to the pump design, the operational performance of a pump depends upon factors such as the downstream load characteristics, pipe friction, and valve performance. Typically, head and flow follow the following relationship:

$$\frac{(Q_2)^2}{(Q_1)^2} = \frac{h_2}{h_1} \tag{10-60}$$

where the subscript 1 refers to the design condition and 2 to the actual conditions. The above equation indicates that head will change as a square of the water flow rate.

Figure 10-42 shows the schematic of a pump, moving a fluid from tank A to tank B, both of which are at the same level. The only force that the pump has to overcome in this case is the pipe friction, variation of which with fluid flow rate is also shown in the figure. On the other for the use shown in Fig. 10-43, the pump in addition to pipe friction should overcome head due to difference in elevation between tanks A and B. In this case, elevation head is constant, whereas the head required to overcome friction depends on the flow rate. Figure 10-44 shows the pump performance requirement of a valve opening and closing.

Pump Selection One of the parameters that is extremely useful in selecting a pump for a particular application is specific speed N_s. Specific speed of a pump can be evaluated based on its design speed, flow, and head:

$$N_s = \frac{NQ^{1/2}}{H^{3/4}} \tag{10-61}$$

where N = rpm, Q is flow rate in gpm, and H is head in ft·lb$_f$/lbm.

Specific speed is a parameter that defines the speed at which impellers of geometrically similar design have to be run to discharge one gallon per minute against a one-foot head. In general, pumps with a low specific speed have a low capacity and high specific speed, high capacity. Specific speeds of different types of pumps are shown in Table 10-14 for comparison.

Another parameter that helps in evaluating the pump suction limitations, such as cavitation, is suction-specific speed.

$$S = \frac{NQ^{1/2}}{(\text{NPSH})^{3/4}} \tag{10-62}$$

Typically, for single-suction pumps, suction-specific speed above 11,000 is considered excellent. Below 7000 is poor and 7000–9000 is of an average design. Similarly, for double-suction pumps, suction-specific speed above 14,000 is considered excellent, below 7000 is poor, and 9000–11,000 is average.

Figure 10-45 shows the schematic of specific-speed variation for different types of pumps. The figure clearly indicates that, as the specific speed increases, the ratio of the impeller outer diameter D_1 to inlet or eye diameter D_2 decreases, tending to become unity for pumps of axial-flow type.

Typically, axial flow pumps are of high flow and low head type and have a high specific speed. On the other hand, purely radial pumps are of high head and low flow rate capability and have a low specific speed. Obviously, a pump with a moderate flow and head has an average specific speed.

A typical pump selection chart such as shown in Fig. 10-46 calculates the specific speed for a given flow, head, and speed requirements. Based on the calculated specific speed, the optimal pump design is indicated.

Process Pumps This term is usually applied to single-stage pedestal-mounted units with single-suction overhung impellers and with a single packing box. These pumps are ruggedly designed for

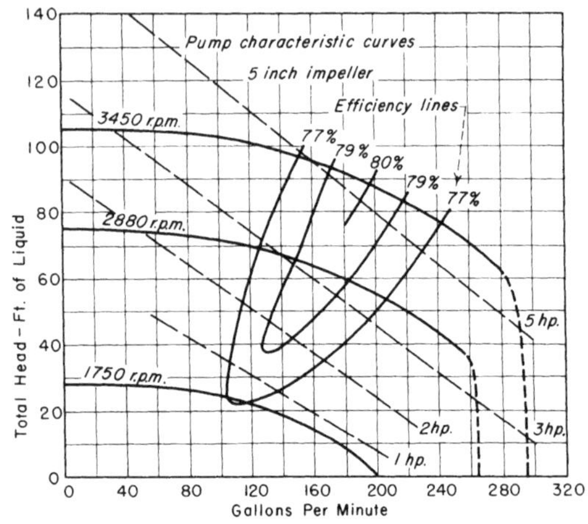

FIG. 10-39 Characteristic curve of a centrifugal pump at various speeds. To convert gallons per minute to cubic meters per hour, multiply by 0.2271; to convert feet to meters, multiply by 0.3048; to convert horsepower to kilowatts, multiply by 0.746; and to convert inches to centimeters, multiply by 2.54.

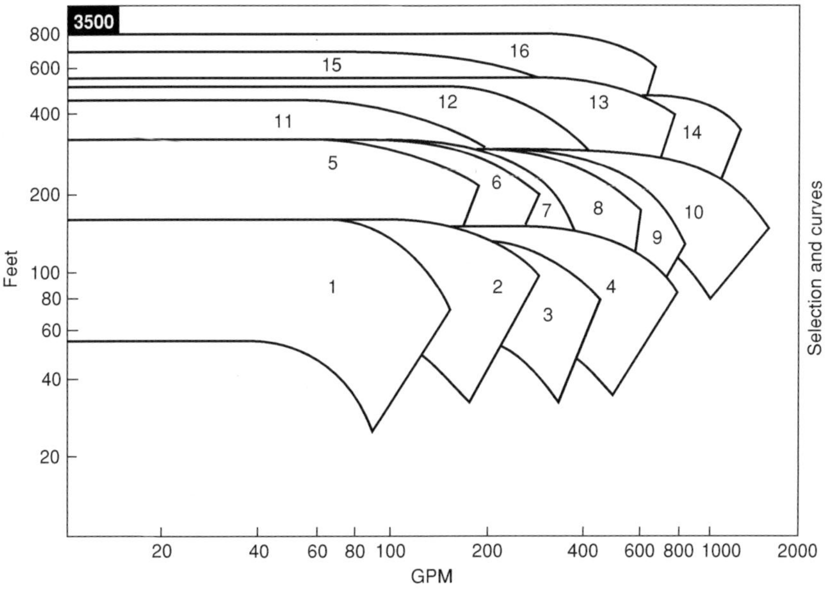

Range No.	Pump	Curve	Range No.	Pump	Curve
1	1.5 × 6 E 731 Plus	A-8475	9	4 × 3 × 8.5 731 Plus	A-8969
2	3 × 1.5 × 6 731 Plus	A-6982	10	6 × 4 × 8.5 731 Plus	A-8547
3	3 × 2 × 6 731 Plus	A-8159	11	2 × 1 × 10 E 731 Plus	A-8496
4	4 × 3 × 6 731 Plus	A-8551	12	3 × 1.5 × 11 E 731 Plus	A-8543
5	1.5 × 1 × 8 731 Plus	A-8153	13	3 × 2 × 11 731 Plus	A-8456
6	3 × 1.5 × 8 731 Plus	A-8155	14	4 × 3 × 11 731 Plus	A-7342
7	3 × 1.5 × 8.5 E 731 Plus	A-8529	15	3 × 1.5 × 13 E 731 Plus	A-8492
8	3 × 2 × 8.5 E 731 Plus	A-8506	16	3 × 2 × 13 731 Plus	A-7338

FIG. 10-40 Performance curves for a range of open impeller pumps.

ease in dismantling and accessibility, with mechanical seals or packing arrangements, and are built especially to handle corrosive or otherwise difficult-to-handle liquids.

Specifically but not exclusively for the chemical industry, most pump manufacturers now build to national standards **horizontal and vertical process pumps.** ASME Standards B73.1—2001 and B73.2—2003 apply to the horizontal (Fig. 10-47) and vertical in-line (Fig. 10-48) pumps, respectively.

The horizontal pumps are available for capacities up to 900 m^3/h (4000 gal/min); the vertical in-line pumps, for capacities up to 320 m^3/h (1400 gal/min). Both horizontal and vertical in-line pumps are available for heads up to 120 m (400 ft). The intent of each ANSI specification is that pumps from all vendors for a given nominal capacity and total dynamic head at a given rotative speed shall be dimensionally interchangeable with respect to mounting, size, and location of suction and discharge nozzles, input shaft, base plate, and foundation bolts.

The vertical in-line pumps, although relatively new additions, are finding considerable use in chemical and petrochemical plants in the United States. An inspection of the two designs will make clear the relative advantages and disadvantages of each.

Chemical pumps are available in a variety of materials. Metal pumps are the most widely used. Although they may be obtained in iron, bronze, and iron with bronze fittings, an increasing number of pumps of ductile-iron, steel, and nickel alloys are being used. Pumps are also available in glass, glass-lined iron, carbon, rubber, rubber-lined metal, ceramics, and a variety of plastics, such units usually being employed for special purposes.

Sealing the Centrifugal Chemical Pump Although detailed treatment of **shaft seals** is presented in the subsection "Sealing of Rotating Shafts," it is appropriate to mention here the special problems of sealing centrifugal chemical pumps. Current practice demands that packing boxes be designed to accommodate both packing and mechanical seals. With either type of seal, one consideration is of paramount importance in chemical service: the liquid present at the sealing surfaces must be free of solids. Consequently, it is necessary to provide a secondary compatible liquid to flush the seal or packing whenever the process liquid is not absolutely clean.

The use of **packing** requires the continuous escape of liquid past the seal to minimize and to carry away the frictional heat developed. If the effluent is toxic or corrosive, quench glands or catch pans are usually employed. Although packing can be adjusted with the pump operating, leaking mechanical seals require shutting down the pump to correct the leak. Properly applied and maintained **mechanical seals** usually show no visible leakage. In general, owing to the more effective performance of mechanical seals, they have gained almost universal acceptance.

Double-Suction Single-Stage Pumps These pumps are used for general water-supply and circulating service and for chemical service when liquids that are noncorrosive to iron or bronze are being handled. They are available for capacities from about 5.7 m^3/h (25 gal/min) up to as high as 1.136×10^4 m^3/h (50,000 gal/min) and heads up to 304 m (1000 ft). Such units are available in iron, bronze, and iron with bronze fittings. Other materials increase the cost; when they are required, a standard chemical pump is usually more economical.

Close-Coupled Pumps (Fig. 10-49) Pumps equipped with a built-in electric motor or sometimes steam-turbine-driven (i.e., with pump

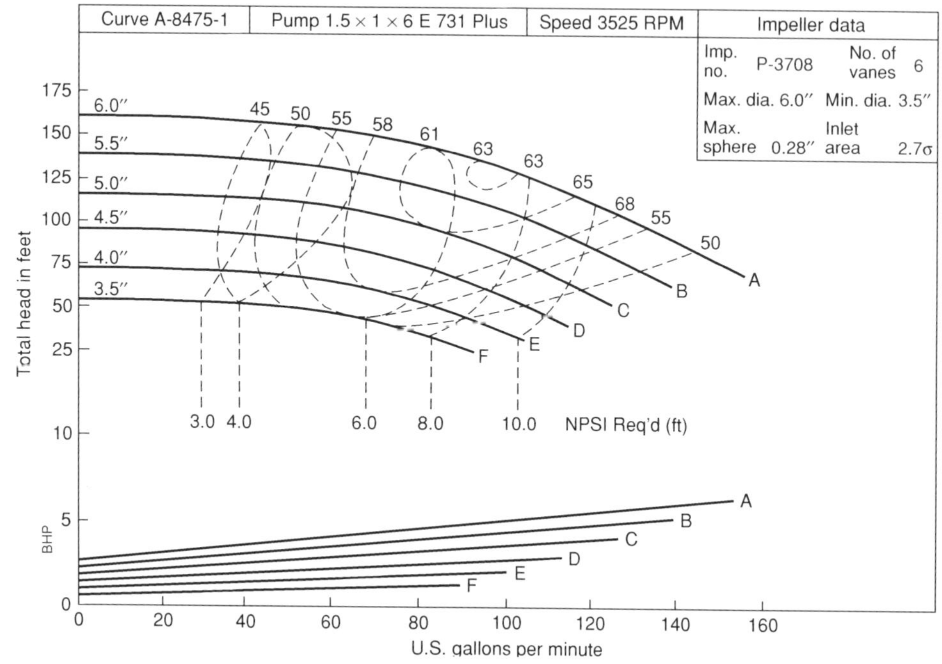

FIG. 10-41 Typical pump performance curve. The curve is shown for water at 85°F. If the specific gravity of the fluid is other than unity, BHP must be corrected.

impeller and driver on the same shaft) are known as close-coupled pumps. Such units are extremely compact and are suitable for a variety of services for which standard iron and bronze materials are satisfactory. They are available in capacities up to about 450 m³/h (2000 gal/min) for heads up to about 73 m (240 ft). Two-stage units in the smaller sizes are available for heads to around 150 m (500 ft).

Canned-Motor Pumps (Fig. 10-50) These pumps command considerable attention in the chemical industry. They are close-coupled units in which the cavity housing the motor rotor and the pump casing are interconnected. As a result, the motor bearings run in the process liquid and all seals are eliminated. Because the process liquid is the bearing lubricant, abrasive solids cannot be tolerated. Standard single-stage canned-motor pumps are available for flows up to 160 m³/h (700 gal/min) and heads up to 76 m (250 ft). Two-stage units are available for heads up to 183 m (600 ft). Canned-motor pumps are being widely used for handling organic solvents, organic heat-transfer liquids, and light oils as well as many clean toxic or hazardous liquids or for installations in which leakage is an economic problem.

Vertical Pumps In the chemical industry, the term **vertical process pump** (Fig. 10-51) generally applies to a pump with a vertical shaft having a length from drive end to impeller of approximately 1 m (3.1 ft) minimum to 20 m (66 ft) or more. Vertical pumps are used as either **wet-pit pumps** (immersed) or **dry-pit pumps** (externally mounted) in conjunction with stationary or mobile tanks containing difficult-to-handle liquids. They have the following advantages: the liquid level is above the impeller, and the pump is thus self-priming; and the shaft seal is above the liquid level and is not wetted by the pumped liquid, which simplifies the sealing task. When no bottom connections are permitted on the tank (a safety consideration for highly corrosive or toxic liquid), the vertical wet-pit pump may be the only logical choice.

These pumps have the following disadvantages: intermediate or line bearings are generally required when the shaft length exceeds about 3 m (10 ft) in order to avoid shaft resonance problems; these

TABLE 10-13 The Affinity Laws

	Constant impeller diameter	Constant impeller speed
Capacity	$\frac{Q_1}{Q_2} = \frac{N_1}{N_2}$	$\frac{Q_1}{Q_2} = \frac{D_1}{D_2}$
Head	$\frac{H_1}{H_2} = \frac{(N_1)^2}{(N_2)^2}$	$\frac{h_1}{h_2} = \frac{(D_1)^2}{(D_2)^2}$
Brake horsepower	$\frac{BHP_1}{BHP_2} = \frac{(N_1)^3}{(N_2)^3}$	$\frac{BHP_1}{BHP_2} = \frac{(P_1)^3}{(P_2)^3}$

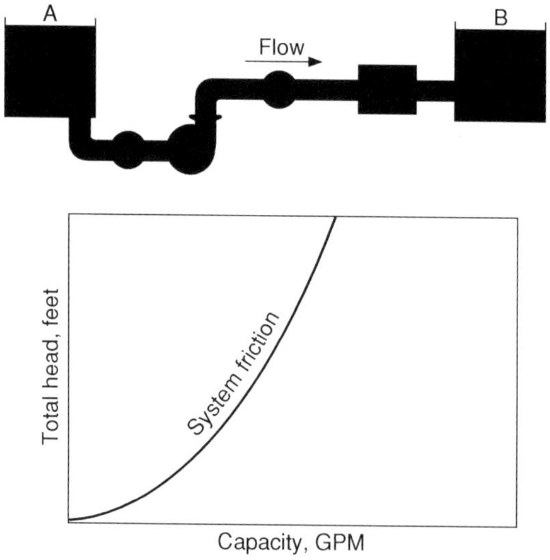

FIG. 10-42 Variation of total head versus flow rate to overcome friction.

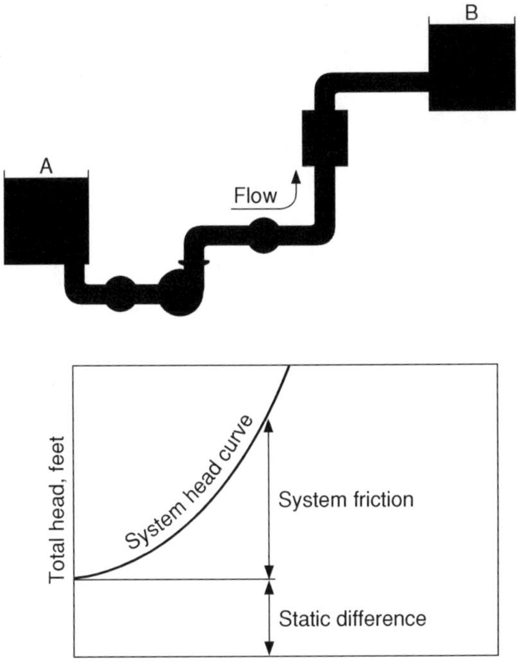

FIG. 10-43 Variation of total head as a function of flow rate to overcome both friction and static head.

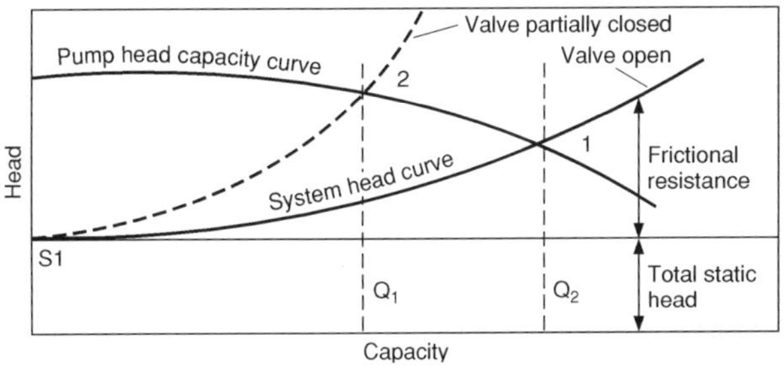

FIG. 10-44 Typical steady-state response of a pump system with a valve fully and partially open.

TABLE 10-14 Specific Speeds of Different Types of Pumps

Pump type	Specific speed range
Below 2,000	Process pumps and feed pumps
2,000–5,000	Turbine pumps
4,000–10,000	Mixed-flow pumps
9,000–15,000	Axial-flow pumps

bearings must be lubricated whenever the shaft is rotating. Since all wetted parts must be corrosion-resistant, low-cost materials may not be suitable for the shaft, column, etc. Maintenance is more costly since the pumps are larger and more difficult to handle.

For abrasive service, vertical cantilever designs requiring no line or foot bearings are available. Generally, these pumps are limited to about a 1-m (3.1-ft) maximum shaft length. Vertical pumps are also used to pump waters to reservoirs. One such application in the Los Angeles water basin has fourteen 4-stage pumps, each pump requiring 80,000 hp to drive them.

Sump Pumps These are small single-stage vertical pumps used to drain shallow pits or sumps. They are of the same general construction as vertical process pumps but are not designed for severe operating conditions.

Multistage Centrifugal Pumps These pumps are used for services requiring heads (pressures) higher than can be generated by a single impeller. All impellers are in series, the liquid passing from one impeller to the next and finally to the pump discharge. The total head then is the summation of the heads of the individual impellers. Deep-well pumps, high-pressure water-supply pumps, boiler-feed pumps, fire pumps, and charge pumps for refinery processes are examples of multistage pumps required for various services.

Multistage pumps may be of the **volute type** (Fig. 10-52), with single- or double-suction impellers (Fig. 10-53), or of the **diffuser type** (Fig. 10-54). They may have horizontally split casings or, for extremely high pressures, 20 to 40 MPa (3000 to 6000 lbf/in^2), vertically split barrel-type exterior casings with inner casings containing diffusers, interstage passages, etc.

PROPELLER AND TURBINE PUMPS

Axial-Flow (Propeller) Pumps (Fig. 10-55) These pumps are essentially very-high-capacity low-head units. Normally they are designed for flows in excess of 450 m^3/h (2000 gal/min) against heads of 15 m (50 ft) or less. They are used to great advantage in closed-loop circulation systems in which the pump casing becomes merely an

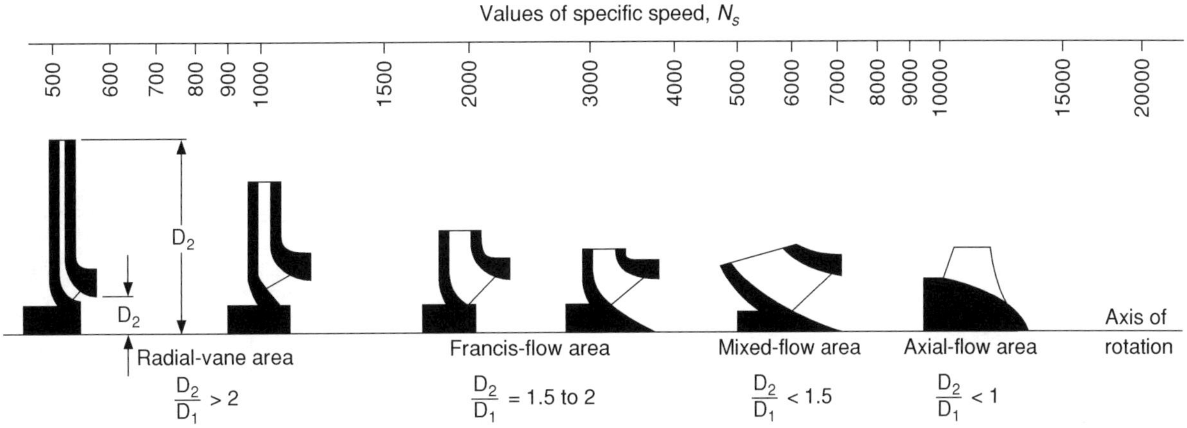

FIG. 10-45 Specific speed variations of different types of pump.

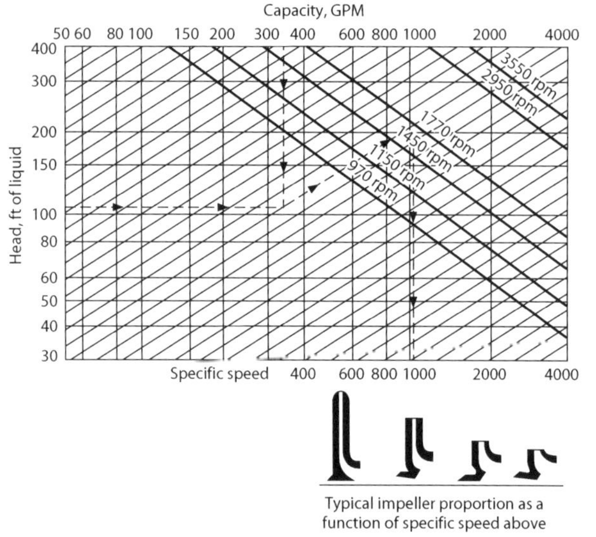

FIG. 10-46 Relationships between specific speed, rotative speed, and impeller proportions (*Worthington Pump Inc.*, Pump World, *vol. 4, no. 2, 1978*).

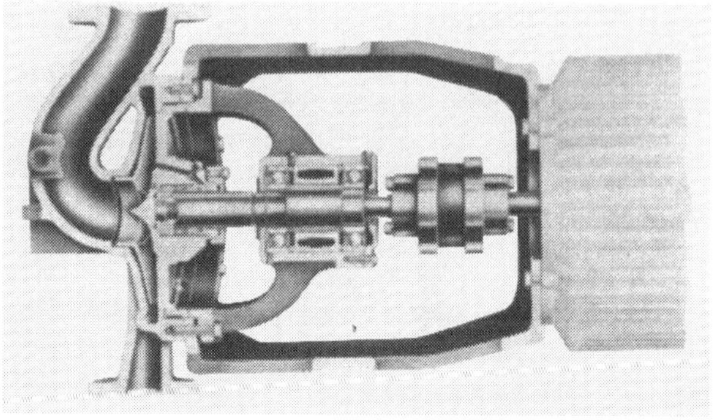

FIG. 10-48 Vertical in-line process pump conforming to ASME standard B73.2—2003. The pump shown is driven by a motor through flexible coupling. Not shown but also conforming to ASME Standard B73.2—2003 are vertical in-line pumps with rigid couplings and with no coupling (impeller-mounted on an extended motor shaft).

elbow in the line. A common installation is for calandria circulation. A characteristic curve of an axial-flow pump is given in Fig. 10-56.

Turbine Pumps The term "turbine pump" is applied to units with mixed-flow (part axial and part centrifugal) impellers. Such units are available in capacities from 20 m^3/h (100 gal/min) upward for heads up to about 30 m (100 ft) per stage. Turbine pumps are usually vertical.

A common form of turbine pump is the vertical pump, which has the pump element mounted at the bottom of a column that serves as the discharge pipe (see Fig. 10-57). Such units are immersed in the liquid to be pumped and are commonly used for wells, condenser circulating water, large-volume drainage, etc. Another form of the pump has a shell surrounding the pumping element which is connected to the intake pipe. In this form, the pump is used on condensate service in power plants and for process work in oil refineries.

Regenerative Pumps Also referred to as turbine pumps because of the shape of the impeller, regenerative pumps employ a combination of mechanical impulse and centrifugal force to produce heads of several hundred meters (feet) at low volumes, usually less than 20 m^3/h (100 gal/min). The impeller, which rotates at high speed with small clearances, has many short radial passages milled on each side at the periphery. Similar channels are milled in the mating surfaces of the casing. Upon entering, the liquid is directed into the impeller passages and proceeds in a spiral pattern around the periphery, passing alternately from the impeller to the casing and receiving successive impulses as it does so. Figure 10-58 illustrates a typical performance-characteristic curve.

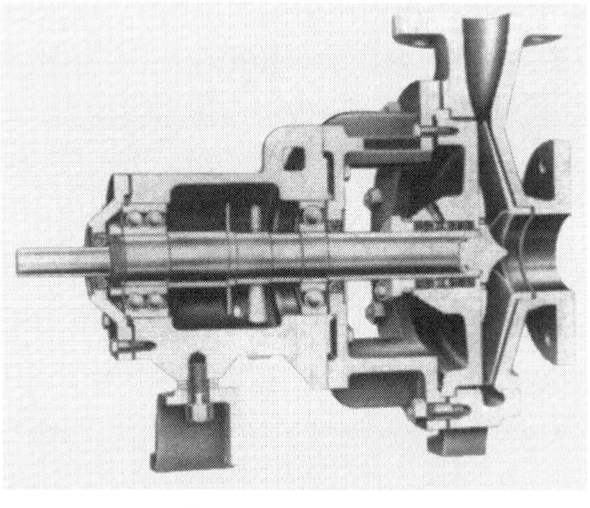

FIG. 10-47 Horizontal process pump conforming to American Society of Mechanical Engineers Standards B73.1—2001.

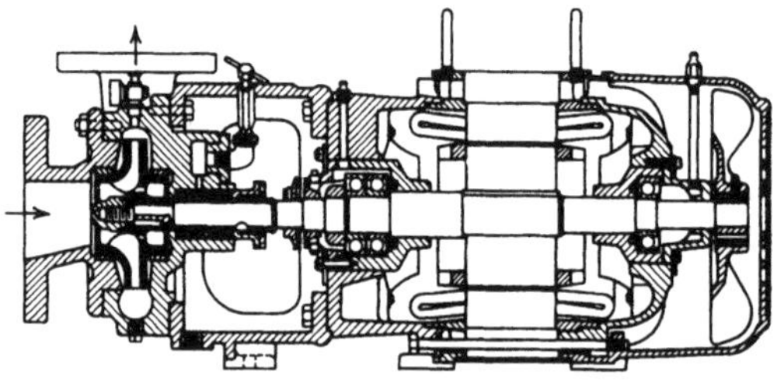

FIG. 10-49 Close-coupled pump.

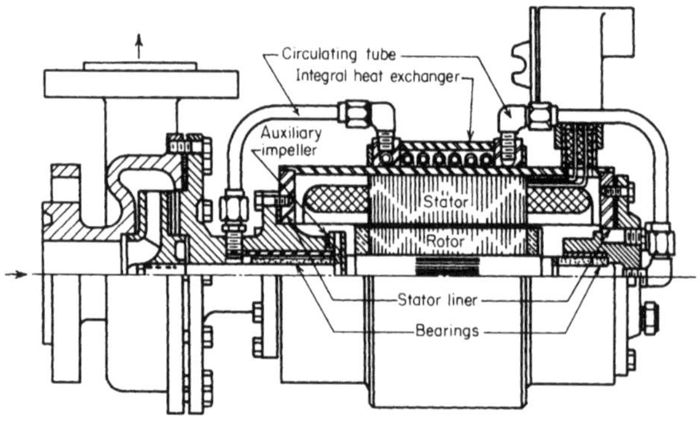

FIG. 10-50 Canned-motor pump. (*Courtesy of Chempump Division, Crane Co.*)

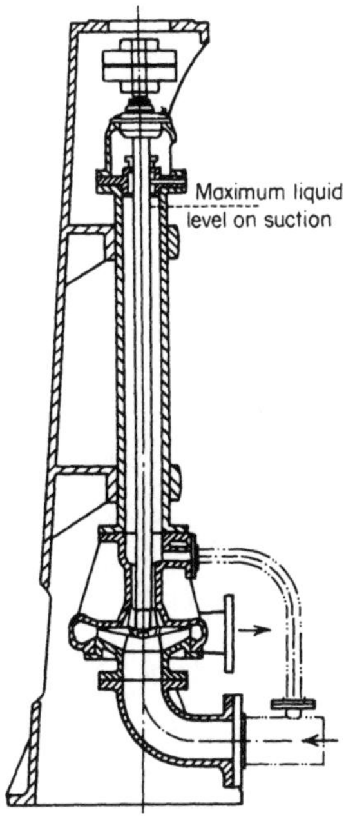

FIG. 10-51 Vertical process pump for dry-pit mounting. (*Courtesy of Lawrence Pumps, Inc.*)

These pumps are particularly useful when low volumes of low-viscosity liquids must be handled at higher pressures than are normally available with centrifugal pumps. Close clearances limit their use to clean liquids. For very high heads, multistage units are available.

JET PUMPS

Jet pumps are a class of liquid-handling device that makes use of the momentum of one fluid to move another.

Ejectors and **injectors** are the two types of jet pumps of interest to chemical engineers. The ejector, also called the siphon, exhauster, or eductor, is designed for use in operations in which the head pumped against is low and is less than the head of the fluid used for pumping. The injector is a special type of jet pump, operated by steam and used for boiler feed and similar services, in which the fluid being pumped is discharged into a space under the same pressure as that of the steam being used to operate the injector.

Figure 10-59 shows a simple design for a jet pump of the ejector type. The pumping fluid enters through the nozzle at the left and passes through the venturi nozzle at the center and out of the discharge opening at the right. As it passes into the venturi nozzle, it develops a suction that causes some of the fluid in the suction chamber to be entrained with the stream and delivered through this discharge.

The efficiency of an ejector or jet pump is low, being only a few percent. The head developed by the ejector is also low except in special types. The device has the disadvantage of diluting the fluid pumped by mixing it with the pumping fluid. In steam injectors for boiler feed and similar services in which the heat of the steam is recovered, efficiency is close to 100 percent.

The simple ejector or siphon is widely used, in spite of its low efficiency, for transferring liquids from one tank to another, for lifting acids, alkalies, or solid-containing liquids of an abrasive nature, and for emptying sumps.

ELECTROMAGNETIC PUMPS

The necessity of circulating liquid-metal heat-transfer media in nuclear-reactor systems has led to development of electromagnetic pumps. All electromagnetic pumps utilize the motor principle: a conductor in a magnetic field, carrying a current which flows at right angles to the direction of the field, has a force exerted on it, the force being mutually perpendicular to both the field and the current. In all electromagnetic pumps, the fluid is the conductor. This force, suitably directed in the fluid, manifests itself as a pressure if the fluid is suitably contained. The field and current can be produced in a number of different ways and the force utilized variously.

Both alternating- and direct-current units are available. While dc pumps (Fig. 10-60) are simpler, their high-current requirement is a definite limitation; ac pumps can readily obtain high currents by making use of transformers. Multipole induction ac pumps have been

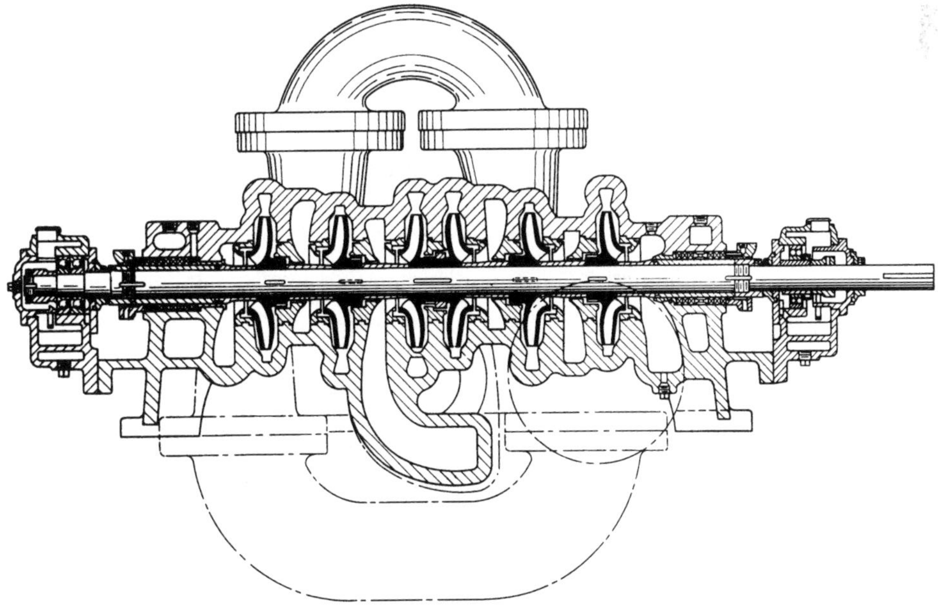

FIG. 10-52 Six-stage volute-type pump.

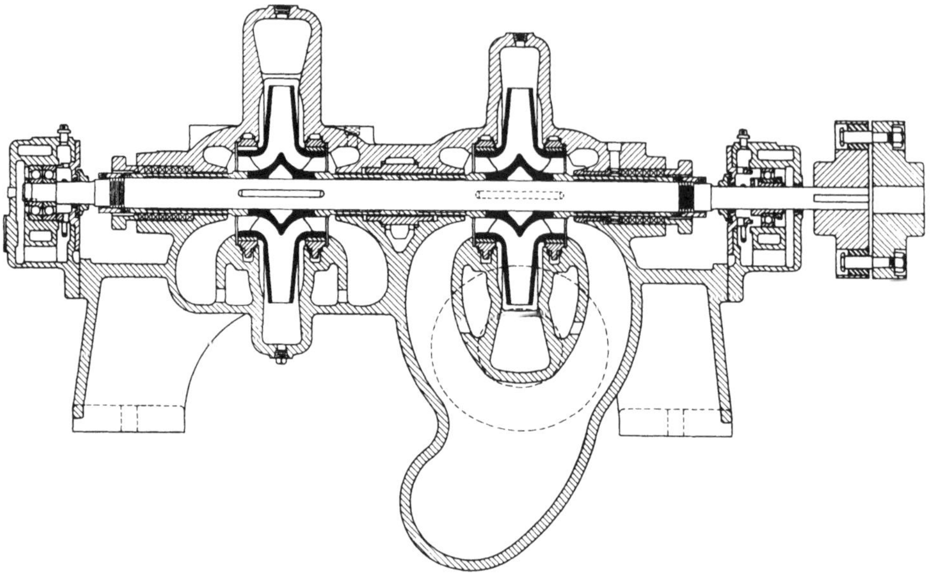

FIG. 10-53 Two-stage pump having double-suction impellers.

built in helical and linear configurations. Helical units are effective for relatively high heads and low flows, while linear induction pumps are best suited to large flows at moderate heads. Electromagnetic pumps are available for flow rates up to 2.271×10^3 m^3/h (10,000 gal/min), and pressures up to 2 MPa (300 lbf/in^2) are practical. Performance characteristics resemble those of centrifugal pumps.

PUMP DIAGNOSTICS

Pump problems vary over a large range depending on the type of pumps and the usage of the pumps. They can be classified in the following manner by the pump type and the service:

1. *Positive-displacement pumps—reciprocating pumps* problems can be classified into the following categories:

a. Compressor valve problems: plate valves, feather valves, concentric disk valves, relief valves

b. Piston and rod assembly: piston rings, cylinder chatter, cylinder cooling, piston-rod packing

c. Lubrication system

2. *Positive displacement pumps—gear-type and roots-type* problems can be classified into the following categories:

a. Rotor dynamic problems: vibration problems, gear problems or roots rotor problems, bearing and seal problems

b. Lubrication systems

3. *Continuous flow pumps such as centrifugal pumps* problems can be classified into the following categories:

a. Cavitation

b. Capacity flow

c. Motor overload

d. Impeller

e. Bearings and seals

f. Lubrication systems

Table 10-15 classifies different types of centrifugal pump-related problems, their possible causes, and corrective actions that can be taken to solve some of the more common issues. These problems in the table are classified into three major categories for these type of pumps:

1. Cavitation
2. Capacity flow
3. Motor overload

The use of vibration monitoring to diagnose pump and compressor problems is discussed at the end of the subsection on compressor problems.

COMPRESSORS

A compressor is a device which pressurizes a working fluid. One of the basic purposes of using a compressor is to compress the fluid and to

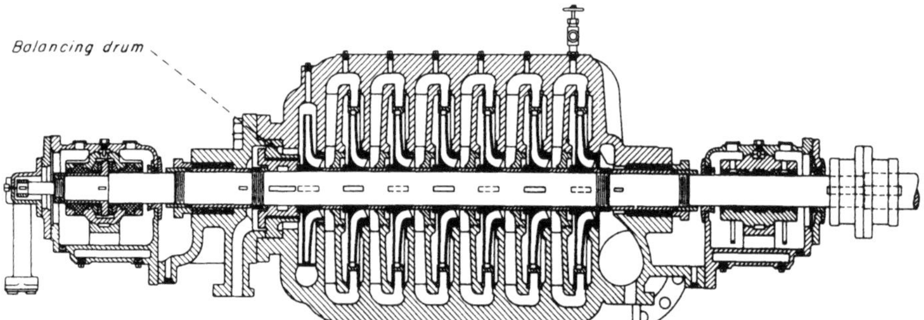

FIG. 10-54 Seven-stage diffuser-type pump.

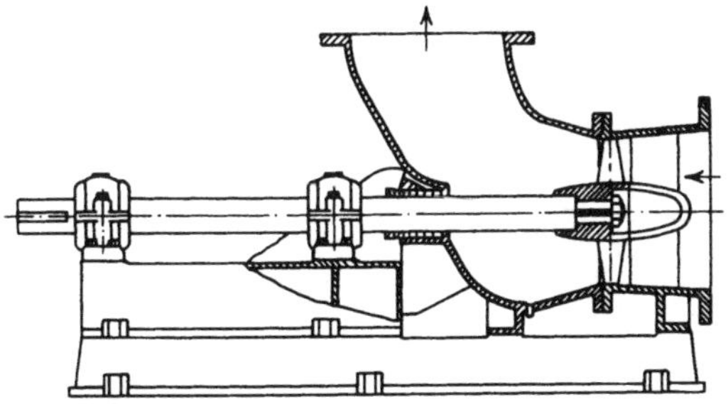

FIG. 10-55 Axial-flow elbow-type propeller pump. (*Courtesy of Lawrence Pumps, Inc.*)

deliver it at a pressure higher than its original pressure. Compression is required for a variety of purposes, some of which are listed below:

1. To provide air for combustion
2. To transport process fluid through pipelines
3. To provide compressed air for driving pneumatic tools
4. To circulate process fluid within a process

Different types of compressors are shown in Fig. 10-61. Positive-displacement compressors are used for intermittent flow in which successive volumes of fluid are confined in a closed space to increase their pressures. Rotary compressors provide continuous flow. In rotary compressors, rapidly rotating parts (impellers) accelerate fluid to a high speed; this velocity is then converted to additional pressure by gradual deceleration in the diffuser or volute which surrounds the impeller. Positive-displacement compressors can be further classified as either reciprocating or rotary type, as shown in Fig. 10-61. The reciprocating compressor has a piston having a reciprocating motion within a cylinder. The rotary positive-displacement compressors have rotating elements whose positive action results in compression and displacement. The rotary positive-displacement compressors can be further subdivided into sliding vane, liquid piston, straight lobe, and helical lobe compressors. The continuous flow compressors (Fig. 10-61) can be classified as either dynamic compressors or ejectors. Ejectors entrain the in-flowing fluid by using a high-velocity gas or steam jet and then convert the velocity of the mixture to pressure in a diffuser. The dynamic compressors have rotating elements, which accelerate the in-flowing fluid, and convert the velocity head to pressure head, partially in the rotating elements and partially in the stationary diffusers or blade. The dynamic compressors can be further subdivided into centrifugal, axial-flow, and mixed-flow compressors. The main flow of gas in the centrifugal compressor is radial. The flow of gas in an axial compressor is axial, and the mixed-flow compressor combines some characteristics of both centrifugal and axial compressors.

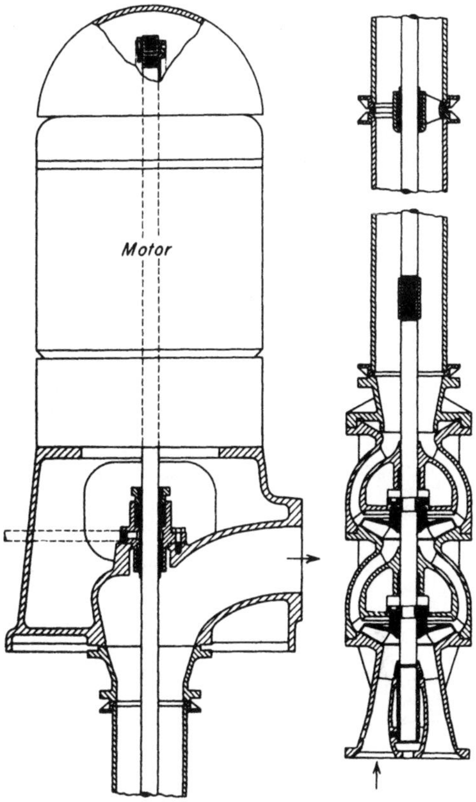

FIG. 10-57 Vertical multistage turbine, or mixed-flow, pump.

It is not always obvious what type of compressor is needed for an application. Of the many types of compressors used in the process industries, some of the more significant are the centrifugal, axial, rotary, and reciprocating compressors. They fall into three categories, as shown in Fig. 10-62.

For very high flows and low pressure ratios, an axial-flow compressor would be best. Axial-flow compressors usually have a higher efficiency, as seen in Fig. 10-63, but a smaller operating region than does

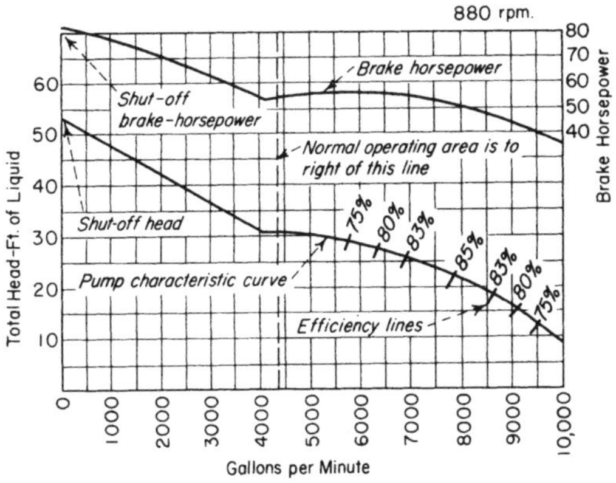

FIG. 10-56 Characteristic curve of an axial-flow pump. To convert gallons per minute to cubic meters per hour, multiply by 0.2271; to convert feet to meters, multiply by 0.3048; and to convert horsepower to kilowatts, multiply by 0.746.

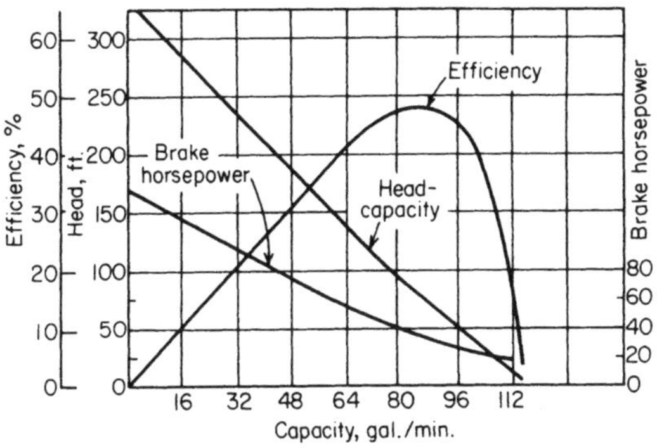

FIG. 10-58 Characteristic curves of a regenerative pump. To convert gallons per minute to cubic meters per hour, multiply by 0.2271; to convert feet to meters, multiply by 0.3048; and to convert horsepower to kilowatts, multiply by 0.746.

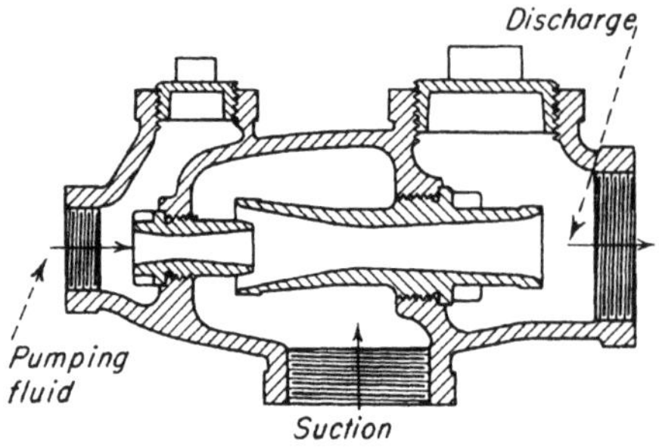

FIG. 10-59 Simple ejector using a liquid-motivating fluid.

a centrifugal machine. Centrifugal compressors operate most efficiently at medium flow rates and high pressure ratios. Rotary and reciprocating compressors (positive-displacement machines) are best used for low flow rates and high pressure ratios. The positive-displacement compressors and, in particular, reciprocating compressors were the most widely used in the process and pipeline industries up to and through the 1960s.

In turbomachinery the centrifugal flow and the axial-flow compressors are the ones used for compressing gases. Positive-displacement compressors such as reciprocating, gear-type, or lobe-type are widely used in the industry for many other applications such as slurry pumping.

The performance characteristics of a single stage of the three main types of compressors are given in Table 10-16. The pressure ratios of the axial and centrifugal compressors have been classified into three groups: industrial, aerospace, and research.

The industrial pressure ratio is low because the operating range needs to be large. The operating range is defined as the range between the surge point and the choke point. The surge point is the point at which the flow is reversed in the compressor. The choke point is the point at which the flow has reached Mach = 1.0, the point where no more flow can get through the unit, a "stone wall." When surge occurs, the flow is reversed, and so are all the forces acting on the compressor, especially the thrust forces. Surge can lead to total destruction of the compressor. Thus surge is a region that must be avoided. Choke conditions cause a large drop in efficiency, but do not lead to destruction of the unit. Note that with the increase in pressure ratio and the number of stages, the operating range is narrowed in axial-flow and centrifugal compressors.

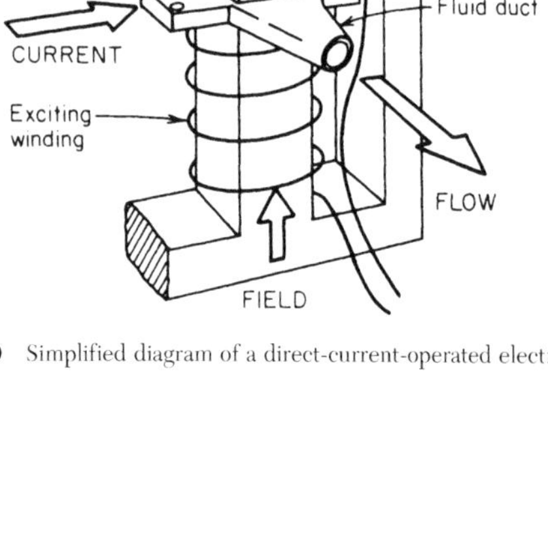

FIG. 10-60 Simplified diagram of a direct-current-operated electromagnetic pump.

Compressor Selection To select the most satisfactory compression equipment, engineers must consider a wide variety of types, each of which offers peculiar advantages for particular applications. Among the major factors to be considered are flow rate, head or pressure, temperature limitations, method of sealing, method of lubrication, power consumption, serviceability, and cost.

To be able to decide which compressor best fits the job, the engineer must analyze the flow characteristics of the units. The following dimensionless numbers describe the flow characteristics.

Reynolds number is the ratio of the inertia forces to the viscous forces

$$N_{\mathrm{Re}} = \frac{\rho V D}{\mu} \tag{10-63}$$

where ρ is the density of the gas, V is the velocity of the gas, D is the diameter of the impeller, and μ is the viscosity of the gas.

The specific speed compares the adiabatic head and flow rate in geometrically similar machines at various speeds.

$$N_s = \frac{N\sqrt{Q}}{H_{\mathrm{ad}}^{3/4}} \tag{10-64}$$

where N is the speed of rotation of the compressor, Q is the volume flow rate, and H is the adiabatic head.

The specific diameter compares head and flow rates in geometrically similar machines at various diameters

$$D_S = \frac{DH^{1/4}}{\sqrt{Q}} \tag{10-65}$$

The flow coefficient is the capacity of the flow rate of the machine

$$\phi = \frac{Q^1}{ND^3} \tag{10-66}$$

The pressure coefficient is the pressure or the pressure rise of the machine

$$\Psi = \frac{H}{N^2D^2} \tag{10-67}$$

In selecting the machines of choice, the use of specific speed and diameter best describe the flow. Figure 10-64 shows the characteristics of the three types of compressors. Other considerations in chemical plant service such as problems with gases which may be corrosive or have abrasive solids in suspension must be dealt with. Gases at elevated temperatures may create a potential explosion hazard, while air at the same temperatures may be handled quite normally; minute amounts of lubricating oil or water may contaminate the process gas and so may not be permissible, and for continuous-process use, a high degree of equipment reliability is required, since frequent shutdowns for inspection or maintenance cannot be tolerated.

COMPRESSION OF GASES

Theory of Compression In any continuous compression process the relation of absolute pressure p to volume V is expressed by the formula

$$pV^n = C = \text{constant} \tag{10-68}$$

The plot of pressure versus volume for each value of exponent n is known as the **polytropic** curve. Since the work W performed in proceeding from p_1 to p_2 along any polytropic curve (Fig. 10-65) is

$$W = \int_1^2 p\, dV \tag{10-69}$$

it follows that the amount of work required is dependent upon the polytropic curve involved and increases with increasing values of n. The path requiring the least amount of input work is $n = 1$, which is equivalent to **isothermal** compression. For **adiabatic** compression (i.e., no heat is being added or taken away during the process), $n = k$ = ratio of specific heat at constant pressure to that at constant volume.

TABLE 10-15 Pump Problems

Possible causes	Corrective action
Cavitating-Type Problems	
Plugged suction screen.	Check for indications of the presence of screen. Remove and clean screen.
Piping gaskets with undersized IDs installed, a very common problem in small pumps.	Install proper-sized gaskets.
Column tray parts or ceramic packing lodged in the impeller eye.	Remove suction piping and debris.
Deteriorated impeller eye due to corrosion.	Replace impeller and overhaul pump.
Flow rate is high enough above design that NPSH for flow rate has increased above NPSH.	Reduce flow rate to that of design.
Lined pipe collapsed at gasket area or ID due to buildup of corrosion products between liner and carbon-steel pipe.	Replace deteriorated piping.
Poor suction piping layout, too many elbows in too many planes, a tee branch almost directly feeding the suction of the other pump, or not enough straight run before the suction flange of the pump.	Redesign piping layout, using fewer elbows and laterals for tees, and have five or more straight pipe diameters before suction flange.
Vertical pumps experience a vortex formation due to loss of submergence required by the pump. Observe the suction surface while the pump is in operation, if possible.	Review causes of vortexing. Consider installation of a vortex breaker such as a bell mouth umbrella or changes to sump design.
Spare pump begins to cavitate when attempt is made to switch it with the running pump. The spare is "backed off" by the running pump because its shutoff head is less than the head produced by the running pump. This is a frequent problem when one pump is turbine-driven and one is motor-driven.	Throttle discharge of running pump until spare can get in system. Slow down running pump if it is a turbine or variable-speed motor.
Suction piping configuration causes adverse fluid rotation when approaching impeller.	Install sufficient straight run of suction piping, or install vanes in piping to break up prerotation.
Velocity of the liquid is too high as it approaches the impeller eye.	Install larger suction piping or reduce flow through pump.
Pump is operating at a low-flow-producing suction recirculation in the impeller eye. This results in a cavitationlike sound.	Install bypass piping back to suction vessel to increase flow through pump. Remember bypass flow may have to be as high as 50 percent of design flow.
Capacity-Type Problems	
Check the discharge block valve opening first. It may be partially closed and thus the problem.	Open block valve completely.
Wear-ring clearances are excessive (closed impeller design).	Overhaul pump. Renew wear rings if clearance is about twice design value for energy and performance reasons.
Impeller-to-case or head clearances are excessive (open impeller design).	Reposition impeller to obtain correct clearance.
Air leaks into the system if the pump suction is below atmospheric pressure.	Take actions as needed to eliminate air leaks.
Increase in piping friction to the discharge vessel due to the following: 1. Gate has fallen off the discharge valve stem. 2. Spring is broken in the spring-type check valve. 3. Check valve flapper pin is worn, and flapper will not swing open. 4. Lined pipe collapsing. 5. Control valve stroke improperly set, causing too much pressure drop.	Take the following actions: 1. Repair or replace gate valve. 2. Repair valve by replacing spring. 3. Overhaul check valve; restore proper clearance to pin and flapper bore. 4. Replace damaged pipe. 5. Adjust control valve stroke as necessary.
Suction and/or discharge vessel levels are not correct, a problem mostly seen in lower-speed pumps.	Calibrate level controllers as necessary.
Motor running backward or impeller of double suction design is mounted backward. Discharge pressure developed in both cases is about one-half design value.	Check for proper rotation and mounting of impeller. Reverse motor leads if necessary.
Entrained gas from the process lowering NPSH available.	Reduce entrained gas in liquid by process changes as needed.
Polymer or scale buildup in discharge nozzle areas.	Shut down pump and remove scale or deposits.
Mechanical seal in suction system under vacuum is leaking air into system, causing pump curve to drop.	Change percentage balance of seal faces or increase spring tension.
The pump may have formed a vortex at high flow rates or low liquid level. Does the vessel have a vortex breaker? Does the incoming flow cause the surface to swirl or be agitated?	Reduce flow to design rates. Raise liquid level in suction vessel. Install vortex breaker in suction vessel.
Variable-speed motor running too slowly.	Adjust motor speed as needed.
Bypassing is occurring between volute channels in a double-volute pump casing due to a casting defect or extreme erosion.	Overhaul pump; repair eroded area.
The positions of impellers are not centered with diffuser vanes. Several impellers will cause vibration and lower head output.	Overhaul pump; reposition individual impellers as needed. Reposition whole rotor by changing thrust collar locator spacer.
When the suction system is under vacuum, the spare pump has difficulty getting into system.	Install a positive-pressure steam (from running pump) to fill the suction line from the block valve through the check valve.
Certain pump designs use an internal bypass orifice port to alter head-flow curve. High liquid velocities often erode the orifice, causing the pump to go farther out on the pump curve. The system head curve increase corrects the flow back up the curve.	Overhaul pump, restore orifice to correct size.
Replacement impeller is not correct casting pattern; therefore NPSH required is different.	Overhaul pump, replace impeller with correct pattern.
Volute and cutwater area of casing is severely eroded.	Overhaul pump; replace casing or repair by welding. Stress-relieve after welding as needed.

TABLE 10-15 Pump Problems *(Concluded)*

Possible causes	Corrective action
Overload Problems	
Polymer buildup between wear surfaces (rings or vanes).	Remove buildup to restore clearances.
Excessive wear ring (closed impeller) or cover-case clearance (open impeller).	Replace wear rings or adjust axial clearance of open impeller. In severe cases, cover or case must be replaced.
Pump circulating excessive liquid back to suction through a breakdown bushing or a diffuser gasket area.	Overhaul pump, replacing parts as needed.
Minimum-flow loop left open at normal rates, or bypass around control valve is open.	Close minimum-flow loop or control valve bypass valve.
Discharge piping leaking under liquid level in sump-type design.	Inspect piping for leakage. Replace as needed.
Electrical switch gear problems cause one phase to have low amperage.	Check out switch gear and repair as necessary.
Specific gravity is higher than design specification.	Change process to adjust specific gravity to design value, or throttle pump to reduce horsepower requirements. This will not correct problem with some vertical turbine pumps that have a flat horsepower-required curve.
Pump motor not sized for end of curve operation.	Replace motor with one of larger size, or reduce flow rate.
Open impeller has slight rub on casing. Most often occurs in operations from 250 to 400°F due to piping strain and differential growth in the pump.	Increase clearance of impeller to casing.
A replacement impeller was not trimmed to the correct diameter.	Remove impeller from pump and turn to correct diameter.

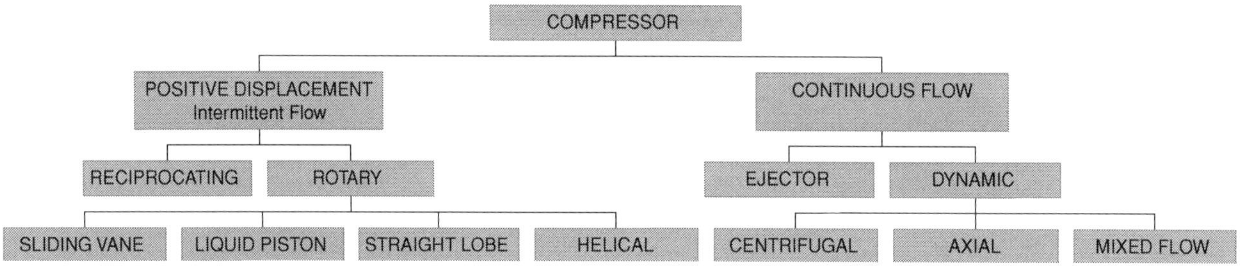

FIG. 10-61 Principal types of compressors.

Since most compressors operate along a polytropic path approaching the adiabatic, compressor calculations are generally based on the adiabatic curve.

Some formulas based upon the adiabatic equation and useful in compressor work are as follows:

Pressure, volume, and **temperature** relations for perfect gases:

$$p_2/p_1 = (V_1/V_2)^k \tag{10-70}$$

$$T_2/T_1 = (V_1/V_2)^{k-1} \tag{10-71}$$

$$p_2/p_1 = (T_2/T_1)^{k/(k-1)} \tag{10-72}$$

Adiabatic Calculations Adiabatic head is expressed as follows: In SI units,

$$H_{ad} = \frac{k \times ZRT_1}{k-1}\left[\left(\frac{p_2}{p_1}\right)^{(k-1)/k} - 1\right] \tag{10-73a}$$

where H_{ad} = adiabatic head, N·m/kg; R = gas constant, J/(kg·K) = 8314/molecular weight; T_1 = inlet gas temperature, K; p_1 = absolute inlet pressure, kPa; and p_2 = absolute discharge pressure, kPa.

In U.S. customary units,

$$H_{ad} = \frac{k}{k-1} ZRT_1\left[\left(\frac{p_2}{p_1}\right)^{(k-1)/k} - 1\right] \tag{10-73b}$$

where H_{ad} = adiabatic head, ft·lbf/lbm; R = gas constant, (ft·lbf)/(lbm·°R) = 1545/molecular weight; T_1 = inlet gas temperature, °R;

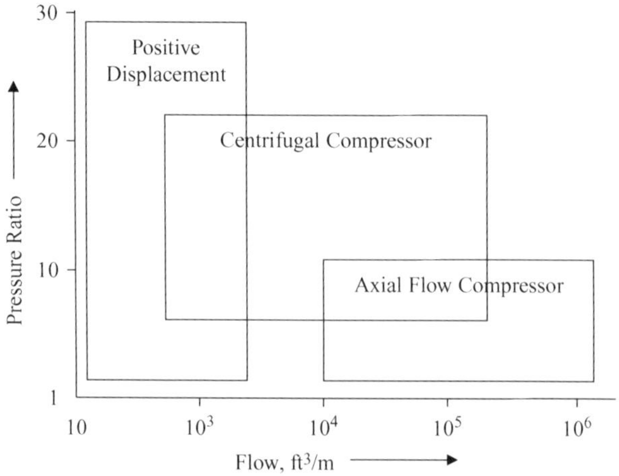

FIG. 10-62 Performance characteristics of different types of compressors.

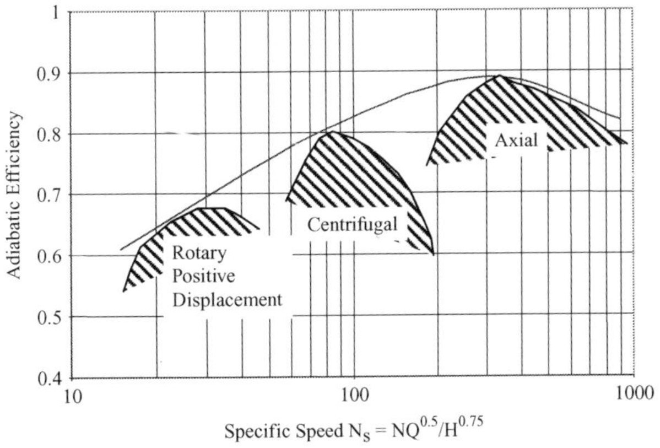

FIG. 10-63 Variation of adiabatic efficiency with specific speed for the three types of compressors.

p_1 = absolute inlet pressure, lbf/in²; and p_2 = absolute discharge pressure, lbf/in².

The **work** expended on the gas during compression is equal to the product of the adiabatic head and the mass flow of gas handled. Therefore, the adiabatic power is as follows:

In SI units,

$$\text{kW}_{ad} = \frac{WH_{ad}}{10^3} = \frac{k \times ZWRT_1}{k-1}\left[\left(\frac{p_2}{p_1}\right)^{(k-1)/k} - 1\right] \quad (10\text{-}74a)$$

or

$$\text{kW}_{ad} = 2.78 \times 10^{-4}\frac{k}{k-1}Q_1p_1\left[\left(\frac{p_2}{p_1}\right)^{(k-1)/k} - 1\right] \quad (10\text{-}75a)$$

where kW_{ad} = power, kW; W = mass flow, kg/s; and Q_1 = volume rate of gas flow, m³/h, at compressor inlet conditions.

In U.S. Customary units,

$$\text{hp}_{ad} = \frac{WH_{ad}}{550} = \frac{k}{k-1}\,\frac{ZWRT_1}{550}\left[\left(\frac{p_2}{p_1}\right)^{(k-1)/k} - 1\right] \quad (10\text{-}74b)$$

or

$$\text{hp}_{ad} = \frac{k}{k-1}\,\frac{Q_1P_1}{3600}\left[\left(\frac{p_2}{p_1}\right)^{(k-1)/k} - 1\right] \quad (10\text{-}75b)$$

where hp_{ad} = power, hp; W = mass flow, lb/s; and Q_1 = volume rate of gas flow, ft³/min.

Adiabatic discharge temperature is

$$T_2 = T_1(p_2/p_1)^{(k-1)/k} \quad (10\text{-}76)$$

The work in a compressor under ideal conditions as previously shown occurs at constant entropy. The actual process is a polytropic process as shown in Fig. 10-65 and given by the equation of state Pv^n = constant.

Adiabatic efficiency is given by the following relationship:

$$\eta_{ad} = \frac{\text{ideal work}}{\text{actual work}} \quad (10\text{-}77)$$

In terms of the change in total temperatures the relationship can be written as:

$$\eta_{ad} = \frac{T_2 - T_1}{T_{2a} - T_1} \quad (10\text{-}78)$$

where T_{2a} is the total actual discharge temperature of the gas. The adiabatic efficiency can be represented in terms of total pressure change:

$$\eta_{ad} = \frac{\left(\frac{P_2}{P_1}\right)^{(k-1)/k} - 1}{\left(\frac{P_2}{P_1}\right)^{(n-1)/n} - 1} \quad (10\text{-}79)$$

Polytropic head can be expressed by the following relationship:

$$H_{\text{poly}} = \frac{n}{n-1}\,\text{ZRT}_1\left\{\left[\left(\frac{P_2}{P_1}\right)^{(n-1)/n} - 1\right]\right\} \quad (10\text{-}80)$$

Likewise for polytropic efficiency, which is often considered as the small stage efficiency, or the hydraulic efficiency:

$$\eta_{pc} = \frac{(k-1)/k}{(n-1)/n} \quad (10\text{-}81)$$

Polytropic efficiency is the limited value of the isentropic efficiency as the pressure ratio approaches 1.0, and the value of the polytropic efficiency is higher than the corresponding adiabatic efficiency.

A characteristic of polytropic efficiency is that the polytropic efficiency of a multistage unit is equal to the stage efficiency if each stage has the same efficiency.

If the compression cycle approaches the isothermal condition, pV = constant, as is the case when several stages with intercoolers are used, a simple approximation of the power is obtained from the following formula:

In SI units,

$$\text{kW} = 2.78 \times 10^{-4} Q_1 p_1 \ln p_2/p_1 \quad (10\text{-}82a)$$

In U.S. customary units,

$$\text{hp} = 4.4 \times 10^{-3} Q_1 p_1 \ln p_2/p_1 \quad (10\text{-}82b)$$

Reciprocating Compressors Reciprocating compressors are used mainly when high-pressure head is required at a low flow. Reciprocating compressors are furnished in either single-stage or multistage types. The number of stages is determined by the required compressor ratio p_2/p_1. The compression ratio per stage is generally limited to 4, although low-capacity units are furnished with compression ratios of 8 and even higher. Generally, the maximum compression ratio is determined by the maximum allowable discharge-gas temperature.

TABLE 10-16 Performance Characteristics of Compressors

Types of compressors	Pressure ratio per stage			Efficiency, %	Operating range surge – choke, %
	Industrial	Aerospace	Research		
Positive displacement	Up to 30	—	—	75–82	—
Centrifugal	1.2–1.9	2.0–7.0	13	75–87 25	Large
Axial	1.05–1.3	1.1–1.45	2.1	80–91	Narrow 3–10

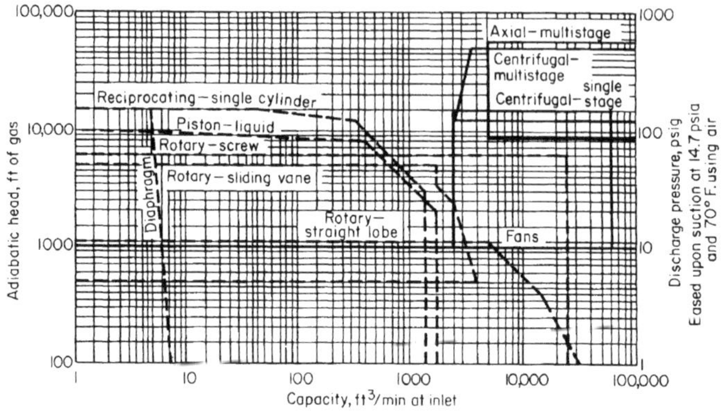

FIG. 10-64 Compressor coverage chart based on the normal range of operation of commercially available types shown. Solid lines: use left ordinate, head. Broken lines: use right ordinate, pressure. To convert cubic feet per minute to cubic meters per hour, multiply by 1.699; to convert feet to meters, multiply by 0.3048; and to convert pounds-force per square inch to kilopascals, multiply by 6.895; (°F − 32)5⁄9 = °C.

Single-acting air-cooled and water-cooled air compressors are available in sizes up to about 75 kW (100 hp). Such units are available in one, two, three, or four stages for pressure as high as 24 MPa (3500 lbf/in²). These machines are seldom used for gas compression because of the difficulty of preventing gas leakage and contamination of the lubricating oil.

The compressors most commonly used for compressing gases have a crosshead to which the connecting rod and piston rod are connected. This provides a straight-line motion for the piston rod and permits simple packing to be used. Figure 10-66 illustrates a simple single-stage machine of this type having a double-acting piston. Either single-acting (Fig. 10-67) or double-acting pistons (Fig. 10-68) may be used, depending on the size of the machine and the number of stages. In some machines double-acting pistons are used in the first stages and single-acting in the later stages.

On multistage machines, intercoolers are provided between stages. These heat exchangers remove the heat of compression from the gas and reduce its temperature to approximately the temperature existing at the compressor intake. Such cooling reduces the volume of gas going to the high-pressure cylinders, reduces the power required for compression, and keeps the temperature within safe operating limits.

Figure 10-69 illustrates a two-stage compressor end such as might be used on the compressor illustrated in Fig. 10-66.

Compressors with horizontal cylinders such as illustrated in Figs. 10-66 to 10-68 are most commonly used because of their accessibility. However, machines are also built with vertical cylinders and other arrangements such as right-angle (one horizontal and one vertical cylinder) and V-angle.

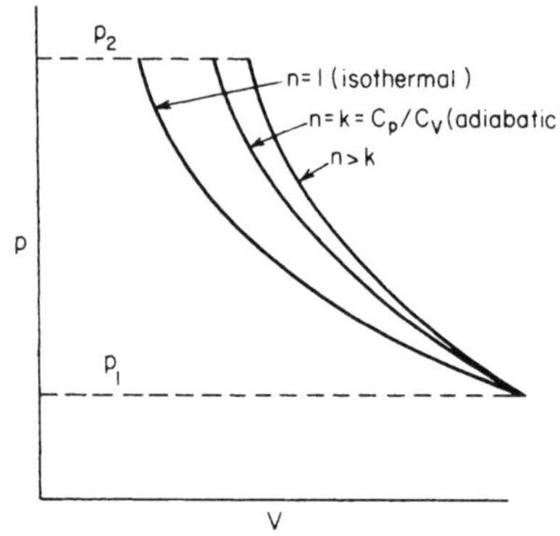

Fig. 10-65 Polytropic compression curves.

Compressors up to around 75 kW (100 hp) usually have a single center-throw crank, as illustrated in Fig. 10-66. In larger sizes compressors are commonly of duplex construction with cranks on each end of the shaft (see Fig. 10-70). Some large synchronous motor-driven units are of four-corner construction; i.e., they are of double-duplex construction with two connecting rods from each of the two crank throws (see Fig. 10-71). Steam-driven compressors have one or more steam cylinders connected directly by piston rod or tie rods to the gas-cylinder piston or crosshead.

Valve Losses Above piston speeds of 2.5 m/s (500 ft/min), suction and discharge valve losses begin to exert significant effects on the actual internal compression ratio of most compressors, depending on the valve port area available. The obvious results are high temperature rise and higher power requirements than might be expected. These effects become more pronounced with higher-molecular-weight gases. Valve problems can be a very major contributor to down time experienced by these machines.

Control Devices In many installations the use of gas is intermittent, and some means of controlling the output of the compressor is therefore necessary. In other cases constant output is required despite variations in discharge pressure, and the control device must operate to maintain a constant compressor speed. Compressor capacity, speed, or pressure may be varied in accordance with requirements. The nature of the control device will depend on the function to be regulated. Regulation of pressure, volume, temperature, or some other factor determines the type of regulation required and the type of the compressor driver.

The most common control requirement is regulation of capacity. Many capacity controls, or unloading devices, as they are usually termed, are actuated by the pressure on the discharge side of the compressor. A falling pressure indicates that gas is being used faster than it is being compressed and that more gas is required. A rising pressure indicates that more gas is being compressed than is being used and that less gas is required.

An obvious method of controlling the capacity of a compressor is to vary the speed. This method is applicable to units driven by variable-speed drivers such as steam pistons, steam turbines, gas engines, diesel engines, etc. In these cases the regulator actuates the steam-admission or fuel-admission valve on the compressor driver and thus controls the speed.

Motor-driven compressors usually operate at constant speed, and other methods of controlling the capacity are necessary. On reciprocating compressors discharging into receivers, up to about 75 kW (100 hp), two types of control are usually available. These are automatic-start-and-stop control and constant-speed control.

Automatic-start-and-stop control, as its name implies, stops or starts the compressor by means of a pressure-actuated switch as the

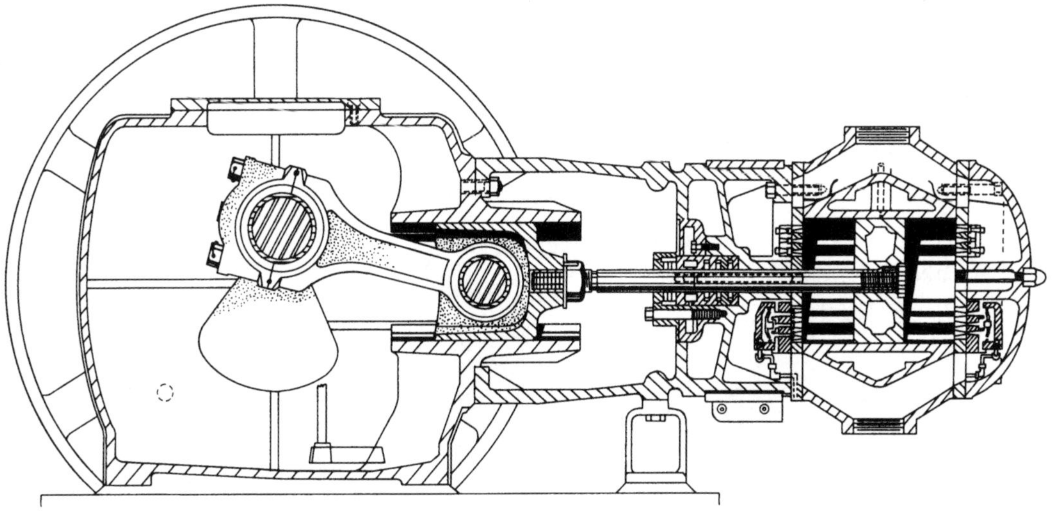

FIG. 10-66 Typical single-stage, double-acting water-cooled compressor.

gas demand varies. It should be used only when the demand for gas will be intermittent.

Constant-speed control should be used when gas demand is fairly constant. With this type of control, the compressor runs continuously but compresses only when gas is needed. Three methods of unloading the compressor with this type of control are in common use: (1) **closed suction unloaders,** (2) **open inlet-valve unloaders,** and (3) **clearance unloaders.** The closed suction unloader consists of a pressure-actuated valve which shuts off the compressor intake. Open inlet-valve unloaders (see Fig. 10-72) operate to hold the compressor inlet valves open and thereby prevent compression. Clearance unloaders (see Fig. 10-73) consist of pockets or small reservoirs which are opened when unloading is desired. The gas is compressed into them on the compression stroke and reexpands into the cylinder on the return stroke, thus preventing the compression of additional gas.

It is sometimes desirable to have a compressor equipped with both constant-speed and automatic-start-and-stop control. When this is done, a switch allows immediate selection of either type.

Motor-driven reciprocating compressors above about 75 kW (100 hp) in size are usually equipped with a step control. This is in reality a variation of constant-speed control in which unloading is accomplished in a series of steps, varying from full load down to no load.

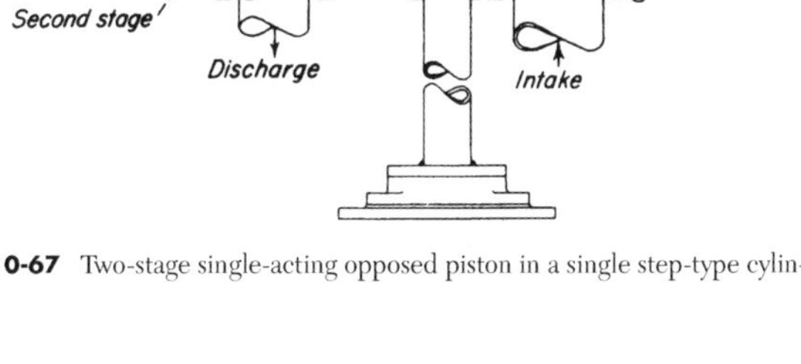

FIG. 10-67 Two-stage single-acting opposed piston in a single step-type cylinder.

Three-step control (full load, one-half load, and no load) is usually accomplished with inlet-valve unloaders. **Five-step control** (full load, three-fourths load, one-half load, one-fourth load, and no load) is accomplished by means of clearance pockets (see Fig. 10-74). On some machines, inlet-valve and clearance-control unloading are used in combination.

Although such control devices are usually automatically operated, manual operation is satisfactory for some services. When manual operation is provided, it often consists of a valve or valves to open and close clearance pockets. In some cases, a movable cylinder head is provided for variable clearance in the cylinder (see Fig. 10-75).

When no capacity control or unloading device is provided, it is necessary to provide bypasses between the inlet and discharge in order that the compressor can be started against no load (see Fig. 10-76).

Nonlubricated Cylinders Most compressors use oil to lubricate the cylinder. In some processes, however, the slightest oil contamination

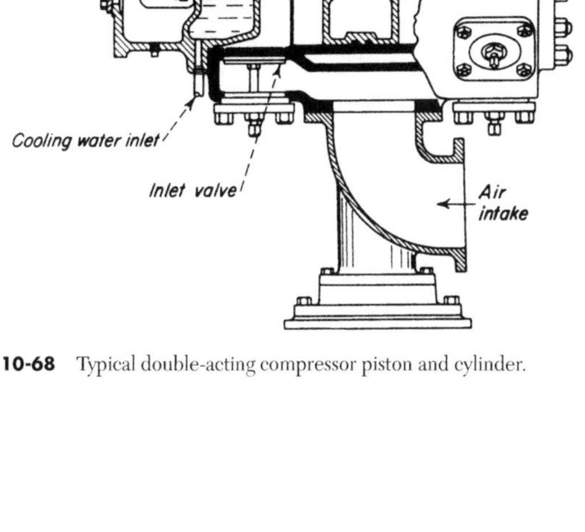

FIG. 10-68 Typical double-acting compressor piston and cylinder.

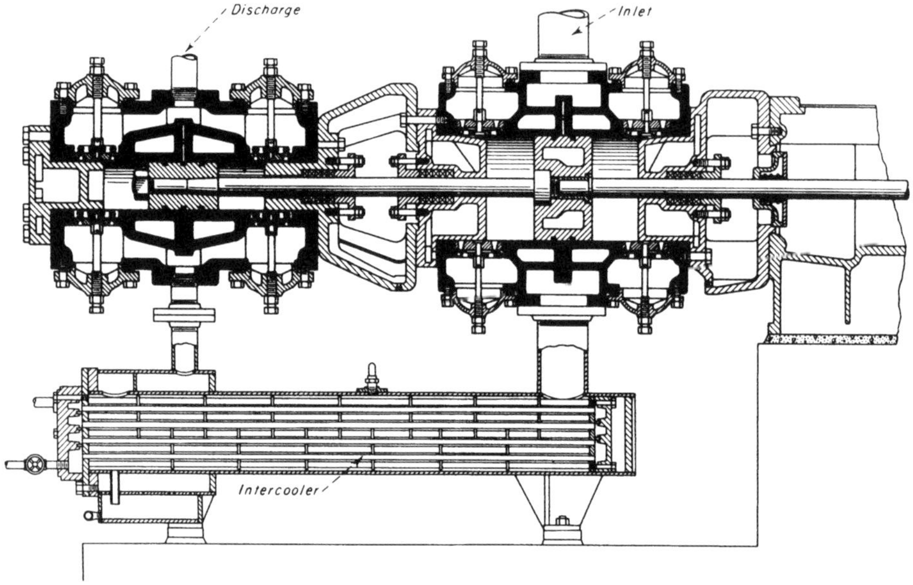

FIG. 10-69 Two-stage double-acting compressor cylinders with intercooler.

is objectionable. For such cases a number of manufacturers furnish a "nonlubricated" cylinder (see Fig. 10-77). The piston on these cylinders is equipped with piston rings of graphitic carbon or Teflon° as well as pads or rings of the same material to maintain proper clearance between the piston and the cylinder. Plastic packing of a type that requires no lubricant is used on the stuffing box. Although oil-wiper rings are used on the piston rod where it leaves the compressor frame, minute quantities of oil might conceivably enter the cylinder on the rod. If even such small amounts of oil are objectionable, an extended cylinder connecting piece can be furnished. This simply lengthens the piston rod enough so that no portion of the rod can alternately enter the frame and the cylinder.

In many cases, a small amount of gas leaking through the packing is objectionable. Special connecting pieces are furnished between the cylinder and the frame, which may be either single-compartment or double-compartment. These may be furnished gastight and vented back to the suction or filled with a sealing gas or fluid and held under a slight pressure.

°®Du Pont tetrafluoroethylene fluorocarbon resin.

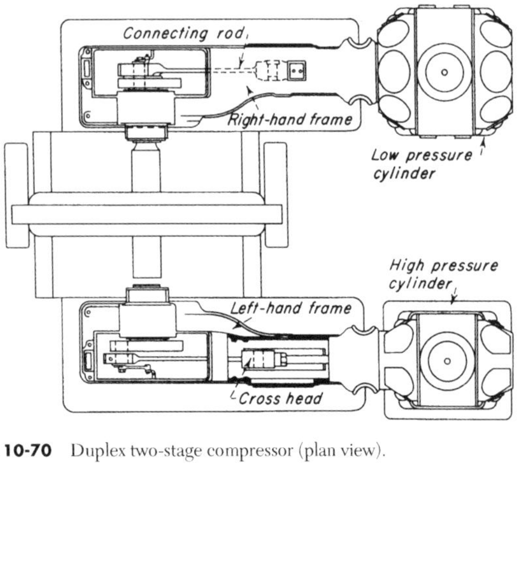

FIG. 10-70 Duplex two-stage compressor (plan view).

High-Pressure Compressors There is a definite trend in the chemical industry toward the use of high-pressure compressors with discharge pressures of from 34.5 to 172 MPa (5000 to 25,000 lbf/in^2) and with capacities from 8.5 × 103 to 42.5 × 10^3 m^3/h (5000 to 25,000 ft^3/min). These require special design, and a complete knowledge of the characteristics of the gas is necessary. In most cases, these types of applications use the barrel-type centrifugal compressor.

The gas usually deviates considerably from the perfect-gas laws, and in many cases temperature or other limitations necessitate a thorough engineering study of the problem. These compressors usually have five, six, seven, or eight stages, and the cylinders must be properly proportioned to meet the various limitations involved and also to balance the load among the various stages. In many cases, scrubbing or other processing is carried on between stages. High-pressure cylinders are steel forgings with single-acting plungers (see Fig. 10-78). The compressors are usually designed so that the pressure load against the plunger is opposed by one or more single-acting pistons of the lower pressure stages. Piston-rod packing is usually of the segmental-ring metallic type. Accurate fitting and correct lubrication are very important. High-pressure compressor valves are designed for the conditions involved. Extremely high-grade engineering and skill are necessary.

Piston-Rod Packing Proper piston-rod packing is important. Many types are available, and the most suitable is determined by the gas handled and the operating conditions for a particular unit.

There are many types and compositions of soft packing, semimetallic packing, and metallic packing. In many cases, metallic packing is to be recommended. A typical low-pressure packing arrangement is shown in Fig. 10-79. A high-pressure packing arrangement is shown in Fig. 10-80.

When wet, volatile, or hazardous gases are handled or when the service is intermittent, an auxiliary packing gland and soft packing are usually employed (see Fig. 10-81).

Metallic Diaphragm Compressors (Fig. 10-82) These are available for small quantities [up to about 17 m^3/h (10 ft^3/min)] for compression ratios as high as 10:1 per stage. Temperature rise is not a serious problem, as the large wall area relative to the gas volume permits sufficient heat transfer to approach isothermal compression. These compressors possess the advantage of having no seals for the process gas. The diaphragm is actuated hydraulically by a plunger pump.

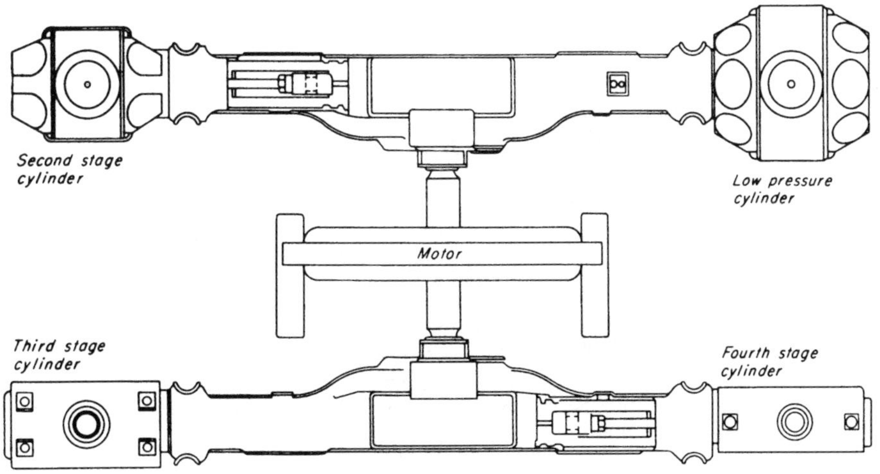

FIG. 10-71 Four-corner four-stage compressor (plan view).

FANS AND BLOWERS

Fans are used for low pressures where generally the delivery pressure is less than 3.447 kPa (0.5 lb/in^2), and blowers are used for higher pressures. However, they are usually below delivery pressures of 10.32 kPa (1.5 lbf/in^2). These units can either be centrifugal or the axial-flow type.

Fans and blowers are used for many types of ventilating work such as air-conditioning systems. In large buildings, blowers are often used due to the high delivery pressures needed to overcome the pressure drop in the ventilation system. Most of these blowers are of the centrifugal type. Blowers are also used to supply draft air to boilers and furnaces. Fans are used to move large volumes of air or gas through ducts, supplying air for drying, conveying material suspended in the gas stream, removing fumes, condensing towers and other high-flow, low-pressure applications.

Axial-Flow Fans These are designed to handle very high flow rates and low pressure. The disc-type fans are similar to those of a household fan. They are usually for general circulation or exhaust work without ducts.

The so-called propeller-type fans with blades that are aerodynamically designed (as seen in Fig. 10-83) can consist of two or more stages. The air in these fans enters in an axial direction and leaves in an axial direction. The fans usually have inlet guide vanes followed by a rotating blade, followed by a stationary (stator) blade.

Centrifugal Blowers These blowers have air or gases entering in the axial direction and being discharged in the radial direction. These blowers have three types of blades: radial or straight blades, forward-curved blades, and backward-curved blades (Figs. 10-84 to 10-86).

Radial blade blowers as seen in Fig. 10-84 are usually used in large-diameter or high-temperature applications. The blades being radial in direction have very low stresses as compared to the backward or forward

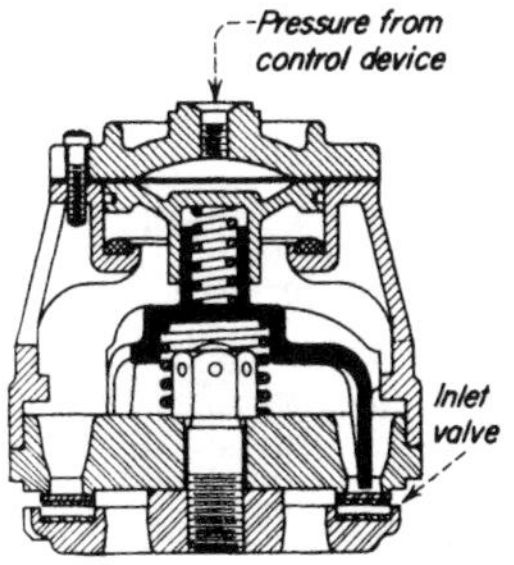

FIG. 10-72 Inlet-valve unloader.

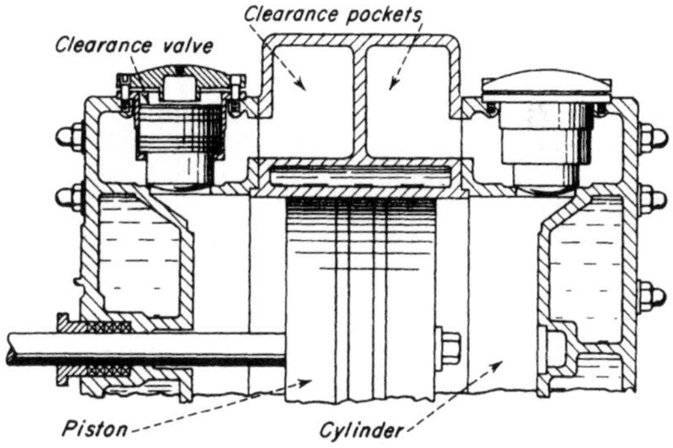

FIG. 10-73 Clearance-control cylinder. *(Courtesy of Ingersoll-Rand.)*

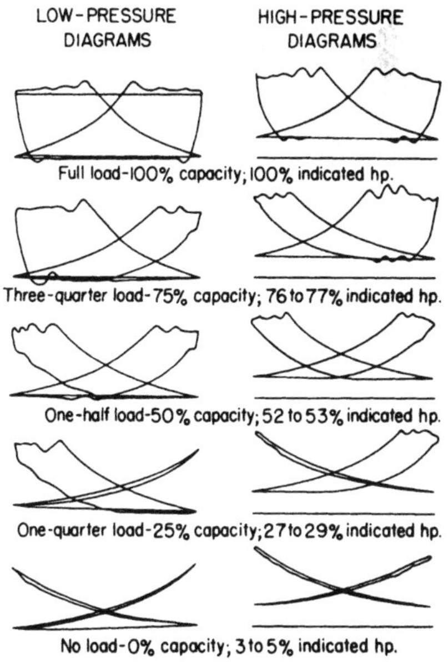

FIG. 10-74 Actual indicator diagram of a two-stage compressor showing the operation of clearance control at five load points.

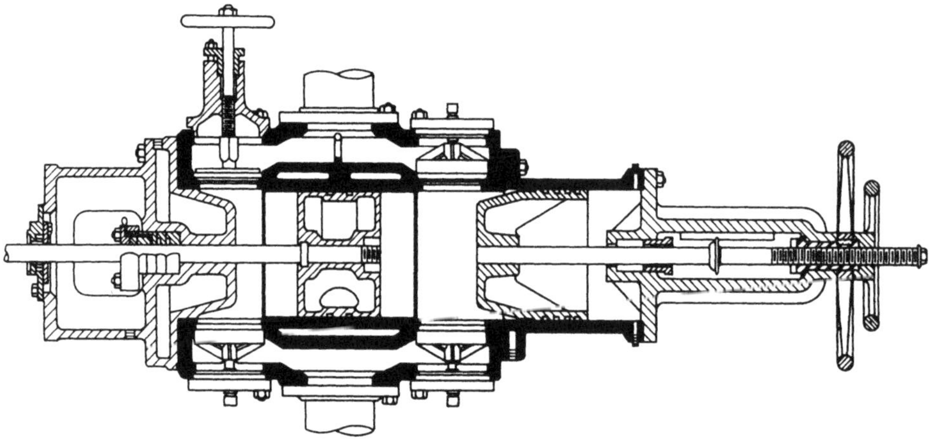

FIG. 10-75 Sectional view of a cylinder equipped with a hand-operated valve lifter on one end and a variable-volume clearance pocket at other end.

curve blades. The rotors have anywhere between 4 to 12 blades and usually operate at low speeds. These fans are used in exhaust work especially for gases at high temperature and with suspensions in the flow stream.

Forward-Curved Blade Blowers These blowers discharge the gas at a very high velocity. The pressure supplied by this blower is lower than that produced in the other two blade characteristics. The number of blades in such a rotor can be large—up to 50 blades—and the speed is high—usually 3600–1800 rpm in 60-cycle countries and 3000–1500 rpm in 50-cycle countries.

Backward-Curved Blade Blowers These blowers are used when a higher discharge pressure is needed. It is used over a wide range of applications. Both the forward and backward curved blades do have much higher stresses than the radial bladed blower.

Starting compressor	Stopping compressor
Start with A and D open	Close - - - - C
Close - - - - - - D	Close - - - B
Close - - - - - - - A	Open - - - - A and D
Open - - - - - - - B	
Slowly open - - - C	

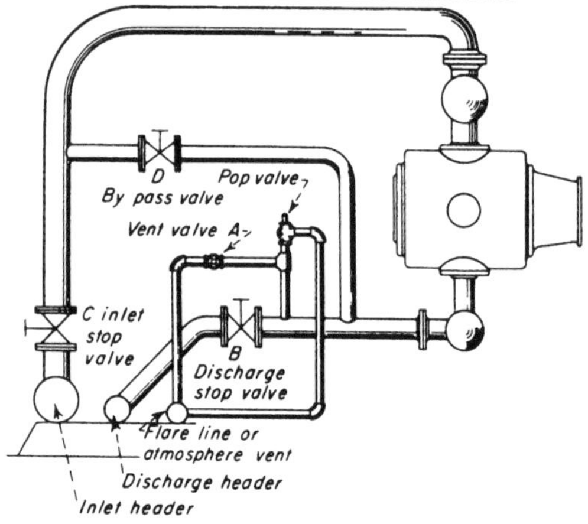

FIG. 10-76 Bypass arrangement for a single-stage compressor. On multistage machines, each stage is bypassed in a similar manner. Such an arrangement is necessary for no-load starting.

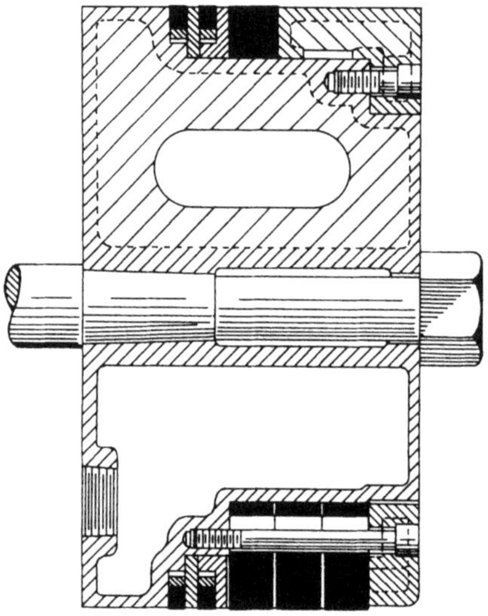

FIG. 10-77 Piston equipped with carbon piston and wearing rings for a nonlubricated cylinder.

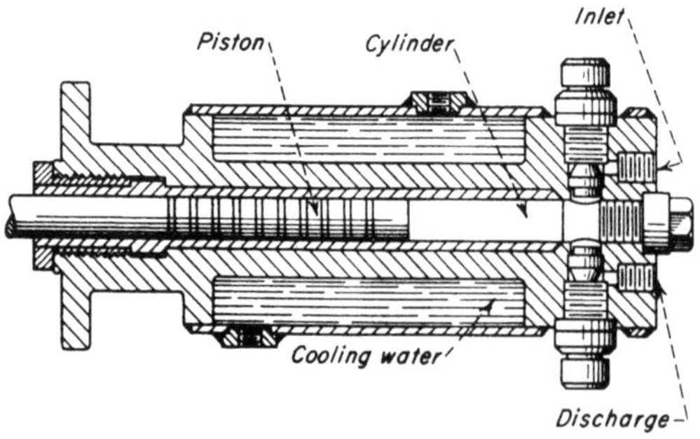

FIG. 10-78 Forged-steel single-acting high-pressure cylinder.

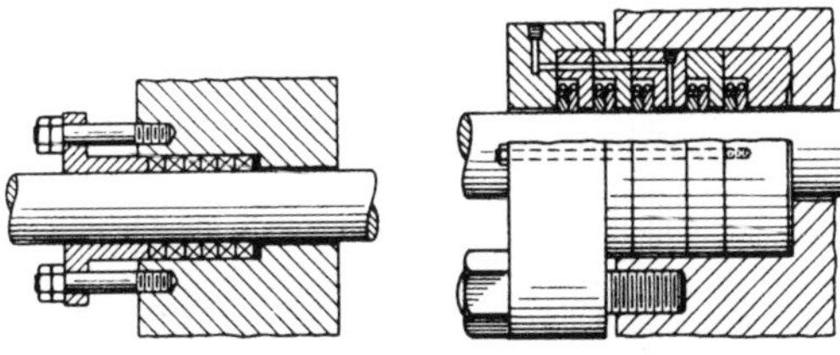

FIG. 10-79 Typical packing arrangements for low-pressure cylinders.

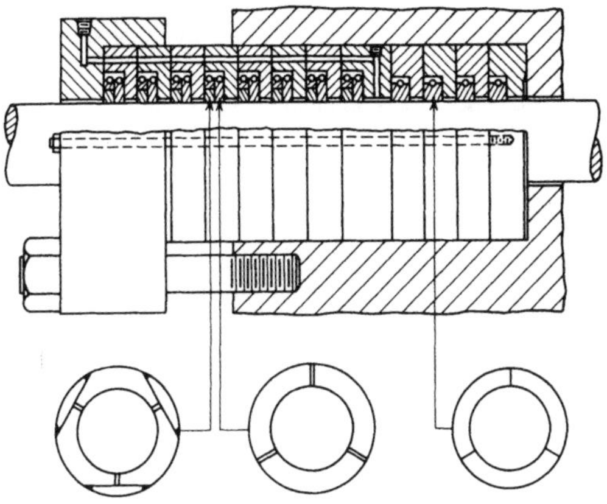

FIG. 10-80 Typical packing arrangement, using metallic packing, for high-pressure cylinders.

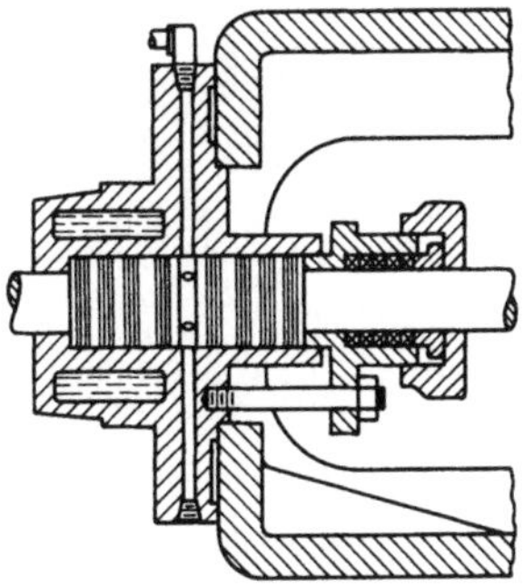

FIG. 10-81 Soft packing in an auxiliary stuffing box for handling gases.

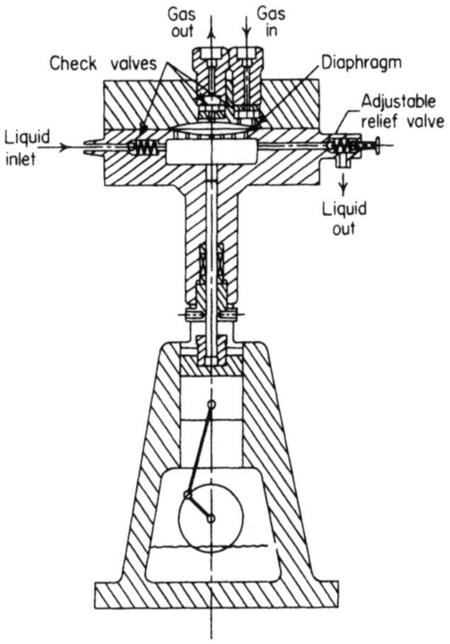

FIG. 10-82 High-pressure, low-capacity compressor having a hydraulically actuated diaphragm. (*Pressure Products Industries.*)

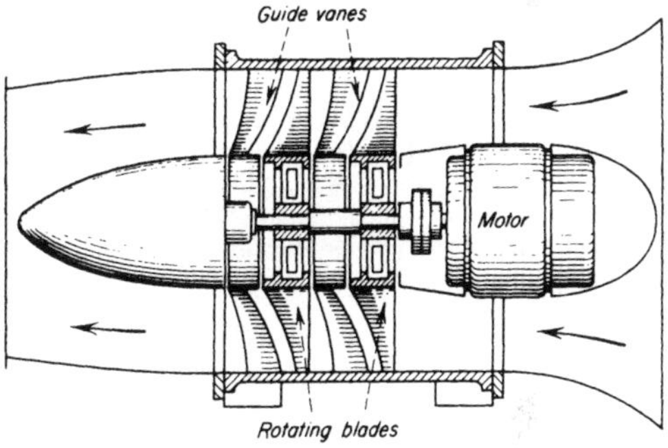

FIG. 10-83 Two-stage axial-flow fan.

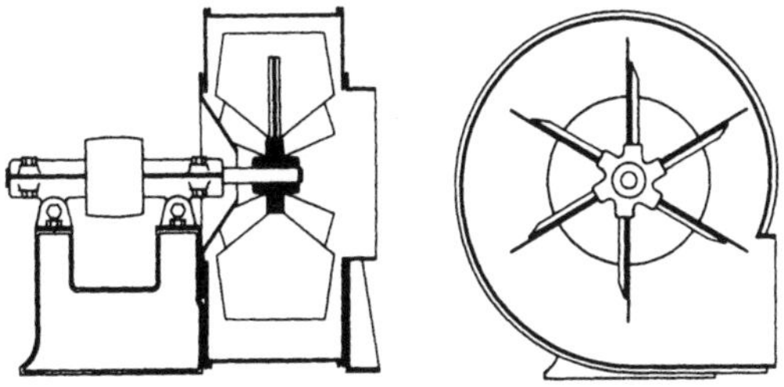

FIG. 10-84 Straight-blade, or steel-plate, fan.

The centrifugal blower produces energy in the air stream by the centrifugal force and imparts a velocity to the gas by the blades. Forward curved blades impart the most velocity to the gas. The scroll-shaped volute diffuses the air and creates an increase in the static pressure by reducing the gas velocity. The change in total pressure occurs in the impeller—this is usually a small change. The static pressure is increased in both the impeller and the diffuser section. Operating efficiencies of the fan range from 40 to 80 percent. The discharge total pressure is the summation of the static pressure and the velocity head.

The power needed to drive the fan can be computed as follows.

$$\text{Power (kW)} = 2.72 \times 10^{-5} QP \tag{10-83}$$

where Q is the fan volume (m^3/h) and P is the total discharge pressure in cm of water column.

In U.S. customary units,

$$\text{hp} = 1.57 \times 10^{-4} Qp \tag{10-84}$$

where hp is the fan power output, hp; Q is the fan volume, ft^3/min; and p is the fan-operating pressure, inches water column.

$$\text{Efficiency} = \frac{\text{air power output}}{\text{shaft power input}} \tag{10-85}$$

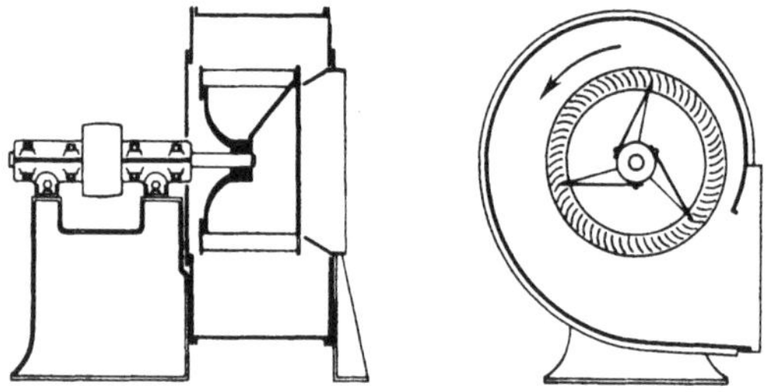

FIG. 10-85 Forward-curved blade, or "scirocco"-type, fan.

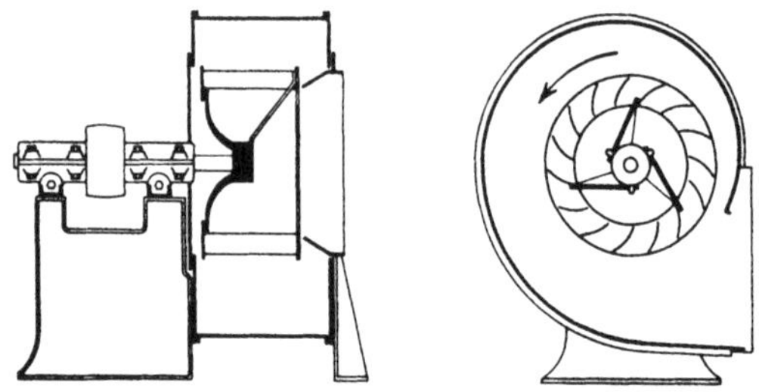

FIG. 10-86 Backward-curved-blade fan.

Fan Performance The performance of a centrifugal fan varies with changes in conditions such as temperature, speed, and density of the gas being handled. It is important to keep this in mind in using the catalog data of various fan manufacturers, since such data are usually based on stated standard conditions. Corrections must be made for variations from these standards. The usual variations are as follows:

When speed varies, (1) capacity varies directly as the speed ratio, (2) pressure varies as the square of the speed ratio, and (3) horsepower varies as the cube of the speed ratio.

When the temperature of air or gas varies, horsepower and pressure vary inversely as the absolute temperature, speed and capacity being constant. See Fig. 10-87.

When the density of air or gas varies, horsepower and pressure vary directly as the density, speed and capacity being constant.

CONTINUOUS-FLOW COMPRESSORS

Continuous-flow compressors are machines where the flow is continuous, unlike positive-displacement machines where the flow is fluctuating. Continuous-flow compressors are also classified as turbomachines. These types of machines are widely used in the chemical and petroleum industry for many services. They are also used extensively in many other industries such as the iron and steel industry, pipeline boosters, and on offshore platforms for reinjection compressors. Continuous-flow machines are usually much smaller in size and produce much less vibration than their counterpart, positive-displacement units.

Centrifugal Compressors The flow in a centrifugal compressor enters the impeller in an axial direction and exits in a radial direction.

In a typical centrifugal compressor, the fluid is forced through the impeller by rapidly rotating impeller blades. The velocity of the fluid is converted to pressure, partially in the impeller and partially in the stationary diffusers. Most of the velocity leaving the impeller is converted into pressure energy in the diffuser as shown in Fig. 10-88. It is normal practice to design the compressor so that half the pressure rise takes place in the impeller and the other half in the diffuser. The diffuser consists of a vaneless space, a vane that is tangential to the impeller, or a combination of both. These vane passages diverge to convert the velocity head into pressure energy.

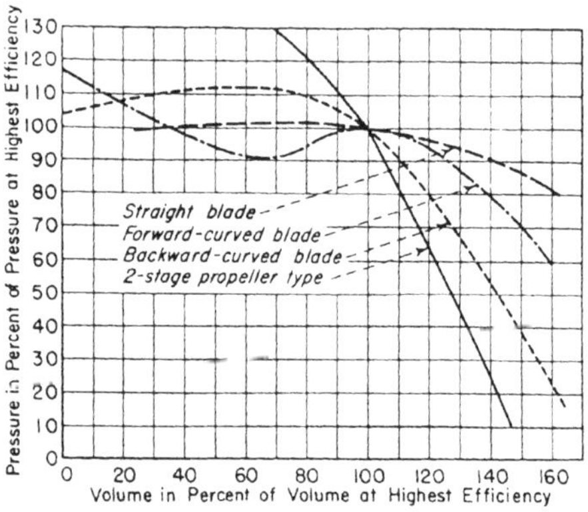

FIG. 10-87 Approximate characteristic curves of various types of fans.

Centrifugal compressors in general are used for higher pressure ratios and lower flow rates compared to lower-stage pressure ratios and higher flow rates in axial compressors. The pressure ratio in a single-stage centrifugal compressor varies depending on the industry and application. In the petrochemical industry the single stage pressure ratio is about 1.2:1. Centrifugal compressors used in the aerospace industry, usually as a compressor of a gas turbine, have pressure ratios between 3:1 to as high as 9:1 per stage.

In the petrochemical industry, the centrifugal compressors consist mainly of casings with multiple stages. In many instances, multiple casings are also used, and to reduce the power required to drive these multiple casings, there are intercoolers between them. Each casing can have up to 9 stages. In some cases, intercoolers are also used between single stages of compressor to reduce the power required for compression. These compressors are usually driven by gas turbines, steam turbines, and electric motors. Speed-increasing gears may be used in conjunction with these drivers to obtain the high speeds at which many of these units operate. Rotative speeds of as high as 50,000 rpm are not uncommon. Most of the petrochemical units run between 9,000 and 15,000 rpm.

The compressor's operating range is between two major regions as seen in Fig. 10-89, which is a performance map of a centrifugal compressor. These two regions are *surge,* which is the lower flow limit of stable operation, and *choke* or *stonewall,* which is the maximum flow through the compressor at a given operating speed. The centrifugal compressor's operating range between surge and choke is reduced as the pressure ratio per stage is increased or the number of stages are added.

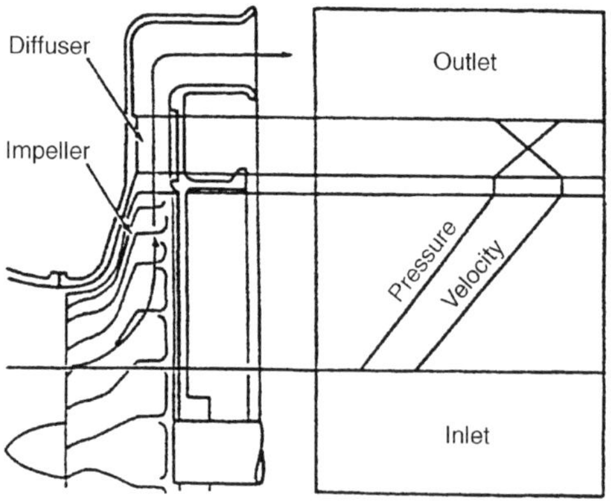

FIG. 10-88 Pressure and velocity through a centrifugal compressor.

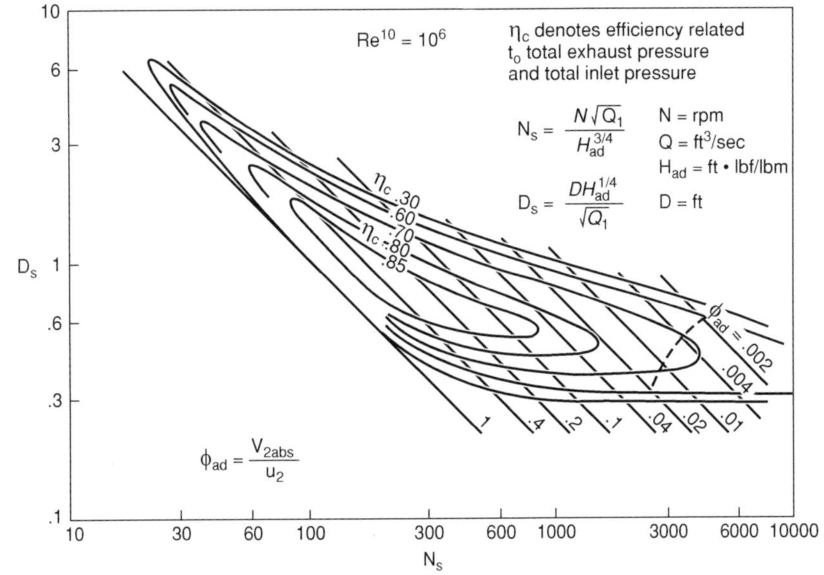

FIG. 10-89 Centrifugal compressor map. (*Balje, O. E., "A Study of Reynolds Number Effects in Turbomachinery,"* Journal of Engineering for Power, *ASME Trans., vol. 86, series A, p. 227.*)

A compressor is said to be in surge when the main flow through the compressor reverses its direction. Surge is often symptomized by excessive vibration and a large audible sound. This flow reversal is accompanied with a very violent change in energy, which causes a reversal of the thrust force. The surge process is cyclic in nature and if allowed to cycle for some time, irreparable damage can occur to the compressor. In a centrifugal compressor, surge is usually initiated at the exit of the impeller or at the diffuser entrance for impellers producing a pressure ratio of less than 3:1. For higher pressure ratios, the initiation of surge can occur in the inducer.

A centrifugal compressor impeller can have three types of blades at the exit of the impeller. These are forward-curved, backward-curved, and radial blades. Forward-curved blades are not often used in a centrifugal compressor's impeller because of the very high-velocity discharge at the compressor that would require conversion of the high velocity to a pressure head in the diffuser, which is accompanied by high losses. Radial blades are used in impellers of high pressure ratio since the stress levels are minimal. Backward-curved blades give the highest efficiency and the largest operating margin of any of the various types of blades in an impeller. Most centrifugal compressors in the petrochemical industry use backward-curved impellers because of the higher efficiency and larger operating range.

Process compressors have impellers with very low pressure ratio impellers and thus large surge-to-choke margins. The common method of classifying process-type centrifugal compressors is based on the number of impellers and the casing design. Sectionalized casing types have impellers that are usually mounted on the extended motor shaft, and similar sections are bolted together to obtain the desired number of stages. Casing material is either steel or cast iron. These machines require minimum supervision and maintenance and are quite economic in their operating range. The sectionalized casing design is used extensively in supplying air for combustion in ovens and furnaces.

The horizontally split type have casings split horizontally at the midsection and the top. The bottom halves are bolted and doweled together. This design type is preferred for large multistage units. The internal parts such as shaft, impellers, bearings, and seals are readily accessible for inspection and repairs by removing the top half. The casing material is cast iron or cast steel.

Barrel casings are used for high pressures in which the horizontally split joint is inadequate. This type of compressor consists of a barrel into which a compressor bundle of multiple stages is inserted. The bundle is itself a horizontally split casing compressor.

Compressor Configuration To properly design a centrifugal compressor, one must know the operating conditions—the type of gas, its pressure, temperature, and molecular weight. One must also know the corrosive properties of the gas so that proper metallurgical selection can be made. Gas fluctuations due to process instabilities must be pinpointed so that the compressor can operate without surging.

Centrifugal compressors for industrial applications have relatively low pressure ratios per stage. This condition is necessary so that the compressors can have a wide operating range while stress levels are kept at a minimum. Because of the low pressure ratios for each stage, a single machine may have a number of stages in one "barrel" to achieve the desired overall pressure ratio. Figure 10-90 shows some of the many configurations. Some factors to be considered when selecting a configuration to meet plant needs are:

1. Intercooling between stages can considerably reduce the power consumed.
2. Back-to-back impellers allow for a balanced rotor thrust and minimize overloading the thrust bearings.
3. Cold inlet or hot discharge at the middle of the case reduces oil-seal and lubrication problems.
4. Single inlet or single discharge reduces external piping problems.
5. Balance planes that are easily accessible in the field can appreciably reduce field-balancing times.
6. Balance piston with no external leakage will greatly reduce wear on the thrust bearings.
7. Hot and cold sections of the case that are adjacent to each other will reduce thermal gradients and thus reduce case distortion.
8. Horizontally split casings are easier to open for inspection than vertically split ones, reducing maintenance time.
9. Overhung rotors present an easier alignment problem because shaft-end alignment is necessary only at the coupling between the compressor and driver.
10. Smaller, high-pressure compressors that do the same job will reduce foundation problems but will have greatly reduced operational range.

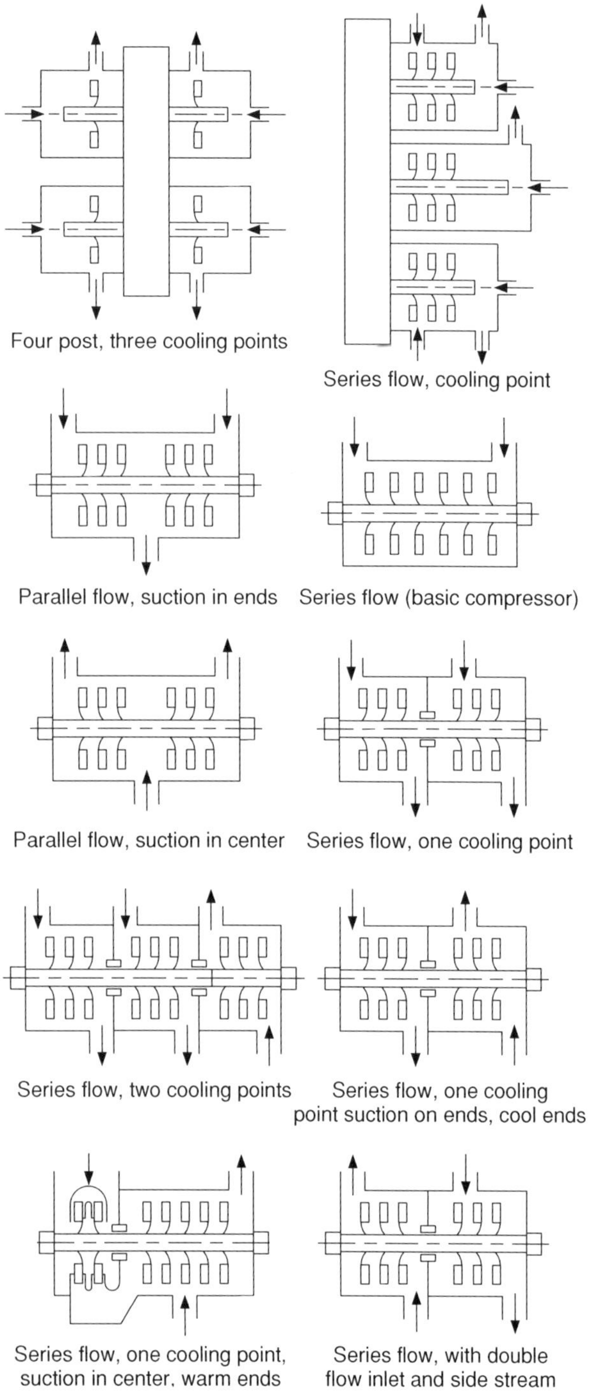

FIG. 10-90 Various configurations of centrifugal compressors.

Impeller Fabrication Centrifugal-compressor impellers are either shrouded or unshrouded. Open, shrouded impellers that are mainly used in single-stage applications are made by investment-casting techniques or by three-dimensional milling. Such impellers are used, in most cases, for the high-pressure-ratio stages. The

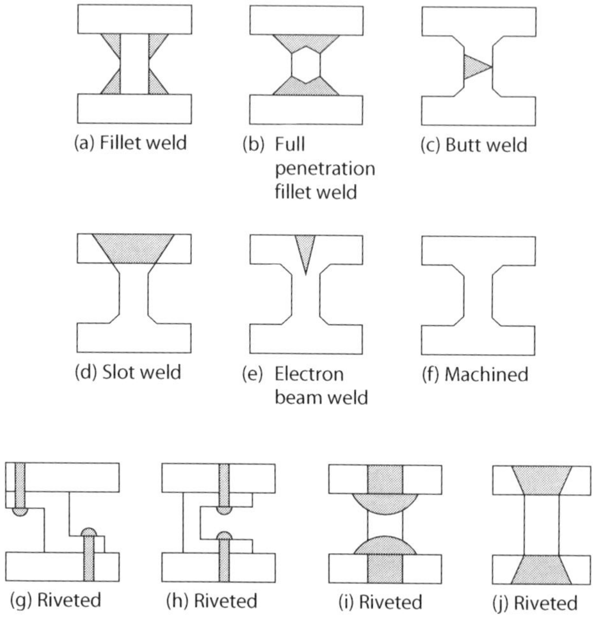

FIG. 10-91 Several fabrication techniques for centrifugal impellers.

shrouded impeller is commonly used in the process compressor because of its low pressure ratio stages. The low tip stresses in this application make it a feasible design. Figure 10-91 shows several fabrication techniques. The most common type of construction is seen in A and B where the blades are fillet-welded to the hub and shroud. In B, the welds are full penetration. The disadvantage in this type of construction is the obstruction of the aerodynamic passage. In C, the blades are partially machined with the covers and then butt-welded down the middle. For backward lean-angled blades, this technique has not been very successful, and there has been difficulty in achieving a smooth contour around the leading edge.

D illustrates a slot-welding technique and is used where blade-passage height is too small (or the backward lean-angle too high) to permit conventional fillet welding. In E, an electron-beam technique is shown. Its major disadvantage is that electron-beam welds should preferably be stressed in tension but, for the configuration of E, they are in shear. The configurations of G through J use rivets. Where the rivet heads protrude into the passage aerodynamic performance is reduced. Riveted impellers were used in the 1960s—they are very rarely used now. Elongation of these rivets occurs at certain critical surge conditions and can lead to major failures.

Materials for fabricating these impellers are usually low-alloy steels, such as AISI 4140 or AISI 4340. AISI 4140 is satisfactory for most applications; AISI 4340 is used for large impellers requiring higher strengths. For corrosive gases, AISI 410 stainless steel (about 12 percent chromium) is used. Monel K-500 is employed in halogen gas atmospheres and oxygen compressors because of its resistance to sparking. Titanium impellers have been applied to chlorine service. Aluminum-alloy impellers have been used in great numbers, especially at lower temperatures (below 300°F). With new developments in aluminum alloys, this range is increasing. Aluminum and titanium are sometimes selected because of their low density. This low density can cause a shift in the critical speed of the rotor, which may be advantageous.

Axial-Flow Compressors Axial-flow compressors are used mainly as compressors for gas turbines. They are also used in the steel industry as blast furnace blowers and in the chemical industry for large nitric acid plants. They are mainly used for applications where the head required is low and the flow large.

Figure 10-92 shows a typical axial-flow compressor. The rotating element consists of a single drum to which are attached several rows

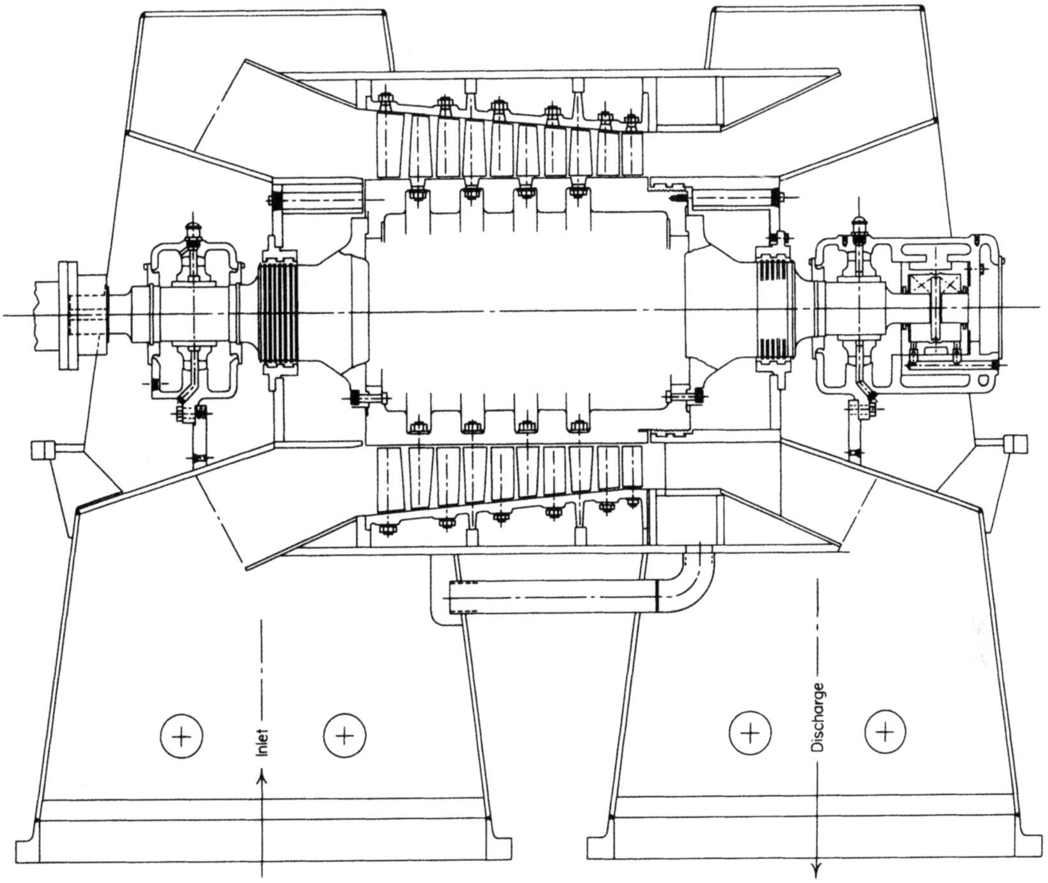

FIG. 10-92 Axial-flow compressor. (*Courtesy of Allis-Chalmers Corporation.*)

of decreasing-height blades having airfoil cross sections. Between each rotating blade row is a stationary blade row. All blade angles and areas are designed precisely for a given performance and high efficiency. The use of multiple stages permits overall pressure increases up to 30:1. The efficiency in an axial flow compressor is higher than the centrifugal compressor.

Pressure ratio per casing can be comparable with those of centrifugal equipment, although flow rates are considerably higher for a given casing diameter because of the greater area of the flow path. The pressure ratio per stage is less than in a centrifugal compressor. The pressure ratio per stage in industrial compressors is between 1:05 and 1:15, and for aeroturbines, 1.1 and 1.2.

The axial-flow compressors used in gas turbines vary depending on the type of turbines. The industrial-type gas turbine has an axial-flow compressor of a rugged construction. These units have blades that have low aspect ratio (R = blade height/blade chord) with minimum streamline curvation, and the shafts are support on sleeve-type bearings. The industrial gas turbine compressor has also a lower pressure ratio per stage (stage = rotor + stationary blade), giving a low blade loading. This also gives a larger operating range than its counterpart the aero axial gas turbine compressor but considerably less than the centrifugal compressor.

The axial-flow compressors in aero gas turbines are heavily loaded. The aspect ratio of the blades, especially the first few stages, can be as high as 4.0, and the effect of streamline curvature is substantial. The streamline configuration is a function of the annular passage area, the camber and thickness distribution of the blade, and the flow angles at the inlet and outlet of the blades. The shafts on these units are supported on antifriction bearings (roller or ball bearings).

The operation of the axial-flow compressor is a function of the rotational speed of the blades and the turning of the flow in the rotor. The stationary blades (stator) are used to diffuse the flow and convert the velocity increased in the rotor to a pressure increase. One rotor and one stator make up a stage in a compressor. One additional row of fixed blades (inlet guide vanes) is frequently used at the compressor inlet to ensure that air enters the first stage rotors at the desired angle. In addition to the stators, another diffuser at the exit of the compressor further diffuses the gas and, in the case of gas turbines, controls its velocity entering the combustor. The axial-flow compressor has a much smaller operating range "Surge to Choke" than its counterpart in the centrifugal compressor. Because of the steep characteristics of the head/flow capacity curve, the surge point is usually within 10 percent of the design point.

The axial-flow compressor has three distinct stall phenomena. Rotating stall and individual blade stall are aerodynamic phenomena. Stall flutter is an aeroelastic phenomenon. Rotating stall (propagating stall) consists of large stall zones covering several blade passages and propagates in the direction of the rotor and at some fraction of rotor speed. The number of stall zones and the propagating rates vary considerably. Rotating stall is the most prevalent type of stall phenomenon. Individual blade stall occurs when all the blades around the compressor annulus stall simultaneously without the occurrence of the stall propagation mechanism. The phenomena of stall flutter is caused by self-excitation of the blade and is aeroelastic. It must be distinguished from classic flutter, since classic flutter is a coupled

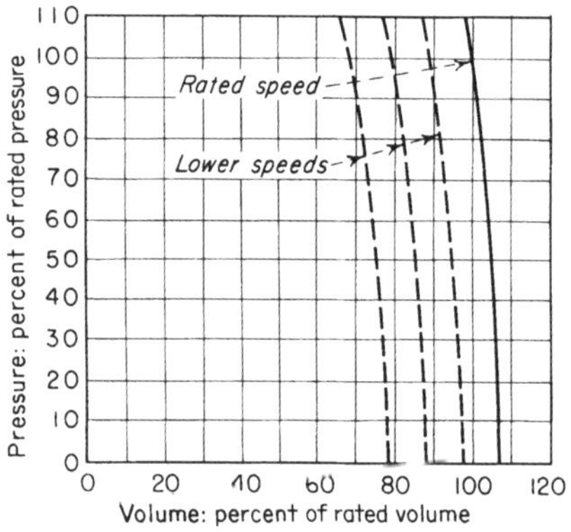

Fig. 10-93 Approximate performance curves for a rotary positive-displacement compressor. The safety valve in discharge line or bypass must be set to operate at a safe value determined by construction.

torsional-flexural vibration that occurs when the freestream velocity over an airfoil section reaches a certain critical velocity. Stall flutter, on the other hand, is a phenomenon that occurs due to the stalling of the flow around a blade. Blade stall causes Karman vortices in the airfoil wake. Whenever the frequency of the vortices coincides with the natural frequency of airfoil, flutter will occur. Stall flutter is a major cause of compressor-blade failure.

Positive-Displacement Compressors Positive-displacement compressors are machines that are essentially constant-volume machines with variable discharge pressures. These machines can be divided into two types:

1. Rotary compressors
2. Reciprocating compressors

Many users consider rotary compressors, such as the "Rootes"-type blower, as turbomachines because their behavior in terms of the rotor dynamics is very close to centrifugal and axial flow machinery. Unlike the reciprocating machines, the rotary machines do not have a very high vibration problem but, like the reciprocating machines, they are positive-displacement machines.

Rotary Compressors Rotary compressors are machines of the positive-displacement type. Such units are essentially constant-volume machines with variable discharge pressure. The volume can be varied only by changing the speed or by bypassing or wasting some of the capacity of the machine. The discharge pressure will vary with the resistance on the discharge side of the system. A characteristic curve typical of the form produced by these rotary units is shown in Fig. 10-93.

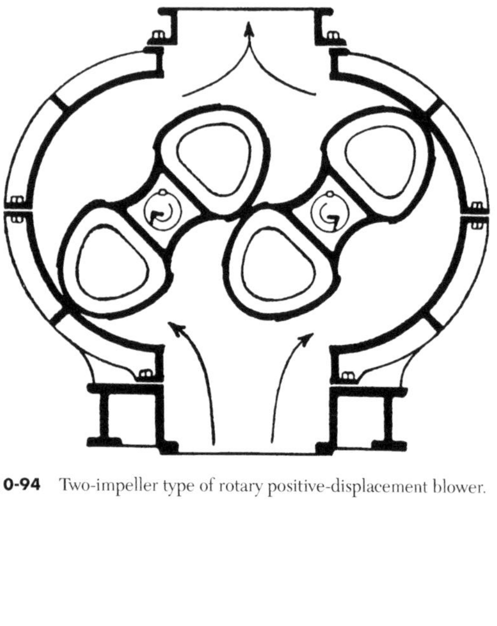

FIG. 10-94 Two-impeller type of rotary positive-displacement blower.

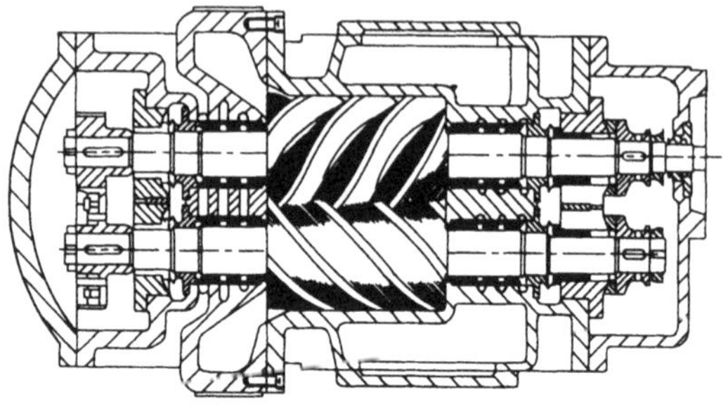

FIG. 10-95 Screw-type rotary compressor.

Rotary compressors are generally classified as of the straight-lobe type, screw type, sliding-vane type, and liquid-piston type.

Straight-Lobe Type This type is illustrated in Fig. 10-94. Such units are available for pressure differentials up to about 83 kPa (12 lbf/in^2) and capacities up to 2.549 × 10^4 m^3/h (15,000 ft^3/min). Sometimes multiple units are operated in series to produce higher pressures; individual-stage pressure differentials are limited by the shaft deflection, which must necessarily be kept small to maintain rotor and casing clearance.

Screw Type This type of rotary compressor, as shown in Fig. 10-95, is capable of handling capacities up to about 4.248 × 10^4 m^3/h (25,000 ft^3/min) at pressure ratios of 4:1 and higher. Relatively small-diameter rotors allow rotative speeds of several thousand rev/min. Unlike the straight-lobe rotary machine, it has male and female rotors whose rotation causes the axial progression of successive sealed cavities. These machines are staged with intercoolers when such an arrangement is advisable. Their high-speed operation usually necessitates the use of suction- and discharge-noise suppressors. The bearings used are sleeve-type bearings. Due to the side pressures experienced, tilting pad bearings are highly recommended.

Sliding-Vane Type This type is illustrated in Fig. 10-96. These units are offered for operating pressures up to 0.86 MPa (125 lbf/in^2) and in capacities up to 3.4 × 10^3 m^3/h (2000 ft^3/min). Generally, pressure ratios per stage are limited to 4:1. Lubrication of the vanes is required, and the air or gas stream therefore contains lubricating oil.

Liquid-Piston Type This type is illustrated in Fig. 10-97. These compressors are offered as single-stage units for pressure differentials up to about 0.52 MPa (75 lbf/in^2) in the smaller sizes and capacities up to 6.8 × 10^3 m^3/h (4000 ft^3/min) when used with a lower pressure differential. Staging is employed for higher pressure differentials. These units have found wide application as vacuum pumps on wet-vacuum service. Inlet and discharge ports are located in the impeller hub. As the vaned impeller rotates, centrifugal force drives the sealing liquid

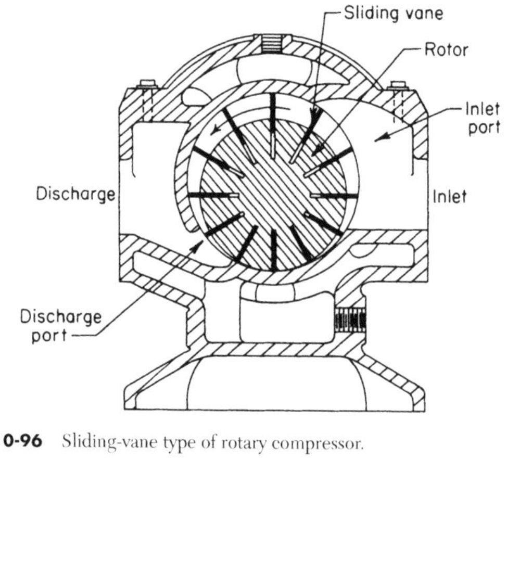

FIG. 10-96 Sliding-vane type of rotary compressor.

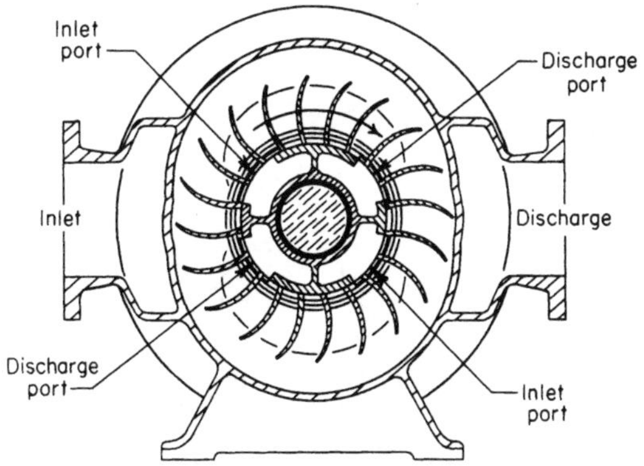

FIG. 10-97 Liquid-piston type of rotary compressor.

against the walls of the elliptical housing, causing the air to be successively drawn into the vane cavities and expelled against discharge pressure. The sealing liquid must be externally cooled unless it is used in a once-through system. A separator is usually employed in the discharge line to minimize carryover of entrained liquid. Compressor capacity can be considerably reduced if the gas is highly soluble in the sealing liquid.

The liquid-piston type of compressor has been of particular advantage when hazardous gases are being handled. Because of the gas-liquid contact and because of the much greater liquid specific heat, the gas-temperature rise is very small.

EJECTORS

An ejector is a simplified type of vacuum pump or compressor which has no pistons, valves, rotors, or other moving parts. Figure 10-98 illustrates a steam-jet ejector. It consists essentially of a nozzle which discharges a high-velocity jet across a suction chamber that is connected to the equipment to be evacuated. The gas is entrained by the steam and carried into a venturi-shaped diffuser which converts the velocity energy into pressure energy. Figure 10-99 shows a large-sized ejector, sometimes called a booster ejector, with multiple nozzles. Nozzles are devices in subsonic flow that have a decreasing area and accelerate the flow. They convert pressure energy to velocity energy. A minimum area is reached when velocity reaches sonic flow. In supersonic flow, the nozzle is an increasing area device. A diffuser in subsonic flow has an increasing area and converts velocity energy into pressure energy. A diffuser in supersonic flow has a decreasing area.

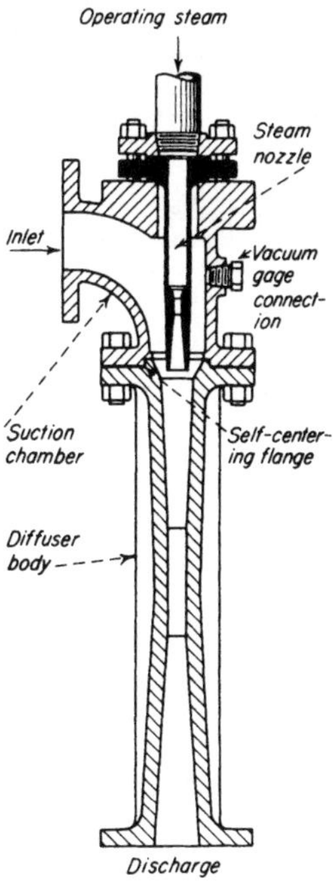

FIG. 10-98 Typical steam-jet ejector.

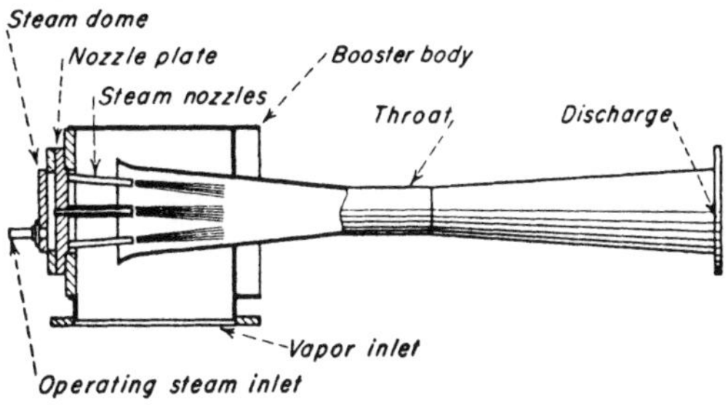

FIG. 10-99 Booster ejector with multiple steam nozzles.

Two or more ejectors may be connected in series or stages. Also, a number of ejectors may be connected in parallel to handle larger quantities of gas or vapor.

Liquid- or air-cooled condensers are usually used between stages. Liquid-cooled condensers may be of either the direct-contact (barometric) or the surface type. By condensing vapor the load on the following stage is reduced, thus minimizing its size and reducing consumption of motive gas. Likewise, a precondenser installed ahead of an ejector reduces its size and consumption if the suction gas contains vapors that are condensable at the temperature condition available. An **aftercondenser** is frequently used to condense vapors from the final stage, although this does not affect ejector performance.

Ejector Performance The performance of any ejector is a function of the area of the motive-gas nozzle and venturi throat, pressure of the motive gas, suction and discharge pressures, and ratios of specific heats, molecular weights, and temperatures. Figure 10-100, based on the assumption of **constant-area mixing,** is useful in evaluating single-stage-ejector performance for compression ratios up to 10 and area ratios up to 100 (see Fig. 10-101 for notation).

For example,* assume that it is desired to evacuate air at 2.94 lbf/in^2 with a steam ejector discharging to 14.7 lbf/in^2 with available steam pressure of 100 lbf/in^2. Entering the chart at $p_{03}/p_{0b} = 5.0$, at $p_{0b}/p_{0a} = 2.94/100 = 0.0294$ the optimum area ratio is 12. Proceeding horizontally to the left, w_b/w_a is approximately 0.15 lb of air per 1 lb of steam. This value must be corrected for the temperature and molecular-weight differences of the two fluids by Eq. (10-86).

$$w/w_a = w_b/w_a \sqrt{T_{0a}M_b/T_{0b}M_a} \qquad (10\text{-}86)$$

In addition, there are empirical correction factors which should be applied. Laboratory tests show that for ejectors with constant-area mixing the actual entrainment and compression ratios will be approximately 90 percent of the calculated values and even less at very small values of p_{0b}/p_{0a}. This compensates for ignoring wall friction in the mixing section and irreversibilities in the nozzle and diffuser. In theory, each point on a given design curve of Fig. 10-100 is associated with an optimum ejector for prevailing operating conditions. Adjacent points on the same curve represent theoretically different ejectors for the new conditions, the difference being that for each ratio of p_{0b}/p_{0a} there is an optimum area for the exit of the motive-gas nozzle. In practice, however, a segment of a given curve for constant A_2/A_t represents the

*All data are given in U.S. customary units since the charts are in these units. Conversion factors to SI units are given on the charts.

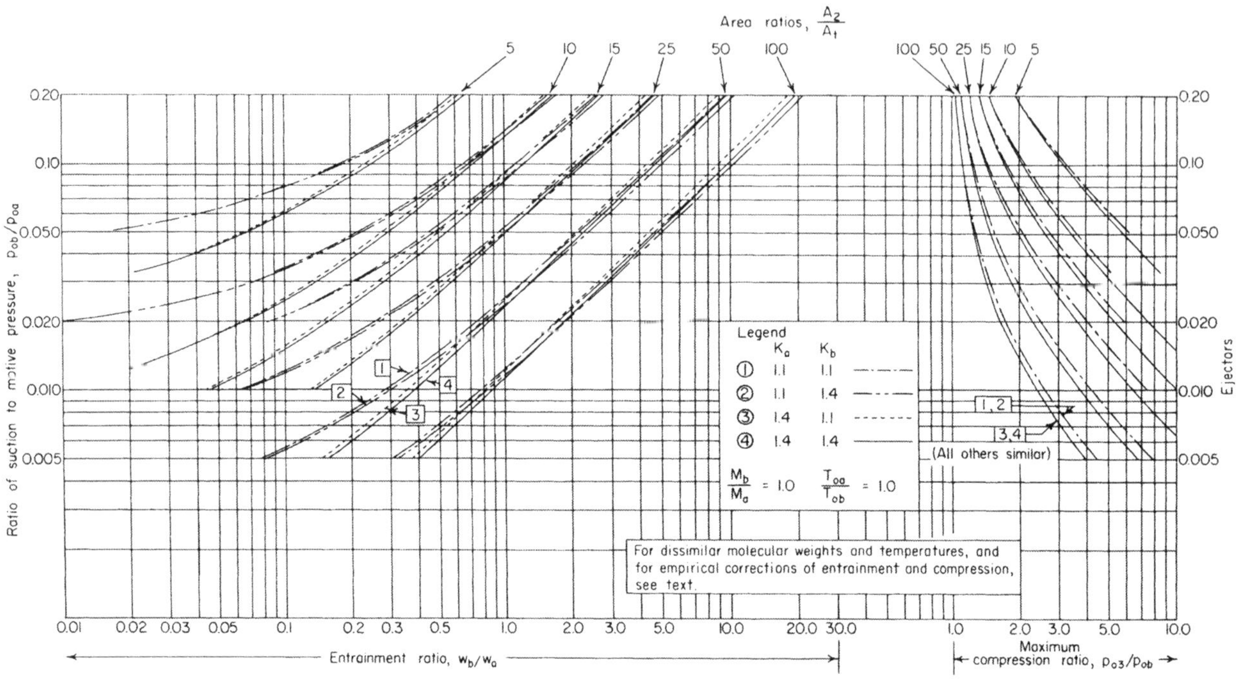

FIG. 10-100 Design curves for optimum single-stage ejectors. [*DeFrate and Hoerl,* Chem. Eng. Prog., **55,** *Symp. Ser. 21, 46 (1959).*]

performance of a single ejector satisfactorily for estimating purposes, provided that the suction pressure lies within 20 to 130 percent of the design suction pressure and the motive pressure within 80 to 120 percent of design motive pressure. Thus the curves can be used to select an optimum ejector for the design point and to estimate its performance at off-design conditions within the limits noted. Final ejector selection should, of course, be made with the assistance of a manufacturer of such equipment.

Uses of Ejectors For the operating range of steam-jet ejectors in vacuum applications, see the subsection "Vacuum Systems."

The choice of the most suitable type of ejector for a given application depends upon the following factors:

1. *Steam pressure.* Ejector selection should be based upon the minimum pressure in the supply line selected to serve the unit.
2. *Water temperature.* Selection is based on the maximum water temperature.
3. *Suction pressure and temperature.* Overall process requirements should be considered. Selection is usually governed by the minimum suction pressure required (the highest vacuum).
4. *Capacity required.* Again overall process requirements should be considered, but selection is usually governed by the capacity required at the minimum process pressure.

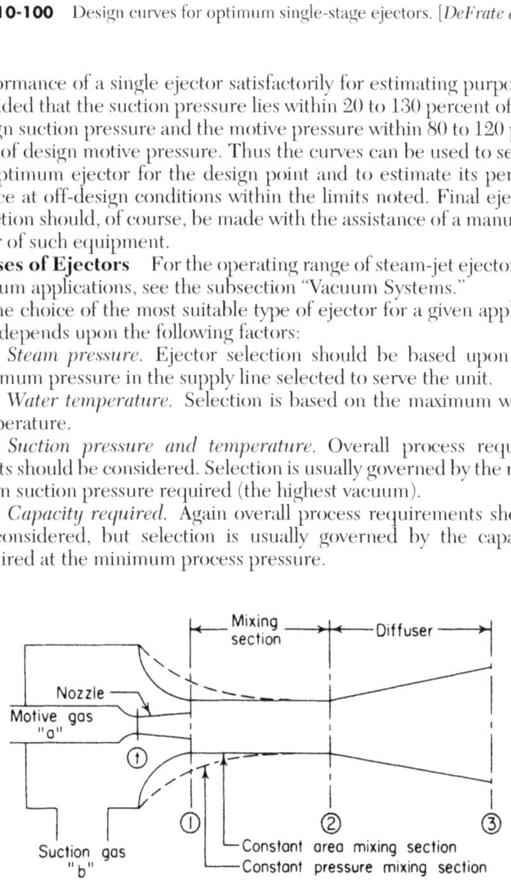

FIG. 10-101 Notation for Fig. 10-100.

Ejectors are easy to operate and require little maintenance. Installation costs are low. Since they have no moving parts, they have long life, sustained efficiency, and low maintenance cost. Ejectors are suitable for handling practically any type of gas or vapor. They are also suitable for handling wet or dry mixtures or gases containing sticky or solid matter such as chaff or dust.

Ejectors are available in many materials of construction to suit process requirements. If the gases or vapors are not corrosive, the diffuser is usually constructed of cast iron and the steam nozzle of stainless steel. For more corrosive gases and vapors, many combinations of materials such as bronze, various stainless-steel alloys, and other corrosion-resistant metals, carbon, and glass can be used.

VACUUM SYSTEMS

Figure 10-102 illustrates the level of vacuum normally required to perform many of the common manufacturing processes. The attainment of various levels is related to available equipment in Fig. 10-103.

Vacuum Equipment The equipment shown in Fig. 10-103 has been discussed elsewhere in this section with the exception of the **diffusion pump.** Figure 10-104 depicts a typical design. A liquid of low absolute vapor pressure is boiled in the reservoir. The vapor is ejected at high velocity in a downward direction through multiple jets and is condensed on the walls, which are cooled by the surrounding coils. Molecules of the gas being pumped enter the vapor stream and are driven downward by collisions with the vapor molecules. The gas molecules are removed through the discharge line by a backing pump such as a rotary oil-sealed unit.

Diffusion pumps operate at very low pressures. The ultimate vacuum attainable depends somewhat upon the vapor pressure of the pump liquid at the temperature of the condensing surfaces. By providing a cold trap between the diffusion pump and the region being evacuated, pressures as low as 10^{-7} mmHg absolute are achieved in this manner. Liquids used for diffusion pumps are mercury and oils of low vapor pressure. Silicone oils have excellent characteristics for this service.

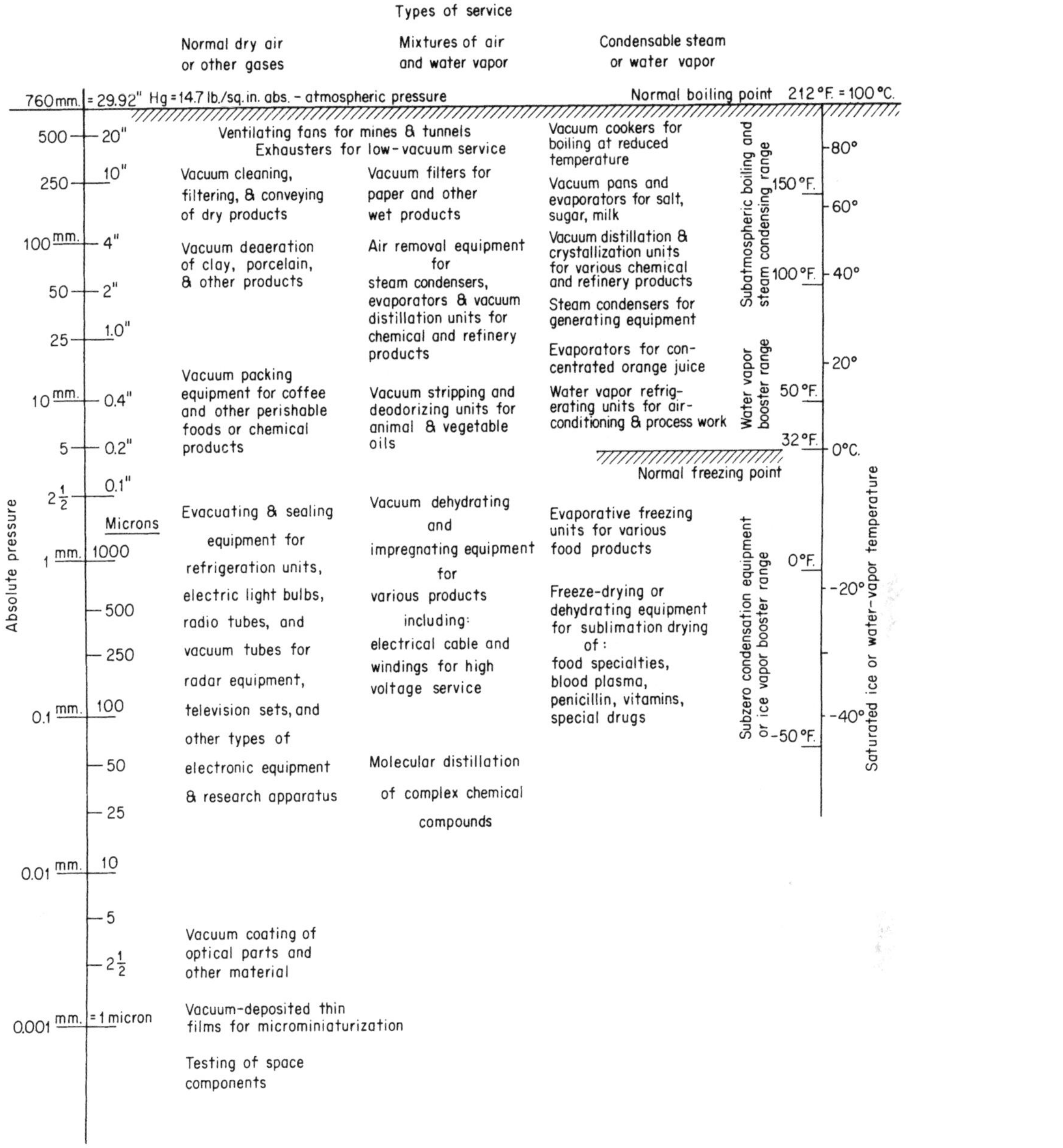

FIG. 10-102 Vacuum levels normally required to perform common manufacturing processes. (*Courtesy of* Compressed Air *magazine.*)

SEALING OF ROTATING SHAFTS

Seals are very important and often critical components in large rotating machinery especially on high-pressure and high-speed equipment. The principal sealing systems used between the rotor and stationary elements fall into two main categories: (1) noncontacting seals and (2) face seals. These seals are an integral part of the rotating system, they affect the dynamic operating characteristics of the machine. The stiffness and damping factors will be changed by the seal geometry and pressures. In operation the rotating shafts have both radial and axial movement. Therefore any seal must be flexible and compact to ensure maximum sealing minimum effect on rotor dynamics.

Noncontact Seals Noncontact seals are used extensively in gas service in high speed rotating equipment. These seals have good mechanical reliability and minimum impact on the rotor dynamics of the system. They are not positive sealing. There are two types of noncontact seals: (1) labyrinth seals and (2) ring seals.

Labyrinth Seals The labyrinth is one of the simplest of the many sealing devices. It consists of a series of circumferential strips of metal extending from the shaft or from the bore of the shaft housing to form a cascade of annular orifices. Labyrinth seal leakage is greater than that of clearance bushings, contact seals, or filmriding seals.

The major advantages of labyrinth seals are their simplicity, reliability, tolerance to dirt, system adaptability, very low shaft power

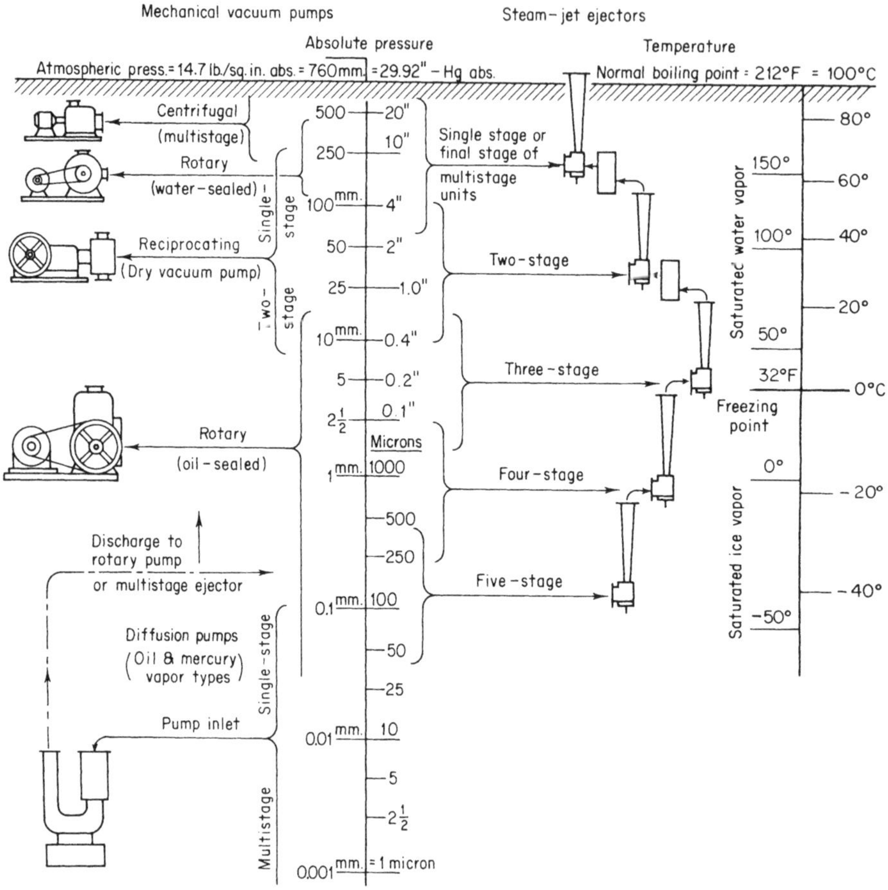

FIG. 10-103 Vacuum levels attainable with various types of equipment. (*Courtesy of* Compressed Air *magazine.*)

consumption, material selection flexibility, minimal effect on rotor dynamics, back diffusion reduction, integration of pressure, lack of pressure limitations, and tolerance to gross thermal variations. The major disadvantages are the high leakage, loss of machine efficiency, increased buffering costs, tolerance to ingestion of particulates with resulting damage to other critical items such as bearings, the possibility of the cavity clogging due to low gas velocities or back diffusion, and the inability to provide a simple seal system that meets OSHA or EPA standards. Because of some of the foregoing disadvantages, many machines are being converted to other types of seals.

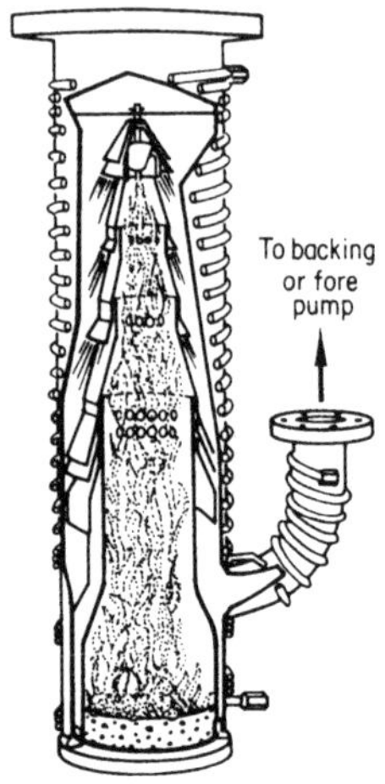

FIG. 10-104 Typical diffusion pump. (*Courtesy of* Compressed Air *magazine.*)

Labyrinth seals are simple to manufacture and can be made from conventional materials. Early designs of labyrinth seals used knife-edge seals and relatively large chambers or pockets between the knives. These relatively long knives are easily subject to damage. The modern, more functional, and more reliable labyrinth seals consist of sturdy, closely spaced lands. Some labyrinth seals are shown in Fig. 10-105. Figure 10-105*a* is the simplest form of the seal. Figure 10-105*b* shows a grooved seal; it is more difficult to manufacture but produces a tighter seal. Figure 10-105*c* and 10-105*d* is rotating labyrinth-type seals. Figure 10-105*e* shows a simple labyrinth seal with a buffered gas for which pressure must be maintained above the process gas pressure and the outlet pressure (which can be greater than or less than the atmospheric pressure). The buffered gas produces a fluid barrier to the process gas. The eductor sucks gas from the vent near the atmospheric end. Figure 10-105*f* shows a buffered, stepped labyrinth. The step labyrinth gives a tighter seal. The matching stationary seal is usually manufactured from soft materials such as babbitt or bronze, while the stationary or rotating labyrinth lands are made from steel. This composition enables the seal to be assembled with minimal clearance. The lands can therefore cut into the softer materials to provide the necessary running clearances for adjusting to the dynamic excursions of the rotor. To maintain maximum sealing efficiency, it is essential that the labyrinth lands maintain sharp edges in the direction of the flow.

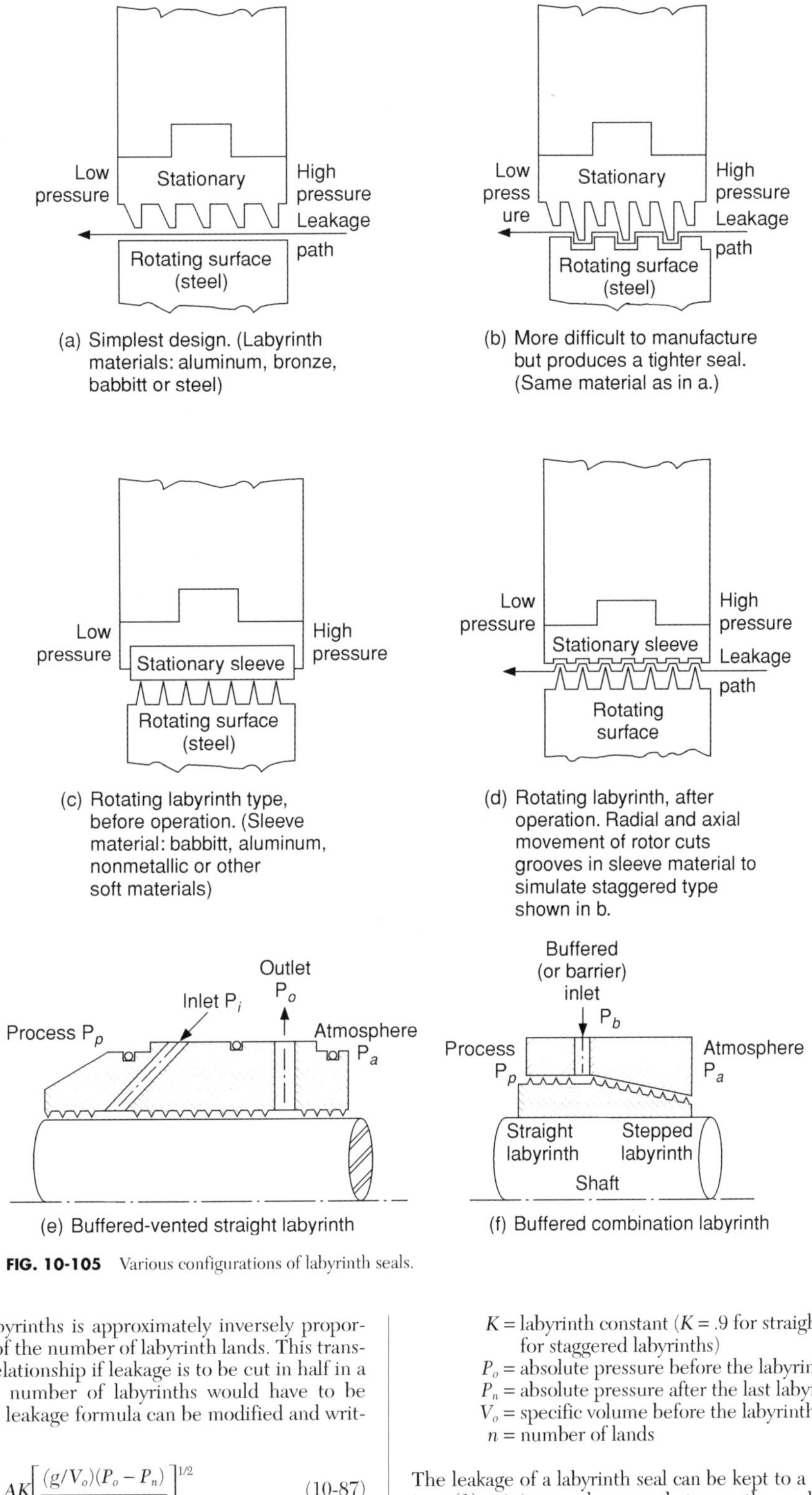

FIG. 10-105 Various configurations of labyrinth seals.

Leakage past these labyrinths is approximately inversely proportional to the square root of the number of labyrinth lands. This translates into the following relationship if leakage is to be cut in half in a four labyrinth seal: The number of labyrinths would have to be increased to 16. The Elgi leakage formula can be modified and written as:

$$m_\ell = AK\left[\frac{(g/V_o)(P_o - P_n)}{n + \ln(P_n/P_o)}\right]^{1/2} \qquad (10\text{-}87)$$

where m_ℓ = leakage
A = leakage area of single throttling
K = labyrinth constant (K = .9 for straight labyrinths, K = .75 for staggered labyrinths)
P_o = absolute pressure before the labyrinth
P_n = absolute pressure after the last labyrinth
V_o = specific volume before the labyrinth
n = number of lands

The leakage of a labyrinth seal can be kept to a minimum by providing: (1) minimum clearance between the seal lands and the seal sleeve, (2) sharp edges on the lands to reduce the flow discharge coefficient, and (3) grooves or steps in the flow path for reducing dynamic head carryover from stage to stage.

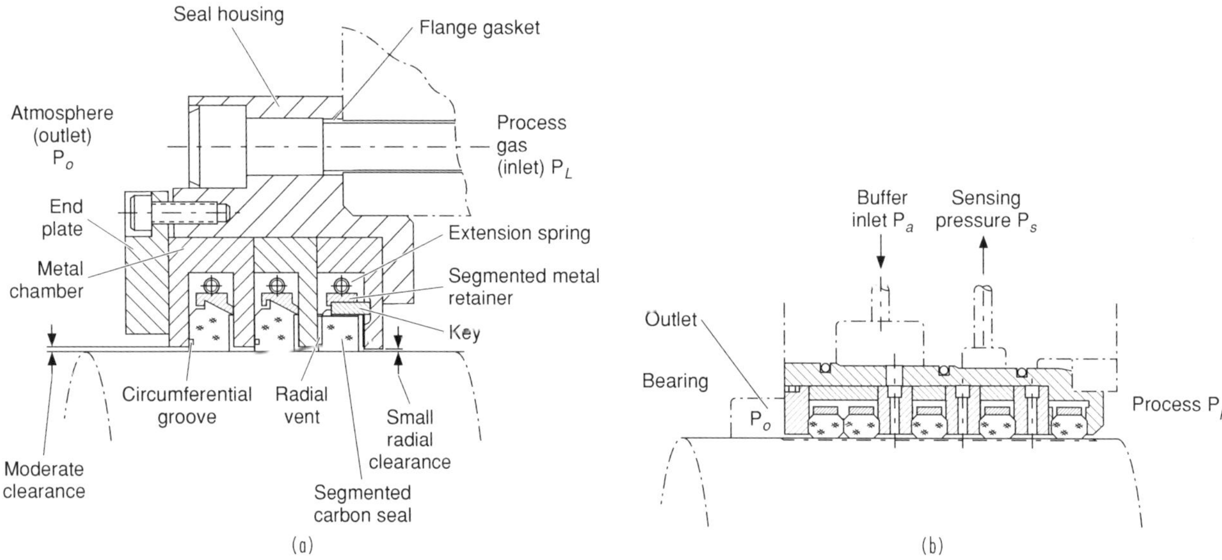

FIG. 10-106 Floating-type restrictive ring seal.

The labyrinth sleeve can be flexibly mounted to permit radial motion for self-aligning effects. In practice, a radial clearance of under 0.008 is difficult to achieve.

Ring Seals The restrictive ring seal is essentially a series of sleeves in which the bores form a small clearance around the shaft. Thus, the leakage is limited by the flow resistance in the restricted area and controlled by the laminar or turbulent friction. There are two types of ring seals: (1) fixed seal rings and (2) floating seal rings. The floating rings permit a much smaller leakage; they can be either the segmented type as shown in Fig. 10-106*a* or the rigid type as shown in Fig. 10-106*b*.

Fixed Seal Rings The fixed-seal ring consists of a long sleeve affixed to a housing in which the shaft rotates with small clearances. Long assemblies must be used to keep leakage within a reasonable limit. Long seal assemblies aggravate alignment and rubbing problems, thus requiring shafts to operate below their capacity. The fixed bushing seal operates with appreciable eccentricity and, combined with large clearances, produces large leakages, thus making this kind of seal impractical where leakage is undesirable.

Floating Seal Rings Clearance seals that are free to move in a radial direction are known as floating seals. The floating characteristics permit them to move freely, thus avoiding severe rubs. Due to differential thermal expansion between the shaft and bushing, the bushings should be made of material with a higher coefficient of thermal expansion. This is achieved by shrinking the carbon into a metallic retaining ring with a coefficient of expansion that equals or exceeds that of the shaft material. It is advisable in high shearing applications to lock the bushings against rotation.

Buildup of dirt and other foreign material lodged between the seal ring and seat will create an excessive spin and damage on the floating seal ring unit. It is therefore improper to use soft material such as babbitt and silver as seal rings.

Packing Seal A common type of rotating shaft seal consists of packing composed of fibers which are first woven, twisted, or braided into strands and then formed into coils, spirals, or rings. To ensure initial lubrication and to facilitate installation, the basic materials are often impregnated. Common materials are braided and twisted rubber and duck, flax, jute, and metallic braids. The so-called plastic packings can be made up with varying amounts of fiber combined with a binder and lubricant for high-speed applications. Maximum temperatures that base materials of packings withstand and still give good service are as follows:

	°C	°F
Flax	38	100
Cotton	93	200
Duck and rubber	149	300
Rubber	177	350
Metallic (lead-based)	218	425
Metallic (aluminum-based)	552	1025
Metallic (copper-based)	829	1525

Packing may not provide a completely leak-free seal. With shaft surface speeds less than approximately 2.5 m/s (500 ft/min), the packing may be adjusted to seal completely. However, for higher speeds some leakage is required for lubrication, friction reduction, and cooling.

Application of Packing Coils and spirals are cut to form closed or nearly closed rings in the stuffing box. Clearance between ends should be sufficient to allow for fitting and possible expansion due to increased temperature or liquid absorption of the packing while in operation.

The correct form of the ring joint depends on materials and service requirements. Braided and flexible metallic packings usually have butt or square joints (Fig. 10-107*a*). With other packing material, service experience indicates that rings cut with bevel or skive joints (Fig. 10-107*b*) are more satisfactory. A slight advantage of the bevel joint over the butt joint is that the bevel permits a certain amount of sliding action, thus absorbing a portion of ring expansion.

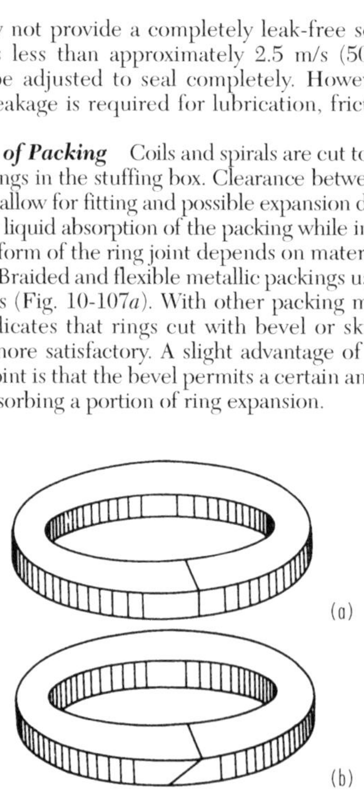

FIG. 10-107 Butt (*a*) and skive (*b*) joints for compression packing rings.

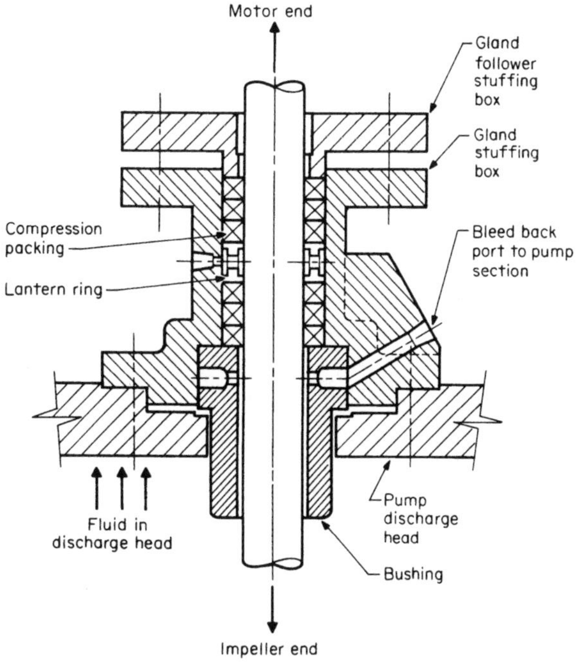

FIG. 10-108 Seal cage or lantern ring. (*Courtesy of Crane Packing Co.*)

In the manufacture of packings, the proper grade and type of **lubricant** is usually impregnated for each service for which the packing is recommended. However, it may be desirable to replenish the lubricant during the normal life of the packing. Lack of lubrication causes packing to become hard and lose its resiliency, thus increasing friction, shortening packing life, and increasing operating costs.

An effective auxiliary device frequently used with packing and rotary shafts is the **seal cage** (or **lantern ring**), shown in Fig. 10-108. The seal cage provides an annulus around the shaft for the introduction of a lubricant, oil, grease, etc. The seal cage is also used to introduce liquid for cooling, to prevent the entrance of atmospheric air, or to prevent the infiltration of abrasives from the process liquid.

The chief advantage of packing over other types of seals is the ease with which it can be adjusted or replaced. Most equipment is designed so that disassembly of major components is not required to remove or add packing rings. The major disadvantages of a packing-type seal are (1) short life, (2) requirement for frequent adjustment, and (3) need for some leakage to provide lubrication and cooling.

Mechanical Face Seals This type of seal forms a running seal between flat precision-finished surfaces. It is an excellent seal against leakages. The sealing surfaces are planes perpendicular to the rotating shaft, and the forces that hold the contact faces are parallel to the shaft axis. For a seal to function properly, there are four sealing points:

1. Stuffing box face
2. Leakage down the shaft
3. Mating ring in the gland plate
4. Dynamic faces

Mechanical Seal Selection There are many factors that govern the selection of seals. These factors apply to any type of seal:

1. Product
2. Seal environment
3. Seal arrangement
4. Equipment
5. Secondary packing
6. Seal face combinations
7. Seal gland plate
8. Main seal body

Product Physical and chemical properties of the liquid or gas being sealed places constraints on the type of material, design, and arrangement of the seal.

Pressure. Pressure affects the choice of material and whether balanced or unbalanced seal design can be used. Most unbalanced seals are good up to 100 psig stuffing box pressure. Over 100 psig, balanced seals should be used.

Temperature. The temperature of the liquid being pumped is important because it affects the seal face material selection as well as the wear life of the seal face.

Lubricity. In any mechanical seal design, there is rubbing motion between the dynamic seal faces. This rubbing motion is often lubricated by the fluid being pumped. Most seal manufacturers limit the speed of their seals to 90 ft/s (30 m/s). This is primarily due to centrifugal forces acting on the seal, which tends to restrict the seal's axial flexibility.

Abrasion. If there are entrained solids in the liquid, it is desirable to have a flushed single inside type with a face combination of very hard material.

Corrosion. This affects the type of seal body: what spring material, what face material, and what type of elastomer or gasket material. The corrosion rate will affect the decision of whether to use a single or multiple spring design because the spring can usually tolerate a greater amount of corrosion without weakening it appreciably.

Seal Environment The design of the seal environment is based on the product and the four general parameters that regulate it:

1. Pressure control
2. Temperature control
3. Fluid replacement
4. Atmospheric air elimination

Seal Arrangement There are four types of seal arrangements:

1. Double seals are standard with toxic and lethal products, but maintenance problems and seal design contribute to poor reliability. The double face-to-face seal may be a better solution.
2. Do not use a double seal in dirty service—the inside seal will hang up.
3. API standards for balanced and unbalanced seals are good guidelines; too low a pressure for a balanced seal may encourage face lift-off.
4. Arrangement of the seal will determine its success more than the vendor. Over 100 arrangements are available.

Equipment The geometry of the pump or compressor is very important in seal effectiveness. Different pumps with the same shaft diameter and the total differential head can present different sealing problems.

Secondary Packing Much more emphasis should be placed on secondary packing especially if Teflon is used. A wide variation in performance is seen between various seal vendors, depending on seal arrangement there can be differences in mating ring packing.

Seal Face Combinations The dynamics of seal faces is better understood today. Seal-face combinations have come a long way in the past 8–10 years. Stellite is being phased out of the petroleum and petrochemical applications. Better grades of ceramic are available, cost of tungsten has come down, and relapping of tungsten are available near most industrial areas. Silicon carbide is being used in abrasive service.

Seal Gland Plate The seal gland plate is caught in between the pump vendor and the seal vendor. Special glands should be furnished by seal vendors, especially if they require heating, quenching, and drain with a floating-throat bushing. Gland designs are complex and may have to be revisited, especially if seals are changed.

Main Seal Body The term *seal body* makes reference to all rotating parts on a pusher seal, excluding shaft packing and seal ring. In many cases it is the chief reason to avoid a particular design for a particular service.

Basically, most mechanical seals have the following components as seen in Fig. 10-109.

1. Rotating seal ring
2. Stationary seal ring
3. Spring devices to provide pressure
4. Static seals

A loading device such as a spring is needed to ensure that in the event of loss or hydraulic pressure the sealing surfaces are kept closed. The amount of the load on the sealing area is determined by the

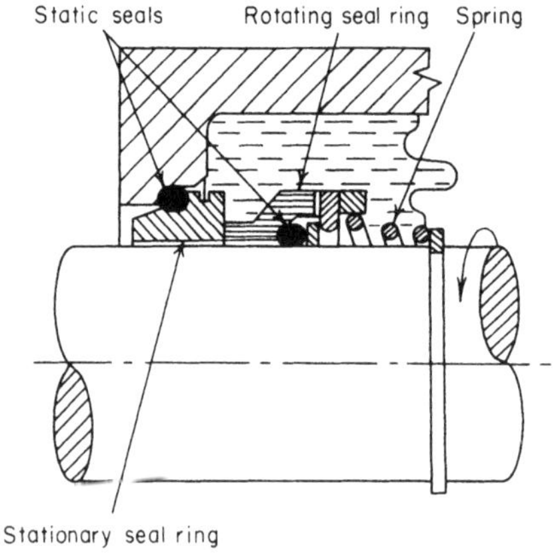

FIG. 10-109 Mechanical-seal components.

degree of "seal balance." Figure 10-110 shows what seal balance means. A completely balanced seal is when the only force exerted on the sealing surfaces is the spring force; i.e., hydraulic pressure does not act on the sealing surface. The type of spring depends on the space available, loading characteristics, and the seal environment. Based on these considerations, either a single or multiple spring can be used. In small axial space, belleville springs, finger washers, or curved washers can be used.

Shaft-sealing elements can be split up into two groups. The first type may be called pusher-type seals and includes the O-ring, V-ring, U-cup, and wedge configurations. Figure 10-111 shows some typical pusher-type seals. The second type is the bellow-type seals, which differ from the pusher-type seals in that they form a static seal between themselves and the shaft.

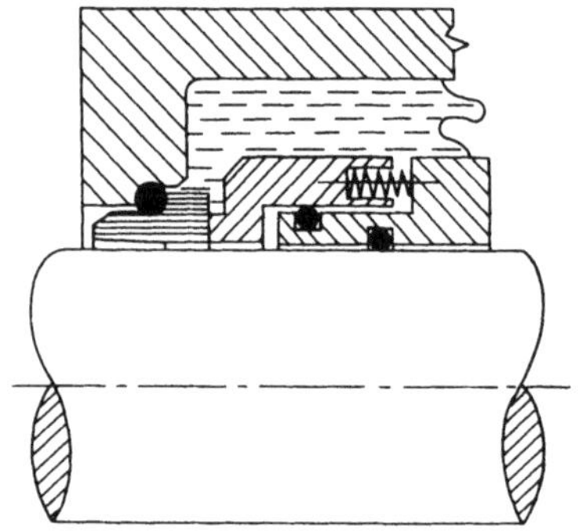

FIG. 10-110 Balanced internal mechanical seal.

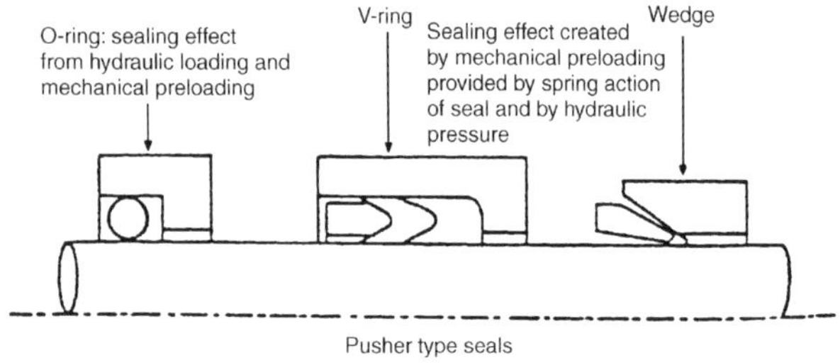

FIG. 10-111 Various types of shaft-sealing elements.

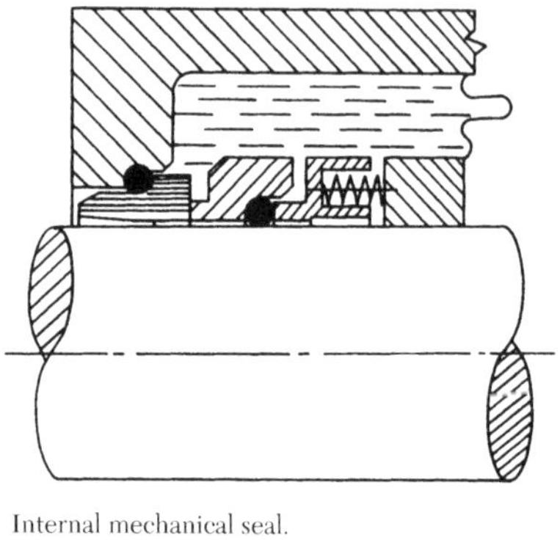

FIG. 10-112 Internal mechanical seal.

Internal and External Seals Mechanical seals are classified broadly as internal or external. **Internal seals** (Fig. 10-112) are installed with all seal components exposed to the fluid sealed. The advantages of this arrangement are (1) the ability to seal against high pressure, since the hydrostatic force is normally in the same direction as the spring force; (2) protection of seal parts from external mechanical damage; and (3) reduction in the shaft length required.

For high-pressure installations, it is possible to balance partially or fully the hydrostatic force on the rotating member of an internal seal by using a stepped shaft or shaft sleeve (Fig. 10-110). This method of relieving face pressure is an effective way of decreasing power consumption and extending seal life.

When abrasive solids are present and it is not permissible to introduce appreciable quantities of a secondary flushing fluid into the process, double internal seals are sometimes used (Fig. 10-113). Both sealing faces are protected by the flushing fluid injected between them even though the inward flow is negligible.

External seals (Fig. 10-114) are installed with all seal components protected from the process fluid. The advantages of this arrangement are that (1) fewer critical materials of construction are required, (2) installation and setting are somewhat simpler because of the exposed position of the parts, and (3) stuffing-box size is not a limiting factor. Hydraulic balancing is accomplished by proper proportioning of the seal face and secondary seal diameters.

Throttle Bushings These bushings (Fig. 10-115) are commonly used with single internal or external seals when solids are present in the fluid and the inflow of a flushing fluid is not objectionable. These close-clearance bushings are intended to serve as flow restrictions through which the maintenance of a small inward flow of flushing fluid prevents the entrance of a process fluid into the stuffing box.

A typical complex seal utilizes both the noncontact and mechanical aspects of sealing. Figure 10-116 shows such a seal with its two major elements. This type of seal will normally have buffering via a labyrinth

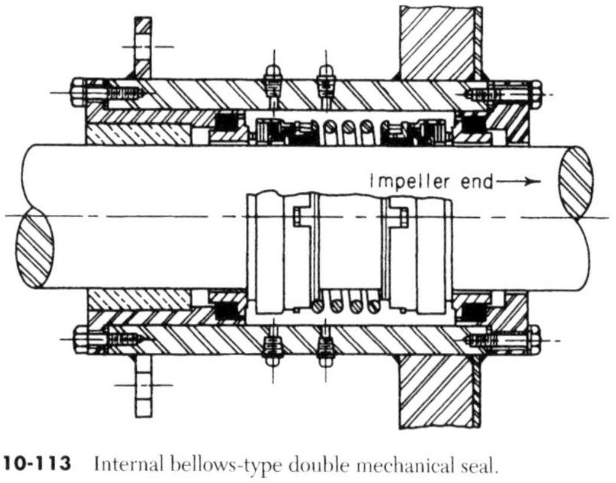

FIG. 10-113 Internal bellows-type double mechanical seal.

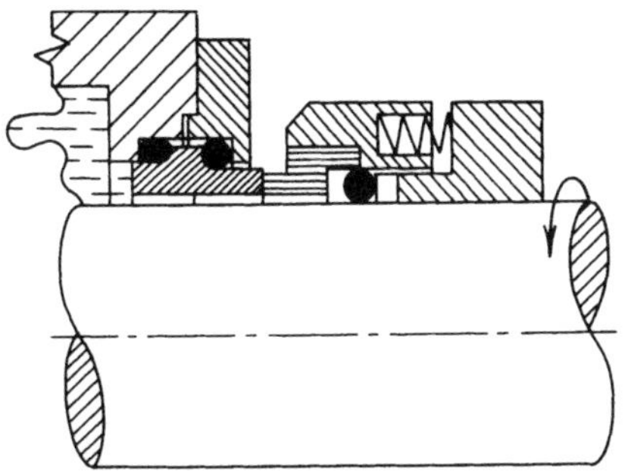

FIG. 10-114 External mechanical seal.

seal and a positive shutdown device. For shutdown, the carbon ring is tightly sandwiched between the rotating seal ring and the stationary sleeve with gas pressure to prevent gas from leaking out when no oil pressure is available.

In operation seal oil pressure is about 30–50 psi over the process gas pressure. The high-pressure oil enters the top and completely fills the seal cavity. A small percentage is forced across the carbon ring seal faces. The rotative speed of the carbon ring can be anywhere between zero and full rotational speed. Oil crossing the seal faces contacts the process gas and therefore is "contaminated oil." The contaminated oil leaves through the contaminated oil drain to a degassifier for purification. The majority of the oil flows through the uncontaminated seal oil drain.

Materials Springs and other metallic components are available in a wide variety of alloys and are usually selected on the basis of temperature and corrosion conditions. The use of a particular mechanical seal is frequently restricted by the temperature limitations of the organic materials used in the static seals. Most elastomers are limited to about 121°C (250°F). Teflon will withstand temperatures of 260°C (500°F) but softens appreciably above 204°C (400°F). Glass-filled Teflon is dimensionally stable up to 232 to 260°C (450 to 500°F).

One of the most common elements used for seal faces is carbon. Although compatible with most process media, carbon is affected by strong oxidizing agents, including fuming nitric acid, hydrogen chloride, and high-temperature air [above 316°C (600°F)]. Normal mating-face materials for carbon are tungsten or chromium carbide, hard steel, stainless steel, or one of the cast irons.

Other sealing-face combinations that have been satisfactory in corrosive service are carbide against carbide, ceramic against ceramic, ceramic against carbon, and carbon against glass. The ceramics have also been mated with the various hard-facing alloys. When selecting seal materials the possibility of galvanic corrosion must also be considered.

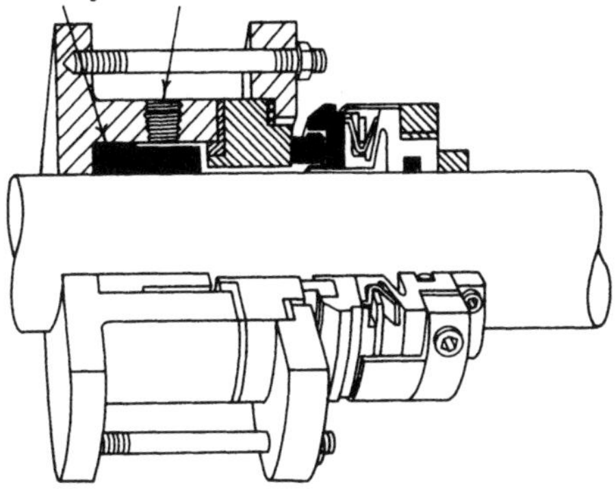

FIG. 10-115 External mechanical seal and throttle bushing.

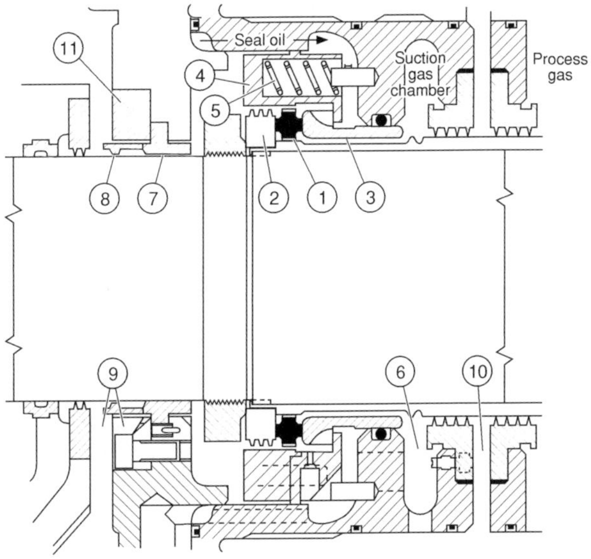

1. Rotating carbon ring
2. Rotating seal ring
3. Stationary sleeve
4. Spring retainer
5. Spring
6. Gas and contaminated oil drain
7. Floating babbitt-faced steel ring
8. Seal wiper ring
9. Seal oil drain line
10. Buffer gas injection port
11. Bypass orifice

FIG. 10-116 Mechanical contact shaft seal.

BEARINGS

Many factors enter into the selection of the proper design for bearings. Some of these factors are:

1. Shaft speed range.
2. Maximum shaft misalignment that can be tolerated.
3. Critical speed analysis and the influence of bearing stiffness on this analysis
4. Loading of the compressor impellers
5. Oil temperatures and viscosity
6. Foundation stiffness
7. Axial movement that can be tolerated
8. Type of lubrication system and its contamination
9. Maximum vibration levels that can be tolerated

Types of Bearings Figure 10-117 shows a number of different types of journal bearings. A description of a few of the pertinent types of journal bearings is given here:

1. *Plain journal.* Bearing is bored with equal amounts of clearance (on the order of one and one-half to two thousands of an inch per inch of journal diameter) between the journal and bearing.

2. *Circumferential grooved bearing.* Normally the oil groove is half the bearing length. This configuration provides better cooling but reduces load capacity by dividing the bearing into two parts.

3. *Cylindrical bore bearings.* Another common bearing type used in turbines. It has a split construction with two axial oil-feed grooves at the split.

4. *Pressure or pressure dam.* Used in many places where bearing stability is required, this bearing is a plain journal bearing with a pressure pocket cut in the unloaded half. This pocket is approximately ¹⁄₃₂ of an inch deep with a width 50 percent of the bearing length. This groove or channel covers an arc of 135° and terminates abruptly in a sharp-edge-edge dam. The direction or rotation is such that the oil is pumped down the channel toward the sharp edge. Pressure dam bearings are for one direction of rotation. They can be used in conjunction with cylindrical bore bearings as shown in Fig. 10-117.

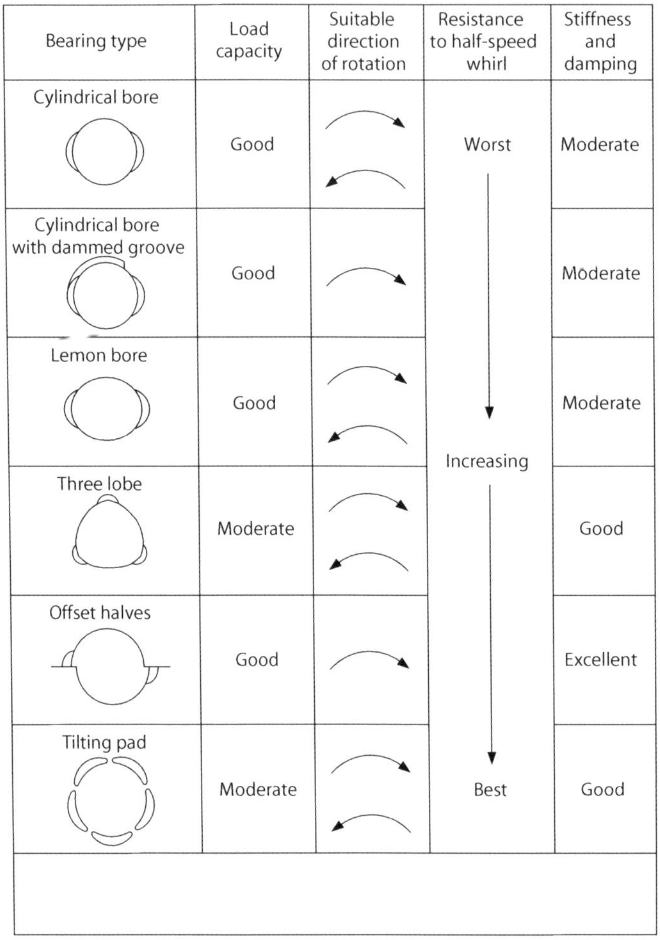

FIG. 10-117 Comparison of general bearing types.

5. *Lemon bore or elliptical.* This bearing is bored with shims split line, which are removed before installation. The resulting shape approximates an ellipse with the major axis clearance approximately twice the minor axis clearance. Elliptical bearings are for both directions of rotation.

6. *Three-lobe bearing.* The three-lobe bearing is not commonly used in turbomachines. It has a moderate load-carrying capacity and can be operated in both directions.

7. *Offset halves.* In principle, this bearing acts very similar to a pressure dam bearing. Its load-carrying capacity is good. It is restricted to one direction of rotation.

8. *Tilt-pad bearings.* This bearing is the most common bearing type in today's machines. It consists of several bearing pads posed around the circumference of the shaft. Each pad is able to tilt to assume the most effective working position. This bearing also offers the greatest increase in fatigue life because of the following advantages:

- Thermal conductive backing material to dissipate heat developed in oil film.
- A thin babbitt layer can be centrifugally cast with a uniform thickness of about 0.005 inch. Thick babbitts greatly reduce bearing life. Babbitt thickness in the neighborhood of .01 reduce the bearing life by more than half.
- Oil film thickness is critical in bearing stiffness calculations. In a tilting-pad bearing, one can change this thickness in a number of ways: (a) change the number of pads; (b) direct the load on or in between the pads; (c) change the axial length of pad.

The previous list contains some of the most common types of journal bearings. They are listed in the order of growing stability. All of the bearings designed for increased stability are obtained at higher manufacturing costs and reduced efficiency. The antiwhirl bearings all impose a parasitic load on the journal, which causes higher-power losses to the bearings and in turn requires higher oil flow to cool the bearing.

Thrust Bearings The most important function of a thrust bearing is to resist the unbalanced force in a machine's working fluid and to maintain the rotor in its position (within prescribed limits). A complete analysis of the thrust load must be conducted. As mentioned earlier, compressors with back-to-back rotors reduce this load greatly on thrust bearings. Figure 10-118 shows a number of thrust-bearing types. Plain, grooved thrust washers are rarely used with any continuous load, and their use tends to be confined to cases where the thrust load is very short duration or possibly occurs at standstill or low speed only. Occasionally, this type of bearing is used for light loads (less than 50 lb/in^2), and in these circumstances the operation is probably hydrodynamic due to small distortions present in the nominally flat bearing surface.

When significant continuous loads have to be taken on a thrust washer, it is necessary to machine into the bearing surface a profile to generate a fluid film. This profile can be either a tapered wedge or occasionally a small step.

Bearing type		Load capacity	Suitable direction of rotation	Tolerance of changing load/speed	Tolerance of misalignment	Space requirement
Plain washer 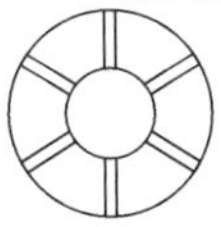		Poor		Good	Moderate	Compact
Taper land 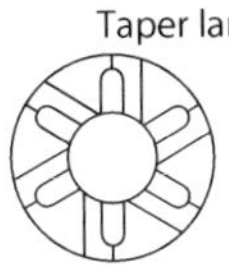	Bidirectional	Moderate		Poor	Poor	Compact
	Unidirectional	Good		Poor	Poor	Compact
Tilting pad 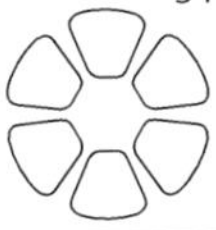	Bidirectional	Good		Good	Good	Greater
	Unidirectional	Good		Good	Good	Greater

FIG. 10-118 Comparison of thrust-bearing types.

The tapered-land thrust bearing, when properly designed, can take and support a load equal to a tilting-pad thrust bearing. With perfect alignment, it can match the load of even a self-equalizing tilting-pad thrust bearing that pivots on the back of the pad along a radial line. For variable-speed operation, tilting-pad thrust bearings as shown in Fig. 10-119 are advantageous when compared to conventional taper-land bearings. The pads are free to pivot to form a proper angle for lubrication over a wide speed range. The self-leveling feature equalizes individual pad loadings and reduces the sensitivity to shaft misalignments that may occur during service. The major drawback of this bearing type is that standard designs require more axial space than a nonequalizing thrust bearing.

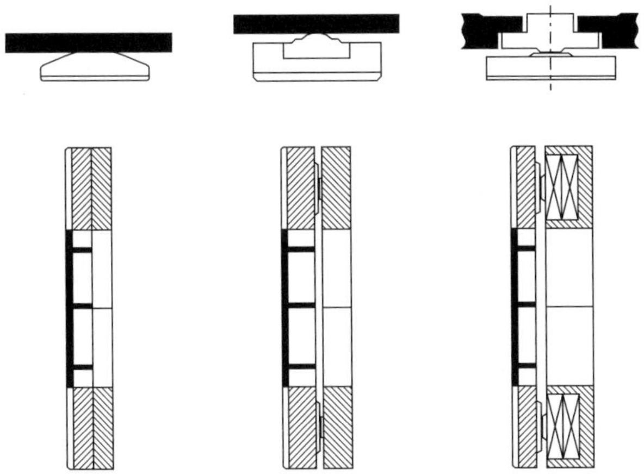

Nonequalizing tilting pad thrust bearing with radial pivot (a) | Nonequalizing tilting pad thrust bearing with ball pivot (b) | Self-equalizing thrust bearing (c)

FIG. 10-119 Various types of thrust bearings.

The thrust-carrying capacity can be greatly improved by maintaining pad flatness and removing heat from the loaded zone. By the use of high thermal conductivity backing materials with proper thickness and proper support, the maximum continuous thrust limit can be increased to 1000 psi or more. This new limit can be used to increase either the factor of safety and improve the surge capacity of a given size bearing or reduce the thrust bearing size and consequently the losses generated for a given load.

Since the higher thermal conductivity material (copper or bronze) is a much better bearing material than the conventional steel backing, it is possible to reduce the babbitt thickness to .010–.030 in. Embedded thermocouples and RTDs will signal distress in the bearing if properly positioned. Temperature-monitoring systems have been found to be more accurate than axial-position indicators, which tend to have linearity problems at high temperatures.

In a change from steel-backing to copper-backing, a different set of temperature limiting criteria should be used. Figure 10-120 shows a typical set of curves for the two backing materials. This chart also shows that drain oil temperature is a poor indicator of bearing operating conditions because there is very little change in drain oil temperature from low load to failure load.

Thrust-Bearing Power Loss The power consumed by various thrust bearing types is an important consideration in any system. Power losses must be accurately predicted so that turbine efficiency can be computed and the oil supply system properly designed.

Figure 10-121 shows a typical power consumption in thrust bearings as a function of unit speed. The total power loss is usually about 0.8–10 percent of the total rate power of the unit. New vector lube bearings reduce the horsepower loss by as much as 30 percent. In large vertical pumps, thrust bearings take not only the load caused by the fluid but also the load caused by the weight of the entire assembly (shaft and impellers). In some large pumps these could be about 60 ft (20 m) high and weigh 16 tons. The thrust bearing for such a pump is over 5 ft (1.7 m) in diameter with each thrust pad weighing over 110 lb (50 kg). In such cases, the entire pump assembly is first floated before the unit is started.

CENTRIFUGAL COMPRESSOR PROBLEMS

Compressors in process gas applications suffer from many problems. The following are some of the major categories in which these problems

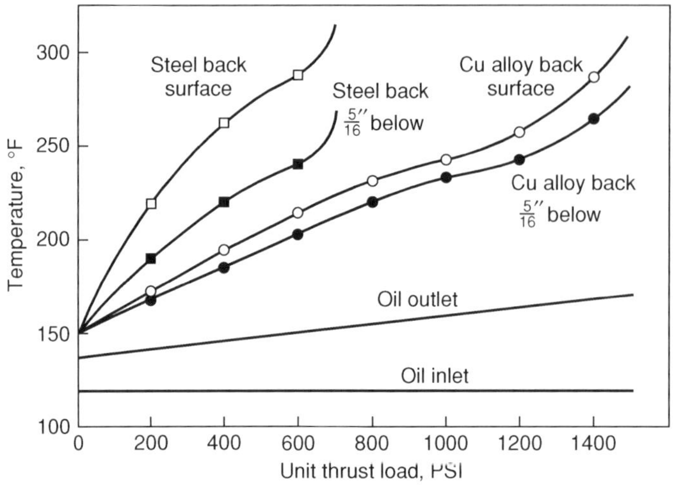

FIG. 10-120 Thrust-bearing temperature characteristics.

fall (see Meherwan P. Boyce, *Centrifugal Compressors: A Basic Guide,* PennWell, 2003):

1. Compressor fouling
2. Compressor failures
3. Impeller problems
4. Rotor thrust problems
5. Seals and bearings problems
6. Bearing maintenance

Compressor Fouling Centrifugal compressors, especially in the process gas applications, suffer greatly from fouling. Fouling is the deposit and nonuniform accumulation of debris in the gas on internal compressor surfaces. Fouling is due to the carryover of liquids and other debris from the suction knockout drums. Debris can roughen compressor surfaces. Polymerization can occur also due to changes in process conditions. In wet gas compressors, ethylene plant cracked gas compressors, and polyethylene recycle compressors, the temperature of the gas must be kept below the threshold temperature that would initiate polymerization. The buildup will usually occur on the hub and the shroud with a larger buildup on the shroud at the elbow of the impeller on closed-face impellers. There is also a buildup on the blades, with the buildup usually more on the pressure side than on the suction side. Often the buildup is the heaviest on the pressure side at the blade exit where there is also flow separation.

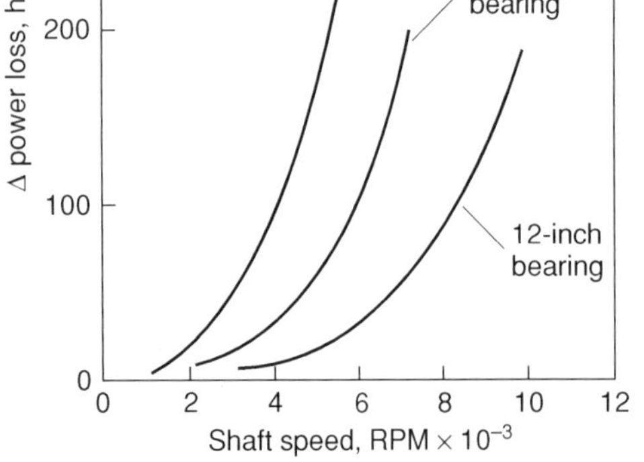

FIG. 10-121 Difference in total-power-loss data test minus catalog frictional losses versus shaft speed for 6 × 6 pad double-element thrust bearings.

Techniques to Prevent Fouling in Process Gas Compressors

1. *Condition monitoring of compressor aerodynamic and mechanical parameters.* Vibration monitoring could also alert the operator to fouling problems.

2. *Process control.* Accurate control of process conditions can prevent fouling in applications where polymers can form. Control of temperature is usually the most important. The following are examples of applications that can be affected by excessive process temperature:

a. Ethylene cracked gas
b. Linear low-density polyethylene
c. High-density propylene
d. Fluid catalytic cracker off-gas (wet gas)
e. Thermal cracker off-gas (wet gas)
f. Coker gas

The temperature below which fouling can be prevented varies with each process, compressor, and application. Monitoring of process conditions is necessary to establish a threshold temperature in each case. In some cases, fouling cannot be prevented with an existing compressor. It may be necessary to modify the aerodynamic design and/or add cooling.

3. *On-line solvent injection.* On-line solvent injection is very successful in various processes. The objective of this measure is to *continuously* inject a small amount of solvent to reduce the friction coefficient of the blade and impeller surface and thus prevent fouling of the surface. The injection should be done from the start; otherwise, the foulant could be dislodged and moved downstream, creating a major problem. The downstream areas are much smaller so foulant lodging there could create a blockage.

Air Compressors

1. *Inlet filter.* In air compressors filter selection is an important factor in preventing fouling. Most high-efficiency air filters have a triple-stage filtration system. Also these filters often have rain shades to prevent water from entering the filters. Site conditions play a very important part in the selection of the filters.

Compressor Blade Coating Coatings protect blades against oxidation, corrosion, and cracking problems. Coatings guard the base metal of the compressor from attack. Other benefits include reduced thermal fatigue from cyclic operation, increased surface smoothness, reduced erosion, and reduced heat flux loading when one is considering thermal barriers. Coatings increase resistance to spalling from light impacts. Coatings also extend compressor life, better endure operational conditions, and serve as a sacrificial layer by allowing the coating to be restripped and recoated on the same base metal.

Compressor Failures In the process industry there are three types of compressors that have very different maintenance problems:

1. Barrel-type compressor
2. Horizontal split casing centrifugal compressor with closed-face impellers
3. Air integral gear-type compressor with open-face impellers

Barrel-Type Compressors Barrel-type compressors are being utilized in the process industry to an increased extent because the barrel design confines gases more effectively than horizontally split cases. This becomes a critical consideration in two areas: high-pressure and low-molecular-weight gas compression. API-617, *Centrifugal Compressors for General Refinery Services,* requires a barrel design based on the molecular percent of hydrogen contained in the process gas and the discharge pressure. API-617 defines high-pressure compressors as units in which the partial pressure of the hydrogen exceeds 200 psig, and specifies that these units be vertically (radially) split. Hydrogen partial pressure is given by the following relationship (in absolute pressures):

$$P_{H2} = P(\%H_2/100) \tag{10-88}$$

Maximum casing working pressure for axially split compressors (psig) is

$$P_{\text{casing}} = \frac{200}{\%H_2/100} \tag{10-89}$$

The casings should include a minimum of a ⅛-in corrosion allowance. Casing strength and rigidity should limit change of alignment to 0.002 in if it is caused by the worst combination of pressure, torque, or allowable piping stress.

The barrel design is essentially a compressor placed inside a pressure vessel. For higher pressures some manufacturers have merely "beefed up" lower-pressure barrel designs, while others, have perfected unique designs such as the "shear ring" head design. All these designs make extensive use of elastomer O-rings as sealing devices. There are several inherent maintenance problems with barrel-type compressors:

Handling. Barrel-type machines must be removed from their foundations for total maintenance. Because barrel machines weigh up to 30 tons, the handling problems can be formidable.

Inner casing alignment. Since this type of compressor consists of a bundle contained within the pressure walls of the barrel, alignment and positive positioning are often very poor and the bundle is free to move to a certain extent. Bundle length is critical. Interstage leakage may occur if the bundle length is not correct. Assembly errors can be particularly detrimental in the case of a stacked diaphragm design, and care must be exercised to maintain proper impeller-diaphragm positioning. Since the bundle is subjected to discharge pressure on one end and suction pressure on the other, a force builds up that is transmitted from diaphragm to diaphragm, causing high loading on the inlet wall.

Internal leakage. The discharge and suction compartments of the inner bundle on a straight-through flow design are normally separated by a single O-ring. Compressors with side nozzles can have several bundles of O-rings, Excessive bundle-to-barrel clearance may cause leakage past the O-rings. O-rings are frequently pinched and cut across the suction nozzle opening in the barrel, a condition that is hard to prevent and doubly hard to detect if it occurs. Pressure differentials in excess of 400 to 500 psi even using good design practice can cause extrusion and failure of the O-rings. In many cases backup rings to the O-rings have been added to prevent failures. Grooves with O-ring ribbons have been added to the horizontal joints of the bundles of almost all the machines to prevent interstage leakage.

Bearing bracket alignment. In contrast to horizontally split compressors where the bearing brackets are normally an integral part of the lower case half, in barrel machines bearing brackets are bolted to the barrel heads. Both the bearing brackets *and* the head are removed during the disassembly operation, thus requiring *all* internal alignment to be reestablished each time maintenance work is performed.

Material problems. To limit the physical size of the case or pressure vessel, to limit the rotor bearing span, and to maximize the number of stages within the heavy barrel, the gas path of a barrel is "squeezed" to a greater extent than in a horizontally split machine. This means the diaphragms and inlet guide vanes are intricate shapes with very small openings. Plain gray cast iron is normally used for these shapes because of casting ease and for other economic reasons. The gray iron is not strong enough in many instances to withstand the pressure differentials imposed on it, resulting in failures. Inlet guide vanes have been especially troublesome. On several occasions inlet guide vanes have been fabricated from wrought stainless and carbon steel materials. Replacement diaphragms and inlet guide vanes cast of nodular iron have also been used to alleviate some of these material problems.

Impeller Problems The high-speed rotation of the impeller of a centrifugal compressor imparts the vital aerodynamic velocity to the flow within the gas path. The buffeting effects of the gas flow can cause fatigue failures in the conventional fabricated shrouded impeller due to vibration-induced alternating stresses. These may be of the following types:

1. Resultant vibration in a principal mode
2. Forced-undamped vibration, associated with aerodynamic buffeting or high acoustic energy levels

The vibratory mode most frequently encountered is of the plate type and involves either the shroud or the disc. Fatigue failure generally originates at the impeller outside diameter, adjacent to a vane often due to the vibratory motion of the shroud or disc. The fatigue crack propagates inward along the nodal line, and finally a section of the shroud or disc tears out.

To eliminate failures of the covered impellers when operating at high density levels, the impellers are frequently scalloped between vanes at the outside diameter. The consequent reduction in disc friction also causes a small increase in impeller efficiency. However, there may be a slight reduction in overall efficiency due to higher losses in the diffuser. The advantages of scalloped impellers from a mechanical point of view are large. Several rotors have been salvaged by scalloping the wheels after a partial failure has occurred.

Rotor Thrust Problems Thrust loads in compressors due to aerodynamic forces are affected by impeller geometry, by pressure rise through the compressor, and by internal leakage due to labyrinth clearances. The impeller thrust is calculated by using correction factors to account for internal leakage, and a balance piston size is selected to compensate for the impeller thrust load. The common assumptions made in the calculation are

1. The radial pressure distribution along the outside of the disc cover is essentially balanced.
2. Only the "eye" area is effective in producing thrust.
3. The pressure differential applied to the "eye" area is equal to the difference between the static pressure at the impeller tip, corrected for the pumping action of the disc, and the total pressure at inlet.

These "common assumptions" are grossly erroneous and can be disastrous when applied to high-pressure, barrel-type compressors where a large part of the impeller-generated thrust is compensated by a balance piston. The actual thrust is about 50 percent more than the calculations indicate. The error is less when the thrust is compensated by opposed impellers, because the mistaken assumptions offset each other.

The magnitude of the thrust is considerably affected by leakage at the impeller labyrinth seals. Increased leakage here produces increased thrust independent of balancing piston labyrinth seal clearance or leakage.

The thrust errors are further compounded in the design of the balancing piston, labyrinths, and bleed line. API-617, *Centrifugal Compressors*, specifies that a separate pressure tap connection be provided to indicate the pressure in the balance chamber. It also specifies that the balance line be sized to handle balance piston labyrinth gas leakage at twice the initial clearance without exceeding the load ratings of the thrust bearing, and that thrust bearings for compressors be selected at no more than 50 percent of the bearing manufacturer's rating. The leaks and the consequential pressure change across the balance piston destabilize the entire rotor system.

Journal Bearing Failures With high-speed machines, simple bearing failures are rare unless they are caused by faulty alignment, distortion, wrong clearance, or dirt. More common are failures caused by vibrations and rotor whirls. Some of these originate in the bearings; others can be amplified or attenuated by the bearings, the bearing cases, and the bearing support structure.

During inspection, all journal bearings should be closely inspected. If the machine has not suffered from excessive vibrations or lubrication problems, the bearings can be reinstalled and utilized.

Four places should be checked for wear during inspection periods:

1. Babbitted shoe surface
2. Pivoting shoe surface and seat in retaining ring
3. Seal ring bore or end plates
4. The shoe thickness at the pivot point or across ball *and* socket (all shoes should be within 0.0005 percent of the same thickness)

Thrust Bearing Failures Tilting pad-type thrust bearings are used in most major pieces of rotating equipment under the general term *Kingsbury type*. A thrust-bearing failure is one of the worst things that can happen to a machine, since it often wrecks the machine. To evaluate the reliability of a thrust bearing arrangement, one must first consider how a failure is initiated and evaluate the merits of the various designs.

Failures caused by bearing overload during normal operation (design error) are rare today, but still far more thrust failures occur than one would expect, considering all the precautions taken by the bearing designer. The causes in the following list are roughly in sequence of importance:

1. *Fluid slugging.* Passing a slug of fluid through a turbine or compressor can increase the thrust to many times its normal level.

2. *Buildup of solids in rotor and/or stator passages ("plugging" of turbine blades).* This problem should be noticed from performance or pressure distribution in the machine (first-stage pressure) long before the failure occurs.

3. *Off-design operation.* This arises especially from backpressure (vacuum), inlet pressure, extraction pressure, or moisture. Many failures are caused by overload, off-design speed, and flow fluctuations.

4. *Compressor surging.* This problem occurs especially in double-flow machines.

5. *Gear coupling thrust.* This is a frequent cause of failure, especially of upstream thrust bearings. The thrust is caused by friction in the loaded teeth that opposes thermal expansion. Therefore, thrust can get very high, since it has no relation to the normal thrust caused by pressure distribution inside the machine. The coupling thrust may act either way, adding to or subtracting from normal thrust. Much depends on tooth geometry and coupling quality.

6. *Dirt in oil.* This is a common cause of failures, especially when combined with other factors. The oil film at the end of the oil wedge is only a small fraction of a thousandth inch thick. If dirt goes through, it can cause the film to rupture, and the bearing may burn out. Therefore, very fine filtering of the oil is required.

7. *Momentary loss of oil pressure.* This type of failure is usually encountered while switching filters or coolers, or in some instances when the dc pump does not come on-line when the main pumps fail.

The thrust bearings must be closely maintained. This type of bearing consists of pivoted segments or pads (usually six) against which the thrust collar revolves, forming a wedge-shaped oil film. This film plus minute misalignment of the thrust collar and the bearing pads causes movement and wear of the various bearing parts. The erroneous thrust calculations discussed earlier cause the bearing to be loaded more heavily than desired. This accelerates the wear problem. There are seven wear points in the bearing. All these points must be checked for wear:

1. The soft babbitted shoe face
2. The hardened steel shoe insert face
3. The face of the hardened steel upper leveling plate
4. The outer edge of the upper leveling plate
5. The upper edge of lower leveling plate
6. The pivot point of the lower leveling plate
7. The inner face of the base ring

To protect thrust bearings, accurate and reliable instrumentation is now available to monitor thrust bearings well enough to ensure safe continuous operation and to prevent catastrophic failure in the event of an upset to the system.

Temperature sensors, such as RTDs (resistance temperature detectors), thermocouples, and thermistors, can be installed directly in the thrust bearing to measure metal temperature.

Axial proximity probes are another means of monitoring rotor position and the integrity of the thrust bearing. This method detects thrust collar runout and rotor movement. In most cases this ideal positioning of the probes is not possible. Many times the probes are indexed to the rotor or other convenient locations and thus do not truly show the movement of the rotor with respect to the thrust bearing.

A critical installation should have the metal temperature sensors in the thrust pad. Axial proximity probes may be used as a backup system. If metal temperatures are high and the rate of change of those temperatures begins to alter rapidly, thrust-bearing failure should be anticipated.

Compressor Seal Problems The extent of the leakage past the seals where the shaft comes through the casing frequently limits the running time of the compressor, yet the seals and the seal systems are not given adequate treatment in the maintenance manuals or in the operating instructions furnished by the compressor manufacturer.

Shaft seals are divided into the following categories by API Standard 617:

- Labyrinth
- Restrictive carbon rings
- Mechanical (contact) type
- Liquid film or floating bushing type
- Liquid film type with pump bushings

The first two seal categories are usually operated dry, and the last three categories require seal oil consoles either separately or as part of the lube system. Each of these seal designs has its own characteristics and maintenance difficulties.

Oil flows and critical clearances are not spelled out well in either the operating instructions or the maintenance manuals. Because of this, several maintenance technique improvements are needed:

1. *Radial clearances.* Radial clearance between the bushing and the shaft and the length of the bushing must be selected to obtain minimum leakage without exceeding fluid temperature limitations.

2. *Quality control.* The flatness, parallelism, and surface finish of the mating sleeve faces must be carefully controlled to obtain maximum seal effectiveness.

3. *Axial clearances.* Axial clearance between the bushing or sleeve and the housing is critical, there should be 12 to 15 mils of clearance per bushing between the bushing or sleeve and the housing; where the sleeves are mounted back-to-back, there will be 25 to 30 mils of clearance total for the seal.

4. *Seal design.* In higher-pressure seals, more than one outboard (i.e., high differential) sleeve may be used. Generally, it is desirable to use a single sleeve because the inboard sleeve operates with up to 80 percent of the total pressure drop across it. The outer sleeve with the lower differential causes lubrication and cooling problems that can shorten the life of one or both sleeves.

5. *Guidelines.* These should be explicit in indicating oil flow rates and the interaction of various components.

6. *Rules of thumb.* There are a few rules of thumb that help in understanding seal operation and maintenance.

a. The oil flow rate will vary: (1) *directly* with the differential pressure and the wetted perimeter of the sleeve; (2) with the *cube* of the radial clearance; (3) with the *square* of the eccentricity of the sleeve and shaft; (4) *inversely* with oil viscosity, temperature, and length of the sleeve.

b. Shear work done on the sealing fluid during its passage through the sleeve raises its temperature to a much higher level than may be expected.

ROTOR DYNAMICS

The rotating elements consist of the impeller and the shaft. The shaft should be made of one-piece, heat-treated forged steel, with the shaft ends tapered for coupling fits. Interstage sleeves should be renewable and made of material which is corrosion-resistant in the specified service. The rotor shaft sensing area observed by the noncontact probes should be concentric with the bearing journals and free of any scratches, marks, or other surface discontinuity. The surface finish should be 16 to 32 μin root mean square, and the area should be demagnetized and treated. Electromechanical runout should not exceed 25 percent of the maximum allowed peak-to-peak vibration amplitude or 0.25 mil, whichever is greater. Although not mentioned in the standard, chrome plating of the shaft in the sensing area is unacceptable. Maximum vibration should not exceed 2.0 mils as given by

$$\text{Vib}_{\text{max}} = \underset{\text{(Vibration)}}{\sqrt{\frac{12{,}000}{\text{rpm}}}} + \underset{\text{(runout)}}{0.25\sqrt{\frac{12{,}000}{\text{rpm}}}} \qquad (10\text{-}90)$$

At the trip speed of the driver (105 percent for a gas turbine), the vibration should not exceed this level by more than 0.5 mil.

The impellers can be an open-faced (stationary shroud) or closed-face (rotating shroud) design. As long as the tip velocities are below 1000 ft/s, closed-face impellers can be used. The standards allow the impellers to be welded, riveted, milled, or cast. Riveted impellers are unacceptable, especially if the impeller loading is high. Impellers are to be assembled on the shaft with a shrink fit with or without key. Shrink fits should be carefully done because excessive shrink fits can cause a problem known as hysteresis whirl. In compressors where the impellers require their thrust to be balanced, a balance drum is acceptable and preferred.

The high-speed pumps or compressors must operate in a region away from any critical speed. The amplification factor used to indicate the severity of the critical speed is given by the relationship

$$\text{AF} = \frac{\text{critical speed}}{\text{peak width of "half-power" point}} \qquad (10\text{-}91)$$

$$\text{AF} = \frac{N_{C1}}{N_2 - N_1} \qquad (10\text{-}92)$$

where $N_2 - N_1$ is the rpm corresponding to the 0.707 peak critical amplitude.

The amplification factor should be below 8 and preferably above 5. A rotor response plot is shown in Fig. 10-122. The operational speed for units operating below their first critical speed should be at least 20 percent below the critical speed. For units operating above their first critical speed, the operational speed must be at least 15 percent above the critical speed and/or 20 percent below any critical speed. The preferred bearings for the various types of installation are tilting-shoe radial bearings and the self-equalizing tilting pad thrust bearings. Radial and thrust bearings should be equipped with embedded temperature sensors to detect pad surface temperatures.

VIBRATION MONITORING

One of the major factors that causes pump failure is vibration, which usually causes seal damage and oil leakage. Vibration in pumps is caused by numerous factors such as cavitation, impeller unbalance, loose bearings, and pipe pulsations.

As the mechanical integrity of the pump system changes, the amplitude of vibration levels change. In some cases, in order to identify the source of vibration, pump speed may have to be varied, as these problems are frequency- or resonance-dependent. Pump impeller imbalance and cavitation are related to this category.

It is advisable in most of these cases to use accelerometers. Displacement probes will not give the high-frequency signals and velocity probes because their mechanical design is very directional and prone to deterioration. Figure 10-123 shows the signal from the various types of probes.

Typically, large-amplitude vibration occurs when the frequency of vibration coincides with that of the natural frequency of the pump system. This results in a catastrophic operating condition that should be avoided. If the natural frequency is close to the upper end of the operating speed range, then the pump system should be stiffened to reduce vibration. On the other hand, if the natural frequency is close to the lower end of the operating range, the unit should be made more flexible. During startup, the pump system may go through its system natural frequency, and vibration can occur. Continuous operation at this operating point should be avoided.

ASME recommends periodic monitoring of all pumps. Pump vibration level should fall within the prescribed limits. The reference vibration level is measured during acceptance testing. This level is specified by the manufacturer.

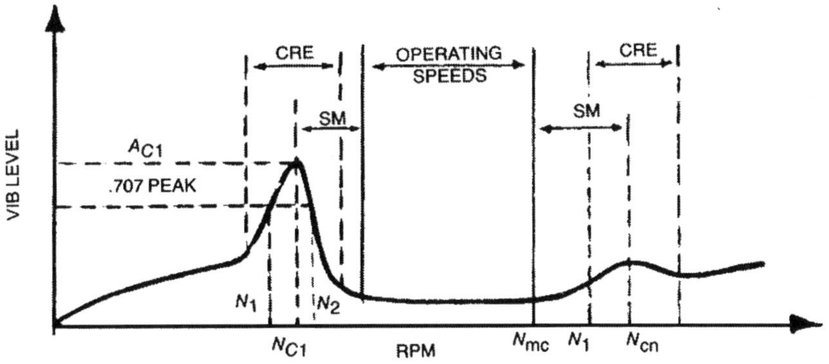

N_{C1} = Rotor 1st critical, center frequency, cycles per minute
N_{cn} = Critical speed, nth
N_+ = Trip speed
N_{mc} = Maximum continuous speed, 105 percent
N_1 = Initial (lesser) speed at .707 × peak amplitude (critical)
N_2 = Final (greater) speed at .707 × peak amplitude (critical)
$N_2 - N_1$ = Peak width at the "half-power" point

AF = Amplification factor $= \frac{N_{C1}}{N_2 - N_1}$
SM = Separation margin
CRE = Critical response envelope
N_{C1} = Amplitude @ N_{C1}
A_{cn} = Amplitude @ N_{cn}

FIG. 10-122 Rotor response plot. This plot is Figure 7 in API Standard 617, *Centrifugal Compressors for General Refinery Services*, 4th ed., 1979. (*Courtesy American Petroleum Institute.*)

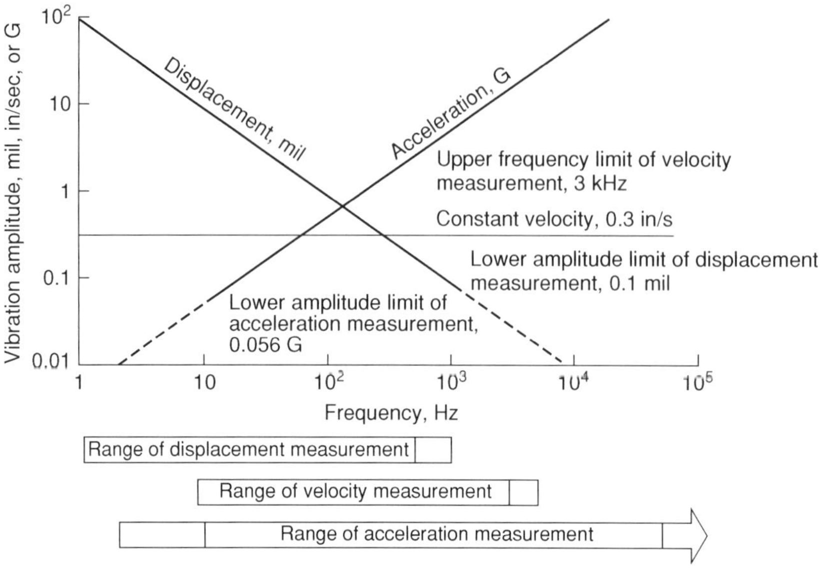

FIG. 10-123 Limitations on machinery vibrations analysis systems and transducers.

TABLE 10-17 Alert Levels

Reference value mils.	Alert mils., μm	Action required mils., μm
$V_r < 0.5$	1.0	1.5
$0.5 < V_r < 2.0$	$2V_r$	$3V_r$
$2.0 < V_r < 5.0$	$2+V_r$	$4+V_r$
$5.0 < V_r$	$1.4V_r$	$1.8V_r$

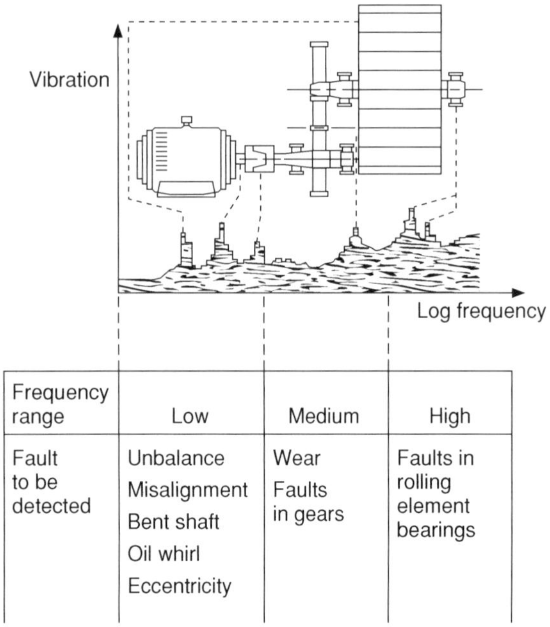

FIG. 10-124 Frequency range of typical machinery faults.

During periodic maintenance, the vibration level should not exceed alert level (see Table 10-17). If the measured level exceeds the alert level then preventive maintenance should be performed, by diagnosing the cause of vibration and reducing the vibration level prior to continued operation.

Typical problems and their vibration frequency ranges are shown in Fig. 10-124.

Collection and analysis of vibration signatures is a complex procedure. By looking at a vibration spectrum, one can identify which components of the pump system are responsible for a particular frequency component. Comparison of vibration signatures at periodic intervals reveals if a particular component is deteriorating. The following example illustrates evaluation of the frequency composition of an electric motor gear pump system.

Example 2: Vibration Consider an electric motor rotating at 1800 rpm driving an 8-vane centrifugal pump rotating at 600 rpm. For this 3:1 speed reduction, assume a gear box having two gears of 100 and 300 teeth. Since 60 Hz is 1 rpm,

Motor frequency = 1800/60 = 30 Hz

Pump frequency = 600/60 = 10 Hz

Gear mesh frequency = 300 teeth × 600 rpm = 3000 Hz

Vane frequency = 8 × 600 rpm = 80 Hz

An ideal vibration spectra for this motor-gear pump assembly would appear as shown in Fig. 10-125.

Figure 10-126 shows an actual pump vibration spectra. In the figure, several amplitude peaks occur at several frequencies.

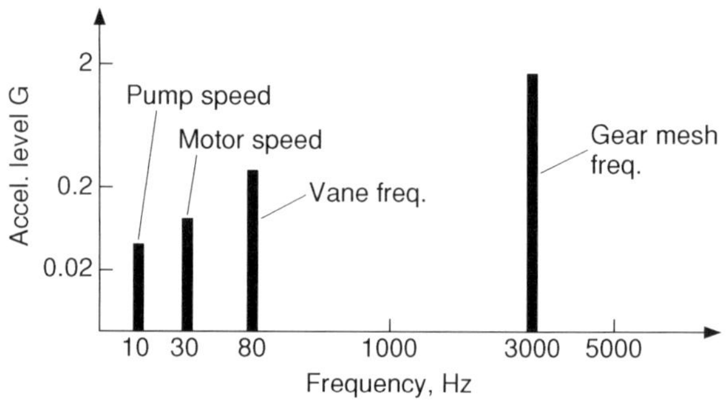

FIG. 10-125 An ideal vibration spectra from an electric motor pump assembly.

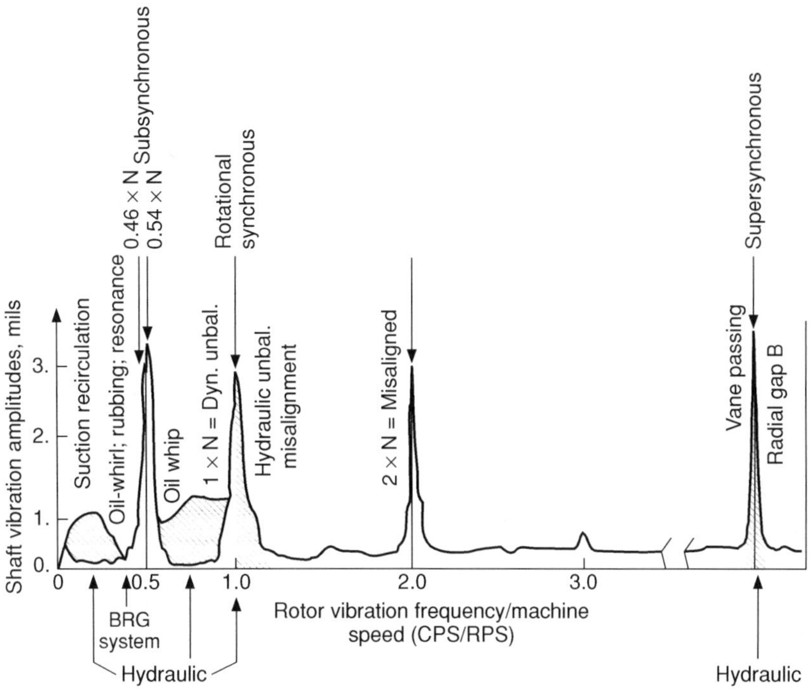

FIG. 10-126 An actual pump vibration spectrum.

PROCESS PLANT PIPING

INTRODUCTION

This section provides general comments that are pertinent to the design of process plant piping. It is intended to provide a convenient summary of commonly used information from various sources. It is not intended to serve as a comprehensive source of requirements or as a substitute for referenced codes, standards, and specifications. It is intended that qualified designers obtain copies of all applicable codes, standards, and specifications and thoroughly review all pertinent requirements of these documents prior to execution of work.

CODES AND STANDARDS

Units: Pipe and Tubing Sizes and Ratings In this subsection pipe and tubing sizes are generally quoted in units of inches. To convert inches to millimeters, multiply by 25.4. Ratings are given in pounds. To convert pounds to kilograms, multiply by 0.454.

Pressure-Piping Codes The code for pressure piping (ASME B31) consists of a number of sections which collectively constitute the code. Table 10-18 shows the status of the B31 code as of July 2005. The sections are published as separate documents for simplicity and convenience. The sections differ extensively.

The Process Piping Code (ASME B31.3) is a subsection of the ASME code for Pressure Piping B31. It was derived from a merging of the code groups for chemical-plant (B31.6) and petroleum-refinery (B31.3) piping into a single committee. Some of the significant requirements of ASME B31.3, Process Piping (2004 edition) are summarized in the following presentation.

Where the word *code* is used in this subsection of the *Handbook* without other identification, it refers to the B31.3 section of ASME B31. The code has been extensively quoted in this subsection of the *Handbook* with the permission of the publisher. The code is published by and copies are available from the American Society of Mechanical Engineers (ASME). Three Park Avenue, New York, New York 10016–5990.

National Standards The American Society of Mechanical Engineers (ASME) and the American Petroleum Institute (API) have established dimensional standards for the most widely used piping components. Lists of these standards as well as specifications for pipe and fitting materials and testing methods of the American Society for Testing and Materials (ASTM), American Welding Society (AWS) specifications, and standards of the Manufacturers Standardization Society of the Valve and Fittings Industry (MSS) can be found in the ASME B31 code sections. Many of these standards contain pressure-temperature ratings which will be of assistance to engineers in their design function. The use of published standards does not eliminate the need for engineering judgment. For example, although the code calculation formulas recognize the need to provide an allowance for corrosion, the standard rating tables for valves, flanges, fittings, etc., do not incorporate a corresponding allowance. Judgments regarding the suitability of these components are left to the designer.

The introduction to the code sets forth engineering requirements deemed necessary for the safe design and construction of piping systems. While safety is the basic consideration of the code, this factor alone will not necessarily govern final specifications for any pressure piping system.

Designers are cautioned that the code is not a design handbook and does not do away with the need for competent engineering judgment.

Government Regulations: OSHA Sections of the ASME B31 code have been adopted with certain reservations or revisions by some state and local authorities as local codes.

TABLE 10-18 Status of ASME B31 Code for Pressure Piping

ASME B31.3 section number, latest issue, and title	General scope and application
B31.1-2004, *Power Piping*	Addresses piping typically found in electric power generating stations, industrial and institutional plants, geothermal heating systems, and central and district heating and cooling systems.
B31.2-1968, *Fuel Gas Piping*	Addresses fuel gas piping systems in and between buildings. The scope includes piping between the outlet of the consumer's meter and the outlet of the first pressure-containing valve upstream of the gas utilization device.
B31.3-2004, *Process Piping*	Addresses piping typically found in petroleum refineries; chemical, pharmaceutical, textile, paper, semiconductor, and cryogenic plants; and related processing plants and terminals.
B31.4-2002, *Pipeline Transportation Systems for Liquid Hydrocarbons and Other Liquids*	Addresses piping transporting products which are predominately liquid between plants and terminals. Included are terminals, tank farms, and pumping, regulating, and metering stations.
B31.5-2001, *Refrigeration Piping and Heat Transfer Components*	Addresses refrigeration piping in packaged units and in commercial and public buildings.
B31.7, *Nuclear Power Piping*	Withdrawn; see *ASME Boiler and Pressure Vessel Code*, Sec. 3.
B31.8-2003, *Gas Transmission and Distribution Piping Systems*	Addresses piping transporting products which are predominately gas between sources and terminals. Included are gas-gathering pipelines and compressor, regulating, and metering stations.
B31.8S-2004, *Managing System Integrity of Gas Pipelines*	Addresses development and implementation of integrity management systems for gas pipelines.
B31.9-1996, *Building Services Piping*	Addresses piping outside the scope of ASME B31.1 that is typically found in multiunit residences and in industrial, institutional, commercial, and public buildings.
B31.11-2002, *Slurry Transportation Piping Systems*	Addresses piping transporting aqueous slurries between plants and terminals and within terminals, pumping, and regulating stations.
B31G-1991, *Manual for Determining the Remaining Strength of Corroded Pipelines: A Supplement to ASME B31 Code for Pressure Piping*	Addresses methods of determining the remaining strength of corroded pipelines that are within the scope of the *ASME B31 Code for Pressure Piping*.

Addenda are issued at intervals between publication of complete editions. Information on the latest issues can be obtained from the American Society of Mechanical Engineers, Three Park Avenue, New York, NY 10016-5990; or on the World Wide Web at ASME.org.

The specific requirements for piping systems in certain services have been promulgated as Occupational Safety and Health Administration (OSHA) regulations. These rules and regulations will presumably be revised and supplemented from time to time and may include specific requirements not addressed by the B31 sections.

International Regulations ASME piping codes have been widely used throughout the world for the design of facilities falling within their defined scopes. Although the use of ASME codes is widely acceptable in areas outside the United States, it is essential to identify additional local or national codes or standards that may apply. Such documents may require qualified third-party review and approval of project specifications, facility design, fabrication, material documentation, inspection, and testing. For example, within the European Community, such requirements are imposed by the Pressure Equipment Directive 97/23/EC (also known as the PED). These requirements must be recognized early in the project to avoid costly error.

CODE CONTENTS AND SCOPE

The code prescribes minimum requirements for materials, design, fabrication, assembly, support, erection, examination, inspection, and testing of piping systems subject to pressure or vacuum. The scope of the piping covered by B31.3 is illustrated in Fig. 10-127. It applies to all fluids including fluidized solids and to all services except as noted in the figure.

The code also excludes piping systems designed for internal gauge pressures at or above zero but less than 0.105 MPa (15 lbf/in^2) provided the fluid handled is nonflammable, nontoxic, and not damaging to human tissues, and its design temperature is from −29°C (−20°F) through 186°C (366°F). Refer to the code for definitions of nonflammable and nontoxic.

Some of the more significant requirements of ASME B31.3 (2004 edition) have been summarized and incorporated in this section of the *Handbook*. For a more comprehensive treatment of code requirements engineers are referred to the B31.3 code and the standards referenced therein.

SELECTION OF PIPE SYSTEM MATERIALS

The selection of material to resist deterioration in service is outside the scope of the B31.3 code (see Sec. 25). Experience has, however, resulted in the following material considerations extracted from the code with the permission of the publisher, the American Society of Mechanical Engineers, New York.

General Considerations* Following are some general considerations which should be evaluated when selecting and applying materials in piping:

1. The possibility of exposure of the piping to fire and the melting point, degradation temperature, loss of strength at elevated temperature, and combustibility of the piping material under such exposure.
2. The susceptibility to brittle failure or failure from thermal shock of the piping material when exposed to fire or to fire-fighting measures, and possible hazards from fragmentation of the material in the event of failure.
3. The ability of thermal insulation to protect piping against failure under fire exposure (e.g., its stability, fire resistance, and ability to remain in place during a fire).
4. The susceptibility of the piping material to crevice corrosion under backing rings, in threaded joints, in socket-welded joints, and in other stagnant, confined areas.
5. The possibility of adverse electrolytic effects if the metal is subject to contact with a dissimilar metal.
6. The compatibility of lubricants or sealants used on threads with the fluid service.

*Extracted from ASME B31.3-2004, Section F323, with permission of the publisher, the American Society of Mechanical Engineers, New York.

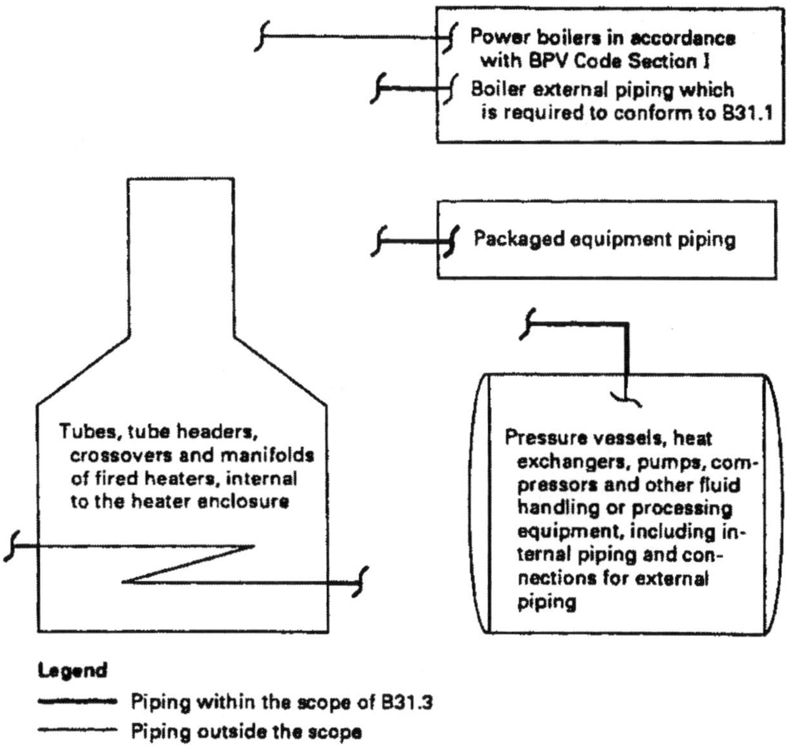

FIG. 10-127 Scope of work covered by process piping code ASME B31.3-2004.

7. The compatibility of packing, seals, and O-rings with the fluid service.

8. The compatibility of materials, such as cements, solvents, solders, and brazing materials, with the fluid service.

9. The chilling effect of sudden loss of pressure on highly volatile fluids as a factor in determining the lowest expected service temperature.

10. The possibility of pipe support failure resulting from exposure to low temperatures (which may embrittle the supports) or high temperatures (which may weaken them).

11. The compatibility of materials, including sealants, gaskets, lubricants, and insulation, used in strong oxidizer fluid service (e.g., oxygen or fluorine).

Specific Material Considerations—Metals* Following are some specific considerations which should be evaluated when applying certain metals in piping.

1. *Irons—cast, malleable, and high silicon (14.5%).* Their lack of ductility and their sensitivity to thermal and mechanical shock.

2. *Carbon steel, and low and intermediate alloy steels.*

a. The possibility of embrittlement when handling alkaline or strong caustic fluids.

b. The possible conversion of carbides to graphite during long time exposure to temperatures above 427°C (800°F) of carbon steels, plain nickel steel, carbon-manganese steel, manganese-vanadium steel, and carbon-silicon steel.

c. The possible conversion of carbides to graphite during long time exposure to temperatures above 468°C (875°F) of carbon-molybdenum steel, manganese-molybdenum-vanadium steel, and chromium-vanadium steel.

d. The advantages of silicon-killed carbon steel (0.1% silicon minimum) for temperatures above 482°C (900°F).

e. The possibility of damage due to hydrogen exposure at elevated temperature (see API RP941); hydrogen damage (blistering) may occur at lower temperatures under exposure to aqueous acid solutions.†

f. The possibility of stress corrosion cracking when exposed to cyanides, acids, acid salts, or wet hydrogen sulfide; a maximum hardness limit is usually specified (see NACE MR0175 and RP0472).†

g. The possibility of sulfidation in the presence of hydrogen sulfide at elevated temperatures.

3. *High-alloy (stainless) steels.*

a. The possibility of stress corrosion cracking of austenitic stainless steels exposed to media such as chlorides and other halides either internally or externally; the latter can result from improper selection or application of thermal insulation, or from use of marking inks, paints, labels, tapes, adhesives, and other accessory materials containing chlorides or other halides.

b. The susceptibility to intergranular corrosion of austenitic stainless steels sensitized by exposure to temperatures between 427 and 871°C (800 and 1600°F); as an example, stress corrosion cracking of sensitized metal at room temperature by polythionic acid (reaction of oxidizable sulfur compound, water, and air); stabilized or low-carbon grades may provide improved resistance (see NACE RP0170).†

c. The susceptibility to intercrystalline attack of austenitic stainless steels on contact with liquid metals (including aluminum, antimony, bismuth, cadmium, gallium, lead, magnesium, tin, and zinc) or their compounds.

*Extracted from ASME B31.3-2004, Section F323, with permission of the publisher, the American Society of Mechanical Engineers, New York.

†Titles of referenced documents are:

API RP941, *Steels for Hydrogen Service at Elevated Temperatures and Pressures in Petroleum Refineries and Petrochemical Plants*

NACE MR0175, *Sulfide Stress-Cracking Resistant Metallic Materials for Oil Field Equipment*

NACE RP0472, *Methods and Controls to Prevent In-Service Cracking of Carbon Steel (P-1) Welds in Corrosive Petroleum Refining Environments*

NACE RP0170, *Protection of Austenitic Stainless Steel in Refineries Against Stress Corrosion Cracking by Use of Neutralizing Solutions During Shutdown d.* The brittleness of ferritic stainless steels at room temperature after service at temperature above 371°C (700°F).

4. *Nickel and nickel-base alloys.*

a. The susceptibility to grain boundary attack of nickel and nickel-base alloys not containing chromium when exposed to small quantities of sulfur at temperatures above 316°C (600°F).

b. The susceptibility to grain boundary attack of nickel-base alloys containing chromium at temperatures above 593°C (1100°F) under reducing conditions and above 760°C (1400°F) under oxidizing conditions.

c. The possibility of stress corrosion cracking of nickel-copper Alloy 400 in hydrofluoric acid vapor in the presence of air, if the alloy is highly stressed (including residual stresses from forming or welding).

5. *Aluminum and aluminum alloys.*

a. The compatibility with aluminum of thread compounds used in aluminum threaded joints to prevent seizing and galling.

b. The possibility of corrosion from concrete, mortar, lime, plaster, or other alkaline materials used in buildings or structures.

c. The susceptibility of Alloy nos. 5083, 5086, 5154, and 5456 to exfoliation or intergranular attack; and the upper temperature limit of 66°C (150°F) shown in Appendix A to avoid such deterioration.

6. *Copper and copper alloys.*

a. The possibility of dezincification of brass alloys.

b. The susceptibility to stress corrosion cracking of copper-based alloys exposed to fluids such as ammonia or ammonium compounds.

c. The possibility of unstable acetylide formation when exposed to acetylene.

7. *Titanium and titanium alloys.* The possibility of deterioration of titanium and its alloys above 316°C (600°F).

8. *Zirconium and zirconium alloys.* The possibility of deterioration of zirconium and zirconium alloys above 316°C (600°F).

9. *Tantalum.* Above 299°C (570°F), the possibility of reactivity of tantalum with all gases except the inert gases. Below 299°C, the possibility of embrittlement of tantalum by nascent (monatomic) hydrogen (but not molecular hydrogen). Nascent hydrogen is produced by galvanic action, or as a product of corrosion by certain chemicals.

10. *Metals with enhanced properties.* The possible loss of strength, in a material whose properties have been enhanced by heat treatment, during long-continued exposure to temperatures above the tempering temperature.

11. The desirability of specifying some degree of production impact testing, in addition to the weld procedure qualification tests, when using materials with limited low-temperature service experience below the minimum temperature stated in ASME B31.3 Table A-1.

Specific Material Considerations—Nonmetals Following are some considerations to be evaluated when applying nonmetals in piping. Refer to Tables 10-19, 10-20, and 10-21 for typical temperature limits.

1. *Static charges.* Because of the possibility of producing hazardous electrostatic charges in nonmetallic piping and metallic piping lined with nonmetals, consideration should be given to grounding the metallic components of such systems conveying nonconductive fluids.

2. *Thermoplastics.* If thermoplastic piping is used aboveground for compressed air or other compressed gases, special precautions should be observed. In determining the needed safeguarding for such services, the energetics and the specific failure mechanism need to be evaluated. Encasement of the plastic piping in shatter-resistant material may be considered.

3. *Borosilicate glass.* Take into account its lack of ductility and its sensitivity to thermal and mechanical shock.

TABLE 10-19 Recommended Temperature Limits for Thermoplastic Pipe*

ASTM spec. no.	Material	Recommended temperature limits †,‡ Minimum °C	Minimum °F	Maximum °C	Maximum °F
...	ABS55535	−40	−40	80	176
...	AP	−18	0	77	170
D 2846, F 441, F 442	CPVC4120	−18	0	99	210
...	ECTFE	−40	−40	149	300
...	ETFE	−40	−40	149	300
D 2513, D 2662, D 2666, D 3000, D 3309	PB2110	−18	0	99	210
D 2104, D 2239, D 2447, D 2513, D 2737, D 3035	PE3408	−34	−30	82	180
...	PEEK	−40	−40	250	450
...	PFA	−40	−40	250	450
...	POP2125	−1	30	99	210
...	PP	−1	30	99	210
D 1785	PVC1120	−18	0	66	150
D 2241	PVC1220	−18	0	66	150
D 2513	PVC2110	−18	0	54	130
D 2672	PVC2120	−18	0	66	150
...	PVDC	4	40	71	160
...	PVDF	−18	0	135	275

*Extracted from ASME B31.3-2004, Table B-1, with permission of the publisher, the American Society of Mechanical Engineers, New York.

†These recommended limits are for low-pressure applications with water and other fluids that do not significantly affect the properties of the thermoplastic. The upper temperature limits are reduced at higher pressures, depending on the combination of fluid and expected service life. Lower temperature limits are affected more by the environment, safeguarding, and installation conditions than by strength.

‡These recommended limits apply only to materials listed. Manufacturers should be consulted for temperature limits on specific types and kinds of materials not listed.

METALLIC PIPING SYSTEM COMPONENTS

Metallic pipe systems comprise the majority of applications. Metallic pipe, tubing, and pipe fittings are divided into two main categories: seamless and welded. Both have advantages and disadvantages in terms of economy and function. Specifications governing the production of these products dictate the permissible mechanical and dimensional variations, and code design calculations account for these variations.

Seamless Pipe and Tubing Seamless pipe and tubing may be formed by various methods. A common technique involves piercing solid round forgings, followed by rolling and drawing. Other techniques include forging and boring, extrusion, and static and centrifugal casting. Piercing frequently produces pipe with a less uniform wall thickness and concentricity of bore than is the case with products produced by other methods. Since seamless products have no weld joints, there is no reduction of strength due to weld joint efficiency.

Welded Pipe and Tubing These products are typically made by forming strips or plate into cylinders and seam-welding by various methods. Manufacturing by welding permits the production of larger diameter pipe than is possible with seamless manufacturing methods, as well as larger diameter/wall thickness ratios. While strip and plate thickness may be more closely controlled than is possible for some seamless products, the specifications governing production are not always more stringent for welded products.

Weld quality has the potential of making the weld weaker than the base material. Depending on the welding method and the degree of nondestructive examination required by the product specification or dictated by the designer, the code assigns a joint efficiency ranging from 60 to 100 percent of the strength of the base material. Although

TABLE 10-20 Recommended Temperature Limits for Thermoplastics Used as Linings*

Materials	Minimum °C	Minimum °F	Maximum °C	Maximum °F
PFA	−198	−325	260	500
PTFE	−198	−325	260	500
FEP	−198	−325	204	400
ECTFE	−198	−325	171	340
ETFE	−198	−325	149	300
PVDF	−18	0	135	275
PP	−18	0	107	225
PVDC	−18	0	79	175

*Extracted from ASME B31.3-2004, Table A323.4.3, with permission of the publisher, the American Society of Mechanical Engineers, New York.

NOTE: These temperature limits are based on material tests and do not necessarily reflect evidence of successful use as piping component linings in specific fluid services at these temperatures. The designer should consult the manufacturer for specific applications, particularly as temperature limits are approached.

Abbreviations for plastics: ABS, acrylonitrile-butadiene-styrene; CPVC, chlorinated poly vinyl chloride; ECTFE, ethylene-chlorotrifluoroethylene; ETFE, ethylene-tetrafluoroethylene; PB, polybutylene; PE, polyethylene; PEEK, poly ether ether ketone; PFA, perfluoroalkoxy copolymer; POP, poly phenylene oxide; PP, polypropylene; PVC, poly vinyl chloride; PVDC, poly vinylidene chloride; PVDF, poly vinylidene fluoride.

some welding methods have the potential of producing short sections of partially fused joints that may develop into small leaks in corrosive conditions, proper matching of the weld method and the type and extent of examination will result in highly reliable joints that are suitable for use in critical services. Welds must be considered when developing specifications for bending, flaring or expanding welded pipe or tubing.

Tubing Tubing sizes typically reflect the actual outside diameter of the product. Pipe is manufactured to nominal diameters, which are not the same as the actual outside diameters for sizes 12 in and smaller. Facilities within the scope of the ASME B31 codes nearly exclusively use pipe, rather than tubing, for applications external to equipment. Tubing is commonly classified as suitable for either mechanical or pressure applications. Tubing is available in size and wall thickness combinations not normally produced as pipe. Tubing wall thickness (gauge) is specified as either average wall or minimum wall. Minimum wall is more costly than average wall, and because of closer tolerances on thickness and diameter, tubing of either gauge system is generally more costly than pipe. Tubing having outside diameters of 2⅜, 2⅞, 3½, and 4½ in are commonly available; however, these sizes are generally considered to be nonstandard for typical piping applications.

Table 10-22 gives some of the more common standard size and wall-thickness combinations together with capacity and weight.

Methods of Joining Pipe Piping joints must be reliably leaktight and provide adequate mechanical strength to resist external loads due to thermal expansion, weight, wind, seismic activity, and other factors. Joints for pipe buried in soil may be subjected to unique external loads resulting from thermal expansion and contraction, settlement, and other factors. Joint designs that permit rotation about an axis perpendicular to the longitudinal axis of the pipe may be advantageous in certain situations.

Disassembly frequency and ease should be considered when selecting joining methods. Ideally the method for joining piping system components provides minimum installed cost, maintains its integrity throughout the lifetime of the facility, provides restraint against axial thrust due to internal pressure, provides strength against external loads equal to that of the pipe, permits unrestricted flow with minimum pressure drop, and is free from crevices that may be detrimental to the product or contribute to corrosion or erosion problems.

Joint design and selection generally involves compromising between the ideal and practical. A number of manufacturers produce patented or "proprietary" joints that embody many ideal characteristics. Some are excellent products and are well suited to special applications. Valves and fittings are often available with proprietary joints that have gained wide acceptance; however, consideration should be given to the possible impact on product delivery time and cost.

Welded Joints The most widely used joint in piping systems is the **butt-weld joint** (Fig. 10-128). In all ductile pipe metals which can be welded, pipe, elbows, tees, laterals, reducers, caps, valves, flanges, and V-clamp joints are available in all sizes and wall thicknesses with ends prepared for butt welding. Joint strength equal to the original pipe (except for work-hardened pipes which are annealed by the welding), unimpaired flow pattern, and generally unimpaired corrosion resistance more than compensate for the necessary careful alignment, skilled labor, and equipment required.

Plain-end pipe used for socket-weld joints (Fig. 10-129) is available in all sizes, but fittings and valves with socket-weld ends are limited to sizes 3 in and smaller, for which the extra cost of the socket is outweighed by much easier alignment and less skill needed in welding.

Socket-welded joints are not as resistant to externally applied bending moments as are butt-welded joints, are not easily examined by volumetric nondestructive examination methods such as radiography and ultrasonic, and should not be used where crevice corrosion has been determined to be of concern. However, they are widely used in sizes 2 in and smaller and are quite satisfactory for most applications when used within the limits established by code restrictions and good engineering judgment. Components with socket-welded ends are generally specified as requiring compliance with ASME B16.11, *Forged-Fittings, Socket-Welding and Threaded.*

Branch Connections Branch connections may be made with manufactured tees, fabricated reinforced and nonreinforced branch connections (Fig. 10-130), or manufactured integrally reinforced branch connections. Butt-welded fittings offer the best opportunity

TABLE 10-21 Recommended Temperature Limits for Reinforced Thermosetting Resin Pipe*

Materials: Resin	Materials: Reinforcing	Recommended temperature limits: Minimum °C	Minimum °F	Maximum °C	Maximum °F
Epoxy	Glass fiber	−29	−20	149	300
Phenolic	Glass fiber	−29	−20	149	300
Furan	Carbon	−29	−20	93	200
Furan	Glass fiber	−29	−20	93	200
Polyester	Glass fiber	−29	−20	93	200
Vinyl ester	Glass fiber	−29	−20	93	200

*Extracted from ASME B31.3-2004, Table A323.4.2C, with permission of the publisher, the American Society of Mechanical Engineers, New York.

NOTE: These temperature limits apply only to materials listed and do not reflect evidence of successful use in specific fluid services at these temperatures. The designer should consult the manufacturer for specific applications, particularly as the temperature limits are approached.

TABLE 10-22 Properties of Steel Pipe

Nominal pipe size, in	Outside diameter, in	Schedule no.	Wall thickness, in	Inside diameter, in	Cross-sectional area		Circumference, ft, or surface, ft²/ft of length		Capacity at 1-ft/s velocity		Weight of plain-end pipe, lb/ft
					Metal, in²	Flow, ft²	Outside	Inside	U.S. gal/min	lb/h water	
⅛	0.405	10S	0.049	0.307	0.055	0.00051	0.106	0.0804	0.231	115.5	0.19
		40ST, 40S	.068	.269	.072	.00040	.106	.0705	.179	89.5	.24
		80XS, 80S	.095	.215	.093	.00025	.106	.0563	.113	56.5	.31
¼	0.540	10S	.065	.410	.097	.00092	.141	.107	.412	206.5	.33
		40ST, 40S	.088	.364	.125	.00072	.141	.095	.323	161.5	.42
		80XS, 80S	.119	.302	.157	.00050	.141	.079	.224	112.0	.54
⅜	0.675	10S	.065	.545	.125	.00162	.177	.143	.727	363.5	.42
		40ST, 40S	.091	.493	.167	.00133	.177	.129	.596	298.0	.57
		80XS, 80S	.126	.423	.217	.00098	.177	.111	.440	220.0	.74
½	0.840	5S	.065	.710	.158	.00275	.220	.186	1.234	617.0	.54
		10S	.083	.674	.197	.00248	.220	.176	1.112	556.0	.67
		40ST, 40S	.109	.622	.250	.00211	.220	.163	0.945	472.0	.85
		80XS, 80S	.147	.546	.320	.00163	.220	.143	0.730	365.0	1.09
		160	.188	.464	.385	.00117	.220	.122	0.527	263.5	1.31
		XX	.294	.252	.504	.00035	.220	.066	0.155	77.5	1.71
¾	1.050	5S	.065	.920	.201	.00461	.275	.241	2.072	1036.0	0.69
		10S	.083	.884	.252	.00426	.275	.231	1.903	951.5	0.86
		40ST, 40S	.113	.824	.333	.00371	.275	.216	1.665	832.5	1.13
		80XS, 80S	.154	.742	.433	.00300	.275	.194	1.345	672.5	1.47
		160	.219	.612	.572	.00204	.275	.160	0.917	458.5	1.94
		XX	.308	.434	.718	.00103	.275	.114	0.461	230.5	2.44
1	1.315	5S	.065	1.185	.255	.00768	.344	.310	3.449	1725	0.87
		10S	.109	1.097	.413	.00656	.344	.287	2.946	1473	1.40
		40ST, 40S	.133	1.049	.494	.00600	.344	.275	2.690	1345	1.68
		80XS, 80S	.179	0.957	.639	.00499	.344	.250	2.240	1120	2.17
		160	.250	0.815	.836	.00362	.344	.213	1.625	812.5	2.84
		XX	.358	0.599	1.076	.00196	.344	.157	0.878	439.0	3.66
1¼	1.660	5S	.065	1.530	0.326	.01277	.435	.401	5.73	2865	1.11
		10S	.109	1.442	0.531	.01134	.435	.378	5.09	2545	1.81
		40ST, 40S	.140	1.380	0.668	.01040	.435	.361	4.57	2285	2.27
		80XS, 80S	.191	1.278	0.881	.00891	.435	.335	3.99	1995	3.00
		160	.250	1.160	1.107	.00734	.435	.304	3.29	1645	3.76
		XX	.382	0.896	1.534	.00438	.435	.235	1.97	985	5.21
1½	1.900	5S	.065	1.770	0.375	.01709	.497	.463	7.67	3835	1.28
		10S	.109	1.682	0.614	.01543	.497	.440	6.94	3465	2.09
		40ST, 40S	.145	1.610	0.800	.01414	.497	.421	6.34	3170	2.72
		80XS, 80S	.200	1.500	1.069	.01225	.497	.393	5.49	2745	3.63
		160	.281	1.338	1.429	.00976	.497	.350	4.38	2190	4.86
		XX	.400	1.100	1.885	.00660	.497	.288	2.96	1480	6.41
2	2.375	5S	.065	2.245	0.472	.02749	.622	.588	12.34	6170	1.61
		10S	.109	2.157	0.776	.02538	.622	.565	11.39	5695	2.64
		40ST, 40S	.154	2.067	1.075	.02330	.622	.541	10.45	5225	3.65
		80ST, 80S	.218	1.939	1.477	.02050	.622	.508	9.20	4600	5.02
		160	.344	1.687	2.195	.01552	.622	.436	6.97	3485	7.46
		XX	.436	1.503	2.656	.01232	.622	.393	5.53	2765	9.03
2½	2.875	5S	.083	2.709	0.728	.04003	.753	.709	17.97	8985	2.48
		10S	.120	2.635	1.039	.03787	.753	.690	17.00	8500	3.53
		40ST, 40S	.203	2.469	1.704	.03322	.753	.647	14.92	7460	5.79
		80XS, 80S	.276	2.323	2.254	.02942	.753	.608	13.20	6600	7.66
		160	.375	2.125	2.945	.02463	.753	.556	11.07	5535	10.01
		XX	.552	1.771	4.028	.01711	.753	.464	7.68	3840	13.69
3	3.500	5S	.083	3.334	0.891	.06063	.916	.873	27.21	13,605	3.03
		10S	.120	3.260	1.274	.05796	.916	.853	26.02	13,010	4.33
		40ST, 40S	.216	3.068	2.228	.05130	.916	.803	23.00	11,500	7.58
		80XS, 80S	.300	2.900	3.016	.04587	.916	.759	20.55	10,275	10.25
		160	.438	2.624	4.213	.03755	.916	.687	16.86	8430	14.32
		XX	.600	2.300	5.466	.02885	.916	.602	12.95	6475	18.58
3½	4.0	5S	.083	3.834	1.021	.08017	1.047	1.004	35.98	17,990	3.48
		10S	.120	3.760	1.463	.07711	1.047	0.984	34.61	17,305	4.97
		40ST, 40S	.226	3.548	2.680	.06870	1.047	0.929	30.80	15,400	9.11
		80XS, 80S	.318	3.364	3.678	.06170	1.047	0.881	27.70	13,850	12.50
4	4.5	5S	.083	4.334	1.152	.10245	1.178	1.135	46.0	23,000	3.92
		10S	.120	4.260	1.651	.09898	1.178	1.115	44.4	22,200	5.61
		40ST, 40S	.237	4.026	3.17	.08840	1.178	1.054	39.6	19,800	10.79

TABLE 10-22 Properties of Steel Pipe *(Continued)*

Nominal pipe size, in	Outside diameter, in	Schedule no.	Wall thickness, in	Inside diameter, in	Cross-sectional area		Circumference, ft, or surface, ft²/ft of length		Capacity at 1-ft/s velocity		Weight of plain-end pipe, lb/ft
					Metal, in²	Flow, ft²	Outside	Inside	U.S. gal/min	lb/h water	
		80XS, 80S	.337	3.826	4.41	.07986	1.178	1.002	35.8	17,900	14.98
		120	0.438	3.624	5.58	0.07170	1.178	0.949	32.2	16,100	19.00
		160	.531	3.438	6.62	.06647	1.178	0.900	28.9	14,450	22.51
		XX	.674	3.152	8.10	.05419	1.178	0.825	24.3	12,150	27.54
5	5.563	5S	.109	5.345	1.87	.1558	1.456	1.399	69.9	34,950	6.36
		10S	.134	5.295	2.29	.1529	1.456	1.386	68.6	34,300	7.77
		40ST, 40S	.258	5.047	4.30	.1390	1.456	1.321	62.3	31,150	14.62
		80XS, 80S	.375	4.813	6.11	.1263	1.456	1.260	57.7	28,850	20.78
		120	.500	4.563	7.95	.1136	1.456	1.195	51.0	25,500	27.04
		160	.625	4.313	9.70	.1015	1.456	1.129	45.5	22,750	32.96
		XX	.750	4.063	11.34	.0900	1.456	1.064	40.4	20,200	38.55
6	6.625	5S	.109	6.407	2.23	.2239	1.734	1.677	100.5	50,250	7.60
		10S	.134	6.357	2.73	.2204	1.734	1.664	98.9	49,450	9.29
		40ST, 40S	.280	6.065	5.58	.2006	1.734	1.588	90.0	45,000	18.97
		80XS, 80S	.432	5.761	8.40	.1810	1.734	1.508	81.1	40,550	28.57
		120	.562	5.501	10.70	.1650	1.734	1.440	73.9	36,950	36.39
		160	.719	5.187	13.34	.1467	1.734	1.358	65.9	32,950	45.34
		XX	.864	4.897	15.64	.1308	1.734	1.282	58.7	29,350	53.16
8	8.625	5S	.109	8.407	2.915	.3855	2.258	2.201	173.0	86,500	9.93
		10S	.148	8.329	3.941	.3784	2.258	2.180	169.8	84,900	13.40
		20	.250	8.125	6.578	.3601	2.258	2.127	161.5	80,750	22.36
		30	.277	8.071	7.265	.3553	2.258	2.113	159.4	79,700	24.70
		40ST, 40S	.322	7.981	8.399	.3474	2.258	2.089	155.7	77,850	28.55
		60	.406	7.813	10.48	.3329	2.258	2.045	149.4	74,700	35.64
		80XS, 80S	.500	7.625	12.76	.3171	2.258	1.996	142.3	71,150	43.39
		100	.594	7.437	14.99	.3017	2.258	1.947	135.4	67,700	50.95
		120	.719	7.187	17.86	.2817	2.258	1.882	126.4	63,200	60.71
		140	.812	7.001	19.93	.2673	2.258	1.833	120.0	60,000	67.76
		XX	.875	6.875	21.30	.2578	2.258	1.800	115.7	57,850	72.42
		160	.906	6.813	21.97	.2532	2.258	1.784	113.5	56,750	74.69
10	10.75	5S	.134	10.482	4.47	.5993	2.814	2.744	269.0	134,500	15.19
		10S	.165	10.420	5.49	.5922	2.814	2.728	265.8	132,900	18.65
		20	.250	10.250	8.25	.5731	2.814	2.685	257.0	128,500	28.04
		30	.307	10.136	10.07	.5603	2.814	2.655	252.0	126,000	34.24
		40ST, 40S	.365	10.020	11.91	.5475	2.814	2.620	246.0	123,000	40.48
		80S, 60XS	.500	9.750	16.10	.5185	2.814	2.550	233.0	116,500	54.74
		80	.594	9.562	18.95	.4987	2.814	2.503	223.4	111,700	64.43
		100	.719	9.312	22.66	.4729	2.814	2.438	212.3	106,150	77.03
		120	.844	9.062	26.27	.4479	2.814	2.372	201.0	100,500	89.29
		140, XX	1.000	8.750	30.63	.4176	2.814	2.291	188.0	94,000	104.13
		160	1.125	8.500	34.02	.3941	2.814	2.225	177.0	88,500	115.64
12	12.75	5S	0.156	12.438	6.17	.8438	3.338	3.26	378.7	189,350	20.98
		10S	0.180	12.390	7.11	.8373	3.338	3.24	375.8	187,900	24.17
		20	0.250	12.250	9.82	.8185	3.338	3.21	367.0	183,500	33.38
		30	0.330	12.090	12.88	.7972	3.338	3.17	358.0	179,000	43.77
		ST, 40S	0.375	12.000	14.58	.7854	3.338	3.14	352.5	176,250	49.56
		40	0.406	11.938	15.74	.7773	3.338	3.13	349.0	174,500	53.52
		XS, 80S	0.500	11.750	19.24	.7530	3.338	3.08	338.0	169,000	65.42
		60	0.562	11.626	21.52	.7372	3.338	3.04	331.0	165,500	73.15
		80	0.688	11.374	26.07	.7056	3.338	2.98	316.7	158,350	88.63
		100	0.844	11.062	31.57	.6674	3.338	2.90	299.6	149,800	107.32
		120, XX	1.000	10.750	36.91	.6303	3.338	2.81	283.0	141,500	125.49
		140	1.125	10.500	41.09	.6013	3.338	2.75	270.0	135,000	139.67
		160	1.312	10.126	47.14	.5592	3.338	2.65	251.0	125,500	160.27
14	14	5S	0.156	13.688	6.78	1.0219	3.665	3.58	459	229,500	23.07
		10S	0.188	13.624	8.16	1.0125	3.665	3.57	454	227,000	27.73
		10	0.250	13.500	10.80	0.9940	3.665	3.53	446	223,000	36.71
		20	0.312	13.376	13.42	0.9750	3.665	3.50	438	219,000	45.61
		30, ST	0.375	13.250	16.05	0.9575	3.665	3.47	430	215,000	54.57
		40	0.438	13.124	18.66	0.9397	3.665	3.44	422	211,000	63.44
		XS	0.500	13.000	21.21	0.9218	3.665	3.40	414	207,000	72.09
		60	0.594	12.812	25.02	0.8957	3.665	3.35	402	201,000	85.05
		80	0.750	12.500	31.22	0.8522	3.665	3.27	382	191,000	106.13
		100	0.938	12.124	38.49	0.8017	3.665	3.17	360	180,000	130.85
		120	1.094	11.812	44.36	0.7610	3.665	3.09	342	171,000	150.79
		140	1.250	11.500	50.07	0.7213	3.665	3.01	324	162,000	170.21
		160	1.406	11.188	55.63	0.6827	3.665	2.93	306	153,000	189.11
16	16	5S	0.165	15.670	8.21	1.3393	4.189	4.10	601	300,500	27.90
		10S	0.188	15.624	9.34	1.3314	4.189	4.09	598	299,000	31.75
		10	0.250	15.500	12.37	1.3104	4.189	4.06	587	293,500	42.05

TABLE 10-22 Properties of Steel Pipe *(Concluded)*

Nominal pipe size, in	Outside diameter, in	Schedule no.	Wall thickness, in	Inside diameter, in	Cross-sectional area		Circumference, ft, or surface, ft²/ft of length		Capacity at 1-ft/s velocity		Weight of plain-end pipe, lb/ft
					Metal, in²	Flow, ft²	Outside	Inside	U.S. gal/min	lb/h water	
		20	0.312	15.376	15.38	1.2985	4.189	4.03	578	289,000	52.27
		30, ST	0.375	15.250	18.41	1.2680	4.189	3.99	568	284,000	62.58
		40, XS	0.500	15.000	24.35	1.2272	4.189	3.93	550	275,000	82.77
		60	0.656	14.688	31.62	1.1766	4.189	3.85	528	264,000	107.50
		80	0.844	14.312	40.19	1.1171	4.189	3.75	501	250,500	136.61
		100	1.031	13.938	48.48	1.0596	4.189	3.65	474	237,000	164.82
		120	1.219	13.562	56.61	1.0032	4.189	3.55	450	225,000	192.43
		140	1.438	13.124	65.79	0.9394	4.189	3.44	422	211,000	223.64
		160	1.594	12.812	72.14	0.8953	4.189	3.35	402	201,000	245.25
18	18	5S	0.165	17.670	9.25	1.7029	4.712	4.63	764	382,000	31.43
		10S	0.188	17.624	10.52	1.6941	4.712	4.61	760	379,400	35.76
		10	0.250	17.500	13.94	1.6703	4.712	4.58	750	375,000	47.39
		20	0.312	17.376	17.34	1.6468	4.712	4.55	739	369,500	58.94
		ST	0.375	17.250	20.76	1.6230	4.712	4.52	728	364,000	70.59
		30	0.438	17.124	24.16	1.5993	4.712	4.48	718	359,000	82.15
		XS	0.500	17.000	27.49	1.5763	4.712	4.45	707	353,500	93.45
		40	0.562	16.876	30.79	1.5533	4.712	4.42	697	348,500	104.67
		60	0.750	16.500	40.64	1.4849	4.712	4.32	666	333,000	138.17
		80	0.938	16.124	50.28	1.4180	4.712	4.22	636	318,000	170.92
		100	1.156	15.688	61.17	1.3423	4.712	4.11	602	301,000	207.96
		120	1.375	15.250	71.82	1.2684	4.712	3.99	569	284,500	244.14
		140	1.562	14.876	80.66	1.2070	4.712	3.89	540	270,000	274.22
		160	1.781	14.438	90.75	1.1370	4.712	3.78	510	255,000	308.50
20	20	5S	0.188	19.624	11.70	2.1004	5.236	5.14	943	471,500	39.78
		10S	0.218	19.564	13.55	2.0878	5.236	5.12	937	467,500	46.06
		10	0.250	19.500	15.51	2.0740	5.236	5.11	930	465,000	52.73
		20, ST	0.375	19.250	23.12	2.0211	5.236	5.04	902	451,000	78.60
		30, XS	0.500	19.000	30.63	1.9689	5.236	4.97	883	441,500	104.13
		40	0.594	18.812	36.21	1.9302	5.236	4.92	866	433,000	123.11
		60	0.812	18.376	48.95	1.8417	5.236	4.81	826	413,000	166.40
		80	1.031	17.938	61.44	1.7550	5.236	4.70	787	393,500	208.87
		100	1.281	17.438	75.33	1.6585	5.236	4.57	744	372,000	256.10
		120	1.500	17.000	87.18	1.5763	5.236	4.45	707	353,500	296.37
		140	1.750	16.500	100.3	1.4849	5.236	4.32	665	332,500	341.09
		160	1.969	16.062	111.5	1.4071	5.236	4.21	632	316,000	397.17
24	24	5S	0.218	23.564	16.29	3.0285	6.283	6.17	1359	679,500	55.37
		10, 10S	0.250	23.500	18.65	3.012	6.283	6.15	1350	675,000	63.41
		20, ST	0.375	23.250	27.83	2.948	6.283	6.09	1325	662,500	94.62
		XS	0.500	23.000	36.90	2.885	6.283	6.02	1295	642,500	125.49
		30	0.562	22.876	41.39	2.854	6.283	5.99	1281	640,500	140.68
		40	0.688	22.624	50.39	2.792	6.283	5.92	1253	626,500	171.29
		60	0.969	22.062	70.11	2.655	6.283	5.78	1192	596,000	238.35
		80	1.219	21.562	87.24	2.536	6.283	5.64	1138	569,000	296.58
		100	1.531	20.938	108.1	2.391	6.283	5.48	1073	536,500	367.39
		120	1.812	20.376	126.3	2.264	6.283	5.33	1016	508,000	429.39
		140	2.062	19.876	142.1	2.155	6.283	5.20	965	482,500	483.12
		160	2.344	19.312	159.5	2.034	6.283	5.06	913	456,500	542.13
30	30	5S	0.250	29.500	23.37	4.746	7.854	7.72	2130	1,065,000	79.43
		10, 10S	0.312	29.376	29.10	4.707	7.854	7.69	2110	1,055,000	98.93
		ST	0.375	29.250	34.90	4.666	7.854	7.66	2094	1,048,000	118.65
		20, XS	0.500	29.000	46.34	4.587	7.854	7.59	2055	1,027,500	157.53
		30	0.625	28.750	57.68	4.508	7.854	7.53	2020	1,010,000	196.08

5S, 10S, 40S, and 80S are extracted from Stainless Steel Pipe, ASME B36.19M-1985, with permission of the publisher, the American Society of Mechanical Engineers, New York. ST = standard wall, XS = extra strong wall, XX = double extra strong wall, and Schedules 10 through 160 are extracted from Welded and Seamless Wrought Steel Pipe, ASME B36.10M-1996, with permission of the same publisher. Refer to these standards for a more comprehensive listing of material sizes and wall thicknesses. Decimal thicknesses for respective pipe sizes represent their nominal or average wall dimensions. Mill tolerances as high as ±12½ percent are permitted.

Plain-end pipe is produced by a square cut. Pipe is also shipped from the mills threaded, with a threaded coupling on one end, or with the ends beveled for welding, or grooved or sized for patented couplings.

To convert inches to millimeters, multiply by 25.4; to convert square inches to square millimeters, multiply by 645; to convert feet to meters, multiply by 0.3048; to convert square feet to square meters, multiply by 0.0929; to convert pounds per foot to kilograms per meter, multiply by 1.49; to convert gallons to cubic meters, multiply by 3.7854×10^{-3}; and to convert pounds to kilograms, multiply by 0.4536.

FIG. 10-128 Butt weld.

for nondestructive examination; however, branch connections are commonly specified for branches smaller than the header, and often best satisfy the design and economic requirements. Design of fabricated branch connections is addressed in the subsection "Pressure Design of Metallic Components: Wall Thickness." Integrally reinforced fittings are generally specified as requiring compliance with Manufacturer's Standardization Society specification MSS SP-97, *Integrally Reinforced Forged Branch Outlet Fittings—Socket Welding, Threaded, and Butt Welding Ends.*

Threaded Joints Pipe with **taper-pipe-thread** ends (Fig. 10-131), per ASME B1.20.1, is available 12 in and smaller, subject to minimum-wall limitations. Fittings and valves with taper-pipe-thread ends are available in most pipe metals.

Principal use of threaded joints is in sizes 2 in and smaller, in metals for which the most economically produced walls are thick enough to withstand pressure and corrosion after reduction in thickness due to threading. For threaded joints over 2 in, assembly difficulty and cost of tools increase rapidly. Careful alignment, required at the start of assembly and during rotation of the components, as well as variation in length produced by diametral tolerances in the threads, severely limits preassembly of the components. Threading is not a precise machining operation, and filler materials known as "pipe dope" are necessary to block the spiral leakage path.

Threads notch the pipe and cause loss of strength and fatigue resistance. Enlargement and contraction of the flow passage at threaded joints creates turbulence; sometimes corrosion and erosion are aggravated at the point where the pipe has already been thinned by threading. The tendency of pipe wrenches to crush pipe and fittings limits the torque available for tightening threaded joints. For low-pressure systems, a slight rotation in the joint may be used to impart flexibility to the system, but this same rotation, unwanted, may cause leaks to develop in higher-pressure systems. In some metals, galling occurs when threaded joints are disassembled.

Straight Pipe Threads These are confined to light-weight couplings in sizes 2 in and smaller (Fig. 10-132). Manufacturers of threaded pipe ship it with such couplings installed on one end of each pipe. The joint obtained is inferior to that obtained with taper threads. The code limits the joint shown in Fig. 10-129 to 1.0 MPa (150 lbf/in^2) gauge maximum, 182°C (360°F) maximum, and to nonflammable, nontoxic fluids.

When both components of a threaded joint are of weldable metal, the joint may be **seal-welded** as shown in Fig. 10-133. Seal welds may be used only to prevent leakage of threaded joints. They are not considered as contributing any strength to the joint. This type of joint is limited to new construction and is not suitable as a repair procedure, since pipe dope in the threads would interfere with welding. Careful consideration should be given to the suitability of threaded joints when joining metals having significantly different coefficients of expansion. Thermal expansion and temperature cycling may eventually result in leakage.

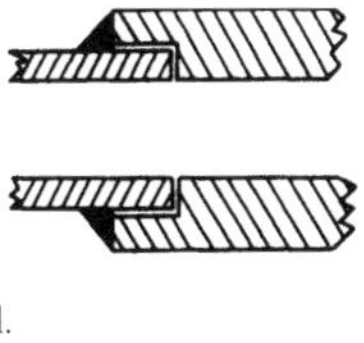

FIG. 10-129 Socket weld.

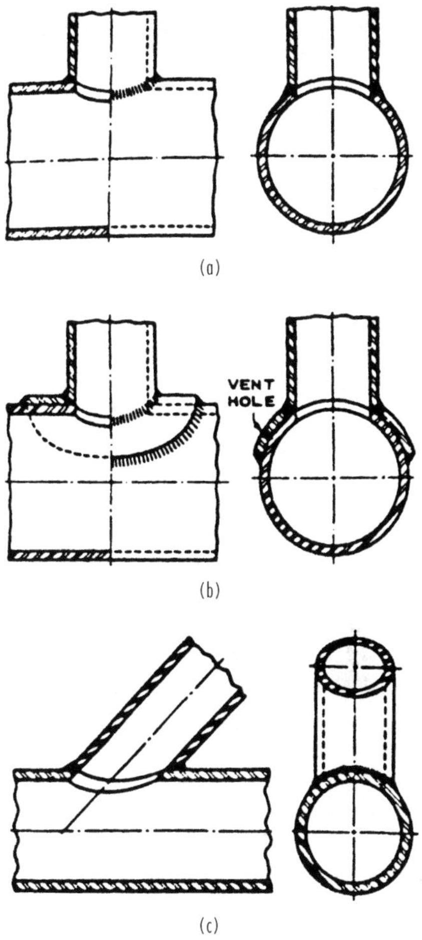

FIG. 10-130 Branch welds. (*a*) Without added reinforcement. (*b*) With added reinforcement. (*c*) Angular branch.

To assist in assembly and disassembly of both threaded and welded systems, **union joints** (Fig. 10-134) are used. They comprise metal-to-metal or gasketed seats drawn together by a shouldered straight thread nut and are available both in couplings for joining two lengths of pipe and on the ends of some fittings. On threaded piping systems in which disassembly is not contemplated, union joints installed at intervals permit future further tightening of threaded joints. Tightening of heavy unions yields tight joints even if the pipe is slightly misaligned at the start of tightening.

Flanged Joints For sizes larger than 2 in when disassembly is contemplated, the flanged joint (Fig. 10-135) is the most widely used. Figures 10-136 and 10-137 illustrate the wide variety of types and facings available. Though flanged joints consume a large volume of metal, precise machining is required only on the facing. Flanged joints do not impose severe diametral tolerances on the pipe. Alignment tolerances required for flanged joints are reasonably achieved with quality construction practices, and in comparison with taper threaded joints, required wrench sizes are smaller and sealing is more easily and reliably obtained.

Manufacturers offer **flanged-end pipe** in only a few metals. Otherwise, flanges are attached to pipe by various types of joints (Fig. 10-136). The lap joint involves a modification of the pipe which may be formed from the pipe itself or by welding a ring or a lap-joint stub end to it. **Flanged-end fittings** and valves are available in all sizes of most pipe metals.

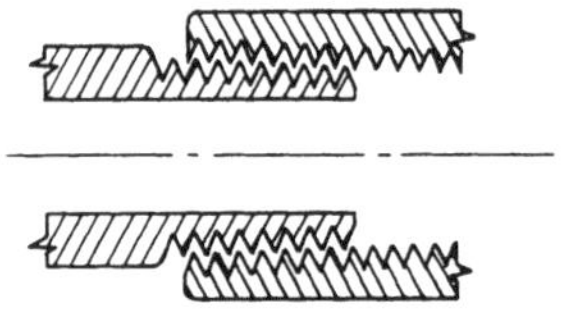

FIG. 10-131 Taper pipe thread.

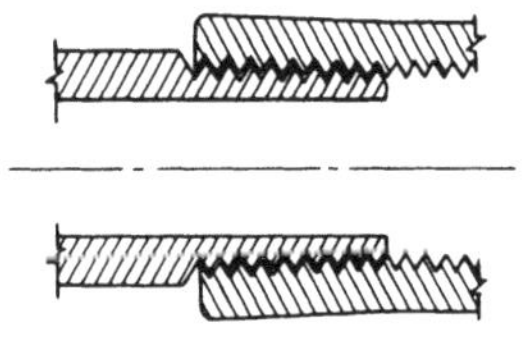

FIG. 10-132 Taper pipe to straight coupling thread.

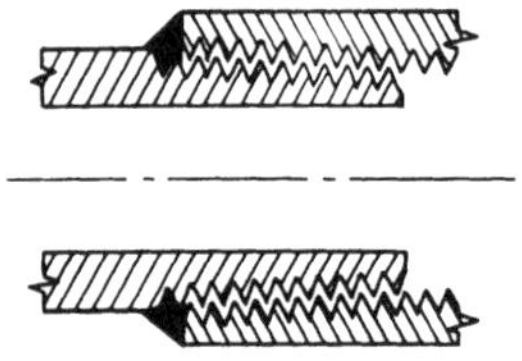

FIG. 10-133 Taper pipe thread seal-welded.

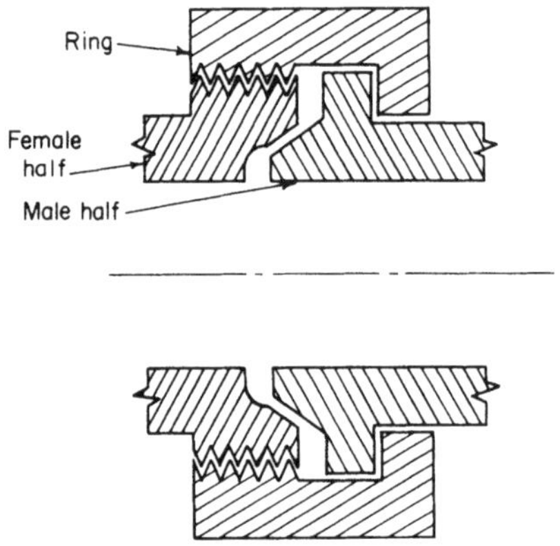

FIG. 10-134 Union.

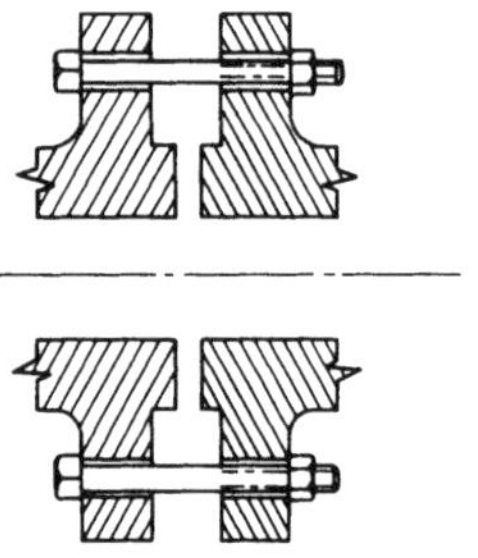

FIG. 10-135 Flanged joint.

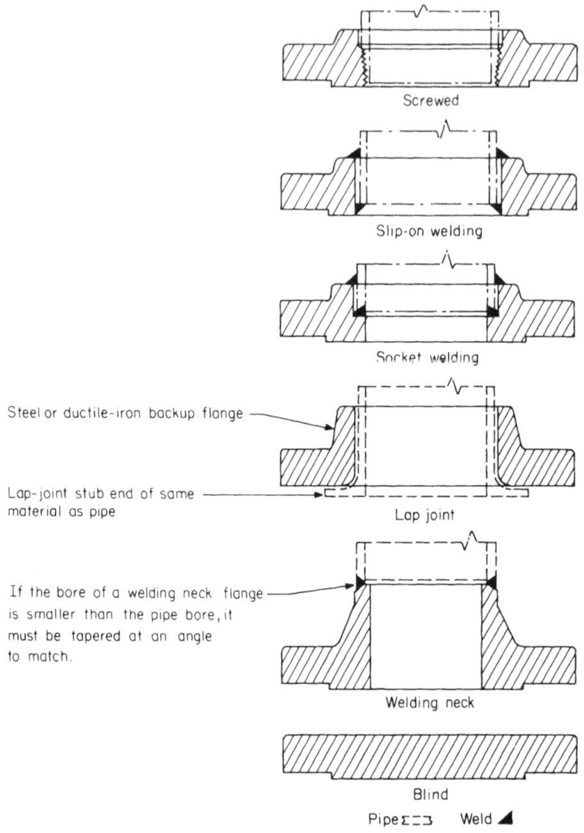

FIG. 10-136 Types of carbon and alloy steel flanges.

Of the flange types shown in Fig. 10-136, welding-neck flanges offer the highest mechanical strength and are the type most suitable for extreme temperatures and cyclic loading. Regardless of the type selected, designers must be aware that the flange's capability to resist external bending moments and maintain its seal does not necessarily match the bending moment capability of the attached pipe.

When selecting the flange type, the designer should review the usage restrictions contained in each section of the ASME B31 code. Each of the other types shown provides significant fabrication and economic advantages and is suitable for many of the routine applications. Lap-joint flanges permit adjustment of the bolt hole orientation and can greatly simplify construction when configurations are complex and bolt hole orientations are difficult to ensure.

Dimensions for alloy and carbon-steel pipe flanges are given in Tables 10-23 through 10-29. The dimensions were extracted from *Pipe Flanges and Flanged Fittings*, ASME B16.5–2003, with permission of the publisher, the American Society of Mechanical Engineers, New York. Dimensions for cast-iron flanges are provided in *Cast Iron Pipe Flanges and Flanged Fittings*, ASME B16.1. Bolt patterns and bolt sizes for Class 125 cast-iron flanges match the ASME B16.5 Class 150 flange dimensions, and bolt patterns for Class 250 cast-iron flanges match the ASME B16.5 Class 300 flange dimensions.

When mating with cast-iron flanged fittings or valves, steel pipe flanges are often preferred to cast-iron flanges because they permit welded rather than screwed assembly to the pipe and because cast-iron pipe flanges, not being reinforced by the pipe, are not so resistant to abuse as flanges cast integrally on cast-iron fittings.

Facing of flanges for alloy and carbon steel pipe and fittings is shown in Fig. 10-137; Class 125 cast-iron pipe and fitting flanges have flat faces, which with full-face gaskets minimize bending stresses; Class 250 cast-iron pipe and fitting flanges have 1.5-mm

TABLE 10-23 Dimensions of ASME B16.5 Class 150 Flanges*

All dimensions in inches

						Length through hub			
Nominal pipe size	Outside diameter of flange	Thickness of flange, minimum	Diameter of bolt circle	Diameter of bolts	No. of bolts	Threaded slip-on socket welding	Lap joint	Welding neck	ANSI B16.1, screwed (125-lb)
½	3.50	0.44	2.38	½	4	0.62	0.62	1.88	
¾	3.88	0.50	2.75	½	4	0.62	0.62	2.06	
1	4.25	0.56	3.12	½	4	0.69	0.69	2.19	0.69
1¼	4.62	0.62	3.50	½	4	0.81	0.81	2.25	0.81
1½	5.00	0.69	3.88	½	4	0.88	0.88	2.44	0.88
2	6.00	0.75	4.75	⅝	4	1.00	1.00	2.50	1.00
2½	7.00	0.88	5.50	⅝	4	1.12	1.12	2.75	1.12
3	7.50	0.94	6.00	⅝	4	1.19	1.19	2.75	1.19
3½	8.50	0.94	7.00	⅝	8	1.25	1.25	2.81	1.25
4	9.00	0.94	7.50	⅝	8	1.31	1.31	3.00	1.31
5	10.00	0.94	8.50	¾	8	1.44	1.44	3.50	1.44
6	11.00	1.00	9.50	¾	8	1.56	1.56	3.50	1.56
8	13.50	1.12	11.75	¾	8	1.75	1.75	4.00	1.75
10	16.00	1.19	14.25	⅞	12	1.94	1.94	4.00	1.94
12	19.00	1.25	17.00	⅞	12	2.19	2.19	4.50	2.19
14	21.00	1.38	18.75	1	12	2.25	3.12	5.00	2.25
16	23.50	1.44	21.25	1	16	2.50	3.44	5.00	2.50
18	25.00	1.56	22.75	1⅛	16	2.69	3.81	5.50	2.69
20	27.50	1.69	25.00	1⅛	20	2.88	4.06	5.69	2.88
24	32.00	1.88	29.50	1¼	20	3.25	4.38	6.00	3.25

*Dimensions from ASME B16.5-2003, unless otherwise noted. To convert inches to millimeters, multiply by 25.4.

(1/16-in) raised faces (wider than on steel flanges) for the same purpose. Carbon steel and ductile- (nodular-) iron lap-joint flanges are widely used as backup flanges with stub ends in piping systems of austenitic stainless steel and other expensive materials to reduce costs (see Fig. 10-136). The code prohibits the use of ductile-iron flanges at temperatures above 343°C (650°F). When the type of facing affects the length through the hub dimension of flanges, correct dimensions for commonly used facings can be determined from the dimensional data in Fig. 10-137.

Gaskets Gaskets must resist corrosion by the fluids handled. The more expensive male-and-female or tongue-and-groove facings may be required to seat hard gaskets adequately. With these facings the gasket generally cannot blow out. Flanged joints, by placing the gasket material under heavy compression and permitting only edge attack by the fluid handled, can use gasket materials which in other joints might not satisfactorily resist the fluid handled.

Standards to which flanges are manufactured (e.g., ASME B16.1, ASME B16.5, ASME B16.47) typically specify a standard surface finish

TABLE 10-24 Dimensions of ASME B16.5 Class 300 Flanges*

All dimensions in inches

						Length through hub			
Nominal pipe size	Outside diameter of flange	Thickness of flange, minimum	Diameter of bolt circle	Diameter of bolts	No. of bolts	Threaded slip-on socket welding	Lap joint	Welding neck	ANSI B16.1, screwed (Class 250)
½	3.75	0.56	2.62	½	4	0.88	0.88	2.06	
¾	4.62	0.62	3.25	⅝	4	1.00	1.00	2.25	
1	4.88	0.69	3.50	⅝	4	1.06	1.06	2.44	0.88
1¼	5.25	0.75	3.88	⅝	4	1.06	1.06	2.56	1.00
1½	6.12	0.81	4.50	¾	4	1.19	1.19	2.69	1.12
2	6.50	0.88	5.00	⅝	8	1.31	1.31	2.75	1.25
2½	7.50	1.00	5.88	¾	8	1.50	1.50	3.00	1.43
3	8.25	1.12	6.62	¾	8	1.69	1.69	3.12	1.56
3½	9.00	1.19	7.25	¾	8	1.75	1.75	3.19	1.62
4	10.00	1.25	7.88	¾	8	1.88	1.88	3.38	1.75
5	11.00	1.38	9.25	¾	8	2.00	2.00	3.88	1.88
6	12.50	1.44	10.62	¾	12	2.06	2.06	3.88	1.94
8	15.00	1.62	13.00	⅞	12	2.44	2.44	4.38	2.19
10	17.50	1.88	15.25	1	16	2.62	3.75	4.62	2.38
12	20.50	2.00	17.75	1⅛	16	2.88	4.00	5.12	2.56
14	23.00	2.12	20.25	1⅛	20	3.00	4.38	5.62	2.69
16	25.50	2.25	22.50	1¼	20	3.25	4.75	5.75	2.88
18	28.00	2.38	24.75	1¼	24	3.50	5.12	6.25	
20	30.50	2.50	27.00	1¼	24	3.75	5.50	6.38	
24	36.00	2.75	32.00	1½	24	4.19	6.00	6.62	

*Dimensions from ASME B16.5-2003, unless otherwise noted. To convert inches to millimeters, multiply by 25.4.

TABLE 10-25 Dimensions of ASME B16.5 Class 400 Flanges*

All dimensions in inches

Nominal pipe size	Outside diameter of flange	Thickness of flange, minimum	Diameter of bolt circle	Diameter of bolts	No. of bolts	Length through hub		
						Threaded slip-on socket welding	Lap joint	Welding neck
½	Use Class 600 dimensions in these sizes.							
¾								
1								
1¼								
1½								
2								
2½								
3								
3½								
4	10.00	1.38	7.88	⅞	8	2.00	2.00	3.50
5	11.00	1.50	9.25	⅞	8	2.12	2.12	4.00
6	12.50	1.62	10.62	⅞	12	2.25	2.25	4.06
8	15.00	1.88	13.00	1	12	2.69	2.69	4.62
10	17.50	2.12	15.25	1¼	16	2.88	4.00	4.88
12	20.50	2.25	17.75	1¼	16	3.12	4.25	5.38
14	23.00	2.38	20.25	1¼	20	3.31	4.62	5.88
16	25.50	2.50	22.50	1⅜	20	3.69	5.00	6.00
18	28.00	2.62	24.75	1⅜	24	3.88	5.38	6.50
20	30.50	2.75	27.00	1½	24	4.00	5.75	6.62
24	36.00	3.00	32.00	1¾	24	4.50	6.25	6.88

*Dimensions from ASME B16.5-2003. To convert inches to millimeters, multiply by 25.4.

for the gasket seal area. Flange rating and type, flange size, flange facing type, gasket style, commodity, design conditions, and bolting must all be considered to ensure proper seating of the gasket and reliable performance. Unless the user is familiar with gasket design and the particular application being considered, it is highly recommended that the gasket manufacturer be consulted regarding gasket selection. Upon request, gasket manufacturers typically provide assistance in determining the proper material selection and the proper gasket style to ensure an economical choice and a reliable system.

When appropriate for the commodity, elastomer sheet gaskets without fillers are generally the least expensive gasket type. They are typically limited to Class 150 and temperatures below 120°C (250°F). Composition sheet gaskets are somewhat more expensive than elastomer sheet gaskets. They are generally composed of an elastomer binder with fiber filler. Their use is generally limited to Class 150 and Class 300, and depending on the filler selected the upper temperature limit typically ranges between 205°C (400°F) and 370°C (700°F). Nonelastomer sheet gaskets, such as graphite sheet gaskets, are generally somewhat more expensive than composition sheet gaskets. Their use is typically limited to Class 150 and Class 300, and the upper temperature limit may be 535°C (1000°F) or higher. Spiral-wound gaskets with graphite, PTFE, or other filler are generally appropriate

TABLE 10-26 Dimensions of ASME B16.5 Class 600 Flanges*

All dimensions in inches

Nominal pipe size	Outside diameter of flange	Thickness of flange, minimum	Diameter of bolt circle	Diameter of bolts	No. of bolts	Length through hub		
						Threaded slip-on socket welding	Lap joint	Welding neck
½	3.75	0.56	2.62	½	4	0.88	0.88	2.06
¾	4.62	0.62	3.25	⅝	4	1.00	1.00	2.25
1	4.88	0.69	3.50	⅝	4	1.06	1.06	2.44
1¼	5.25	0.81	3.88	⅝	4	1.12	1.12	2.62
1½	6.12	0.88	4.50	¾	4	1.25	1.25	2.75
2	6.50	1.00	5.00	⅝	8	1.44	1.44	2.88
2½	7.50	1.12	5.88	¾	8	1.62	1.62	3.12
3	8.25	1.25	6.62	¾	8	1.81	1.81	3.25
3½	9.00	1.38	7.25	⅞	8	1.94	1.94	3.38
4	10.75	1.50	8.50	⅞	8	2.12	2.12	4.00
5	13.00	1.75	10.50	1	8	2.38	2.38	4.50
6	14.00	1.88	11.50	1	12	2.62	2.62	4.62
8	16.50	2.19	13.75	1⅛	12	3.00	3.00	5.25
10	20.00	2.50	17.00	1¼	16	3.38	4.38	6.00
12	22.00	2.62	19.25	1¼	20	3.62	4.62	6.12
14	23.75	2.75	20.75	1¾	20	3.69	5.00	6.50
16	27.00	3.00	23.75	1½	20	4.19	5.50	7.00
18	29.25	3.25	25.75	1⅝	20	4.62	6.00	7.25
20	32.00	3.50	28.50	1⅝	24	5.00	6.50	7.50
24	37.00	4.00	33.00	1⅞	24	5.50	7.25	8.00

*Dimensions from ASME B16.5-2003. To convert inches to millimeters, multiply by 25.4.

TABLE 10-27 Dimensions of ASME B16.5 Class 900 Flanges*

All dimensions in inches

Nominal pipe size	Outside diameter of flange	Thickness of flange, minimum	Diameter of bolt circle	Diameter of bolts	No. of bolts	Length through hub		
						Threaded slip-on socket welding	Lap joint	Welding neck
½	Use Class 1500 dimensions in these sizes.							
¾								
1								
1¼								
1½								
2								
2½								
3	9.50	1.50	7.50	⅞	8	2.12	2.12	4.00
4	11.50	1.75	9.25	1⅛	8	2.75	2.75	4.50
5	13.75	2.00	11.00	1¼	8	3.12	3.12	5.00
6	15.00	2.19	12.50	1⅛	12	3.38	3.38	5.50
8	18.50	2.50	15.50	1⅜	12	4.00	4.50	6.38
10	21.50	2.75	18.50	1⅜	16	4.25	5.00	7.25
12	24.00	3.12	21.00	1⅜	20	4.62	5.62	7.88
14	25.25	3.38	22.00	1½	20	5.12	6.12	8.38
16	27.75	3.50	24.25	1⅝	20	5.25	6.50	8.50
18	31.00	4.00	27.00	1⅞	20	6.00	7.50	9.00
20	33.75	4.25	29.50	2	20	6.25	8.25	9.75
24	41.00	5.50	35.50	2½	20	8.00	10.50	11.50

*Dimensions from ASME B16.5-2003. To convert inches to millimeters, multiply by 25.4.

for applications more demanding than those handled by sheet gaskets. They are generally more expensive than sheet gaskets and are commonly used in Class 150 through Class 1500 services (and higher) class ratings. Because of their breadth of capabilities and the advantages of standardizing, they are often used when less expensive gaskets would suffice. The solid metal outer ring on spiral-wound gaskets serves to center the gasket and provide blowout resistance. With the proper filler, spiral-wound gaskets and some sheet gaskets provide good sealing under fire conditions.

Ring Joint Flanges Ring joint (RTJ) flanges provide sealing capability for pressure-temperature combinations higher than those for which spiral-wound gaskets are typically used. Depending on the service, use of RTJ flanges is often considered in Class 900 and higher applications. RTJ flange facings and gaskets are more expensive than the spiral-wound counterparts. The ring material must be softer than the flange seating surface, and corrosion-resistant to the service. They provide good resistance to leakage under fire conditions. RTJ flanges must be separated in the axial direction to permit insertion and removal of the gasket.

Bolting Bolt strength requirements are addressed to some extent by the code and by code-referenced flange standards. Bolts are categorized by the code as high strength, intermediate strength, and low strength. Bolting materials having allowable stresses meeting or exceeding those of ASTM A193 Grade B7 are categorized as high

TABLE 10-28 Dimensions of ASME B16.5 Class 1500 Flanges*

All dimensions in inches

Nominal pipe size	Outside diameter of flange	Thickness of flange, minimum	Diameter of bolt circle	Diameter of bolts	No. of bolts	Length through hub		
						Threaded slip-on socket welding	Lap joint	Welding neck
½	4.75	0.88	3.25	¾	4	1.25	1.25	2.38
¾	5.12	1.00	3.50	¾	4	1.38	1.38	2.75
1	5.88	1.12	4.00	⅞	4	1.62	1.62	2.88
1¼	6.25	1.12	4.38	⅞	4	1.62	1.62	2.88
1½	7.00	1.25	4.88	1	4	1.75	1.75	3.25
2	8.50	1.50	6.50	⅞	8	2.25	2.25	4.00
2½	9.62	1.62	7.50	1	8	2.50	2.50	4.12
3	10.50	1.88	8.00	1⅛	8	2.88	2.88	4.62
4	12.25	2.12	9.50	1¼	8	3.56	3.56	4.88
5	14.75	2.88	11.50	1½	8	4.12	4.12	6.12
6	15.50	3.25	12.50	1⅜	12	4.69	4.69	6.75
8	19.00	3.62	15.50	1⅝	12	5.62	5.62	8.38
10	23.00	4.25	19.00	1⅞	12	6.25	7.00	10.00
12	26.50	4.88	22.50	2	16	7.12	8.62	11.12
14	29.50	5.25	25.00	2¼	16		9.50	11.75
16	32.50	5.75	27.75	2½	16		10.25	12.25
18	36.00	6.38	30.50	2¾	16		10.88	12.88
20	38.75	7.00	32.75	3	16		11.50	14.00
24	46.00	8.00	39.00	3½	16		13.00	16.00

*Dimensions from ASME B16.5-2003. To convert inches to millimeters, multiply by 25.4.

TABLE 10-29 Dimensions of ASME B16.5 Class 2500*

All dimensions in inches

Nominal pipe size	Outside diameter of flange	Thickness of flange, minimum	Diameter of bolt circle	Diameter of bolts	No. of bolts	Length through hub		
						Threaded	Lap joint	Welding neck
½	5.25	1.19	3.50	¾	4	1.56	1.56	2.88
¾	5.50	1.25	3.75	¾	4	1.69	1.69	3.12
1	6.25	1.38	4.25	⅞	4	1.88	1.88	3.50
1¼	7.25	1.50	5.12	1	4	2.06	2.06	3.75
1½	8.00	1.75	5.75	1⅛	4	2.38	2.38	4.38
2	9.25	2.00	6.75	1	8	2.75	2.75	5.00
2½	10.50	2.25	7.75	1⅛	8	3.12	3.12	5.62
3	12.00	2.62	9.00	1¼	8	3.62	3.62	6.62
4	14.00	3.00	10.75	1½	8	4.25	4.25	7.50
5	16.50	3.62	12.75	1¾	8	5.12	5.12	9.00
6	19.00	4.25	14.50	2	8	6.00	6.00	10.75
8	21.75	5.00	17.25	2	12	7.00	7.00	12.50
10	26.50	6.50	21.25	2½	12	9.00	9.00	16.50
12	30.00	7.25	24.38	2¾	12	10.00	10.00	18.25

*Dimensions from ASME B16.5-2003. To convert inches to millimeters, multiply by 25.4.

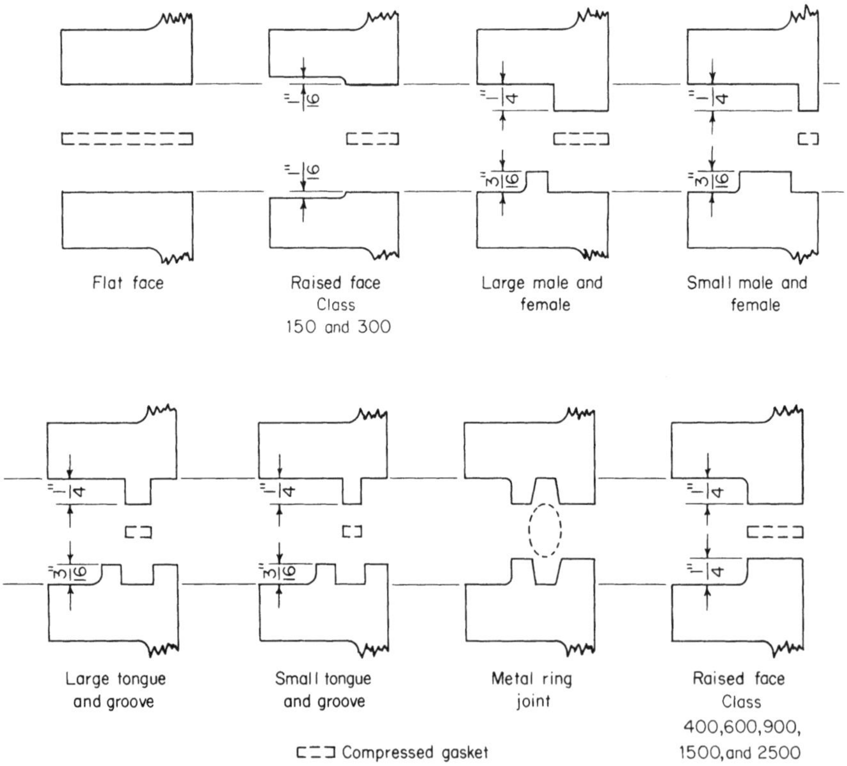

FIG. 10-137 Flange facings, illustrated on welding-neck flanges. (On small male-and-female facings the outside diameter of the male face is less than the outside diameter of the pipe, so this facing does not apply to screwed or slip-on flanges. A similar joint can be made with screwed flanges and threaded pipe by projecting the pipe through one flange and recessing it in the other. However, pipe thicker than Schedule 40 is required to avoid crushing gaskets.) To convert inches to millimeters, multiply by 25.4.

strength. Bolting materials having specified minimum yield strengths of 207 MPa (30 ksi) or less are categorized as low strength. ASTM A307 Grade B is a commonly used specification for low-strength bolting.

The suitability of the strength of any bolting throughout the required temperature range should be verified by the designer. Verification of the suitability of intermediate-strength bolting for the intended joint design is required prior to its use. The code restricts the use of low-strength bolting to nonmetallic gaskets and flanges rated ASME B16.5 Class 300 and lower having bolt temperatures at −29 to 204°C (−20 to 400°F) inclusive. Low-strength bolting is not permitted for use under severe cyclic conditions as defined by the code.

Except when bolting brittle flange materials such as gray cast iron, the code permits the use of high-strength bolting for any style of flanged joint and gasket type combination. Per the code, if either mating flange is specified in accordance with ASME B16.1, ASME B16.24, MSS SP-42, or MSS SP-51, the bolting material shall be no stronger than low-strength unless both mating flanges have a flat face and a full-face gasket is used. Exception to this requirement is permitted if sequence and torque limits for bolt-up are specified, with consideration given to sustained and occasional loads, and displacement strains. When both flanges are flat face and the gasket is full face extending to the outside diameter of the flange, intermediate-strength and high-strength bolts may be used.

Miscellaneous Mechanical Joints

Packed-Gland Joints These joints (Fig. 10-138) require no special end preparation of pipe but do require careful control of the diameter of the pipe. Thus the supplier of the pipe should be notified when packed-gland joints are to be used. Cast- and ductile-iron pipe, fittings, and valves are available with the bell cast on one or more ends. Glands, bolts, and gaskets are shipped with the pipe. Couplings equipped with packed glands at each end, known as Dresser couplings, are available in several metals. The joints can be assembled with small wrenches and unskilled labor, in limited space, and if necessary, under water.

Packed-gland joints are designed to take the same hoop stress as the pipe. They do not resist bending moments or axial forces tending to separate the joints but yield to them to an extent indicated by the manufacturer's allowable-angular-deflection and end movement specifications. Further angular or end movement produces leakage, but end movement can be limited by harnessing or bridling with a combination of rods and welded clips or clamps, or by anchoring to existing or new structures. The crevice between the bell and the spigot may promote corrosion. The joints are widely used in underground lines. They are not affected by limited earth settlement, and friction of the earth is often adequate to prevent end separation. When disassembly by moving pipe axially is not practical, packed-joint couplings which can be slid entirely onto one of the two lengths joined are available.

Poured Joints Figure 10-139 illustrates a poured joint design. With regard to performance and ease of installation, most other joint designs are preferable to poured joints, and their use can generally be avoided.

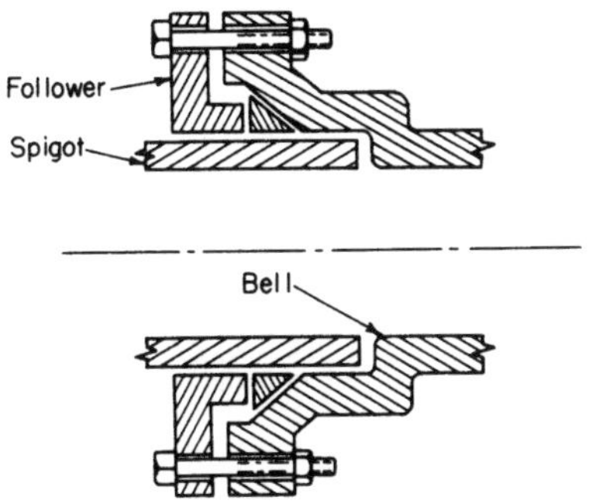

FIG. 10-138 Packed-gland joint.

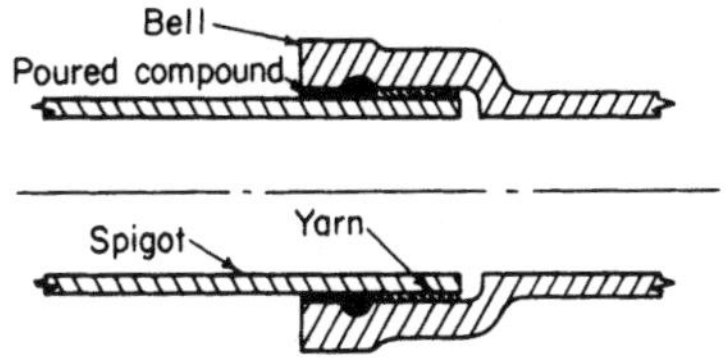

FIG. 10-139 Poured joint.

Push-on Joints These joints (Fig. 10-140) require diametral control of the end of the pipe. They are used for brittle and nonmetallic materials. Pipe, fittings, and valves are furnished with the bells on one or more ends.

Push-on joints do not resist bending moments or axial forces tending to separate the joints but yield to them to an extent limited by the manufacturer's allowable-angular-deflection and end-movement specifications. End movement can be limited by harnessing or bridling with a combination of rods and clamps, or by anchoring to existing or new structures. Some manufacturers offer O-rings with metallic embedments that grip the pipe and prevent axial separation under internal pressure loading. The joints are widely used on underground lines. They are not affected by limited earth settlement, and friction of the earth is often adequate to prevent end separation. A lubricant is used on the O ring during assembly. After this disappears, the O ring bonds somewhat to the spigot and disassembly is very difficult. Disassembly for maintenance is accomplished by cutting the pipe and reassembly by use of a coupling with a packed-gland joint on each end.

Expanded Joints These joints (Fig. 10-141) are confined to the smaller pipe sizes and ductile metals. Various proprietary designs are available in which either the pipe is expanded into the coupling or the

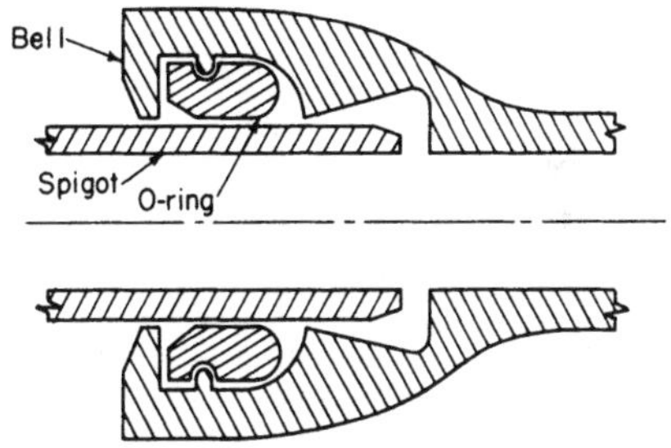

FIG. 10-140 Push-on joint.

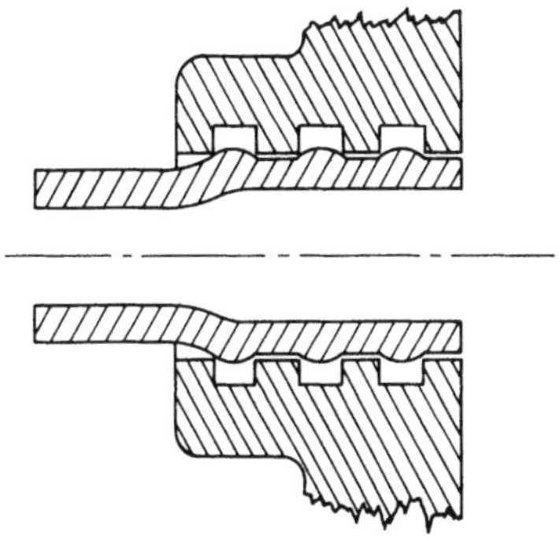

FIG. 10-141 Expanded joint.

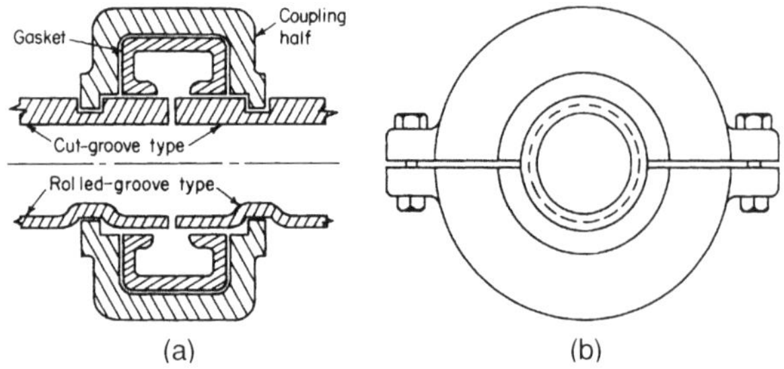

FIG. 10-142 Grooved joint. (*a*) Section. (*b*) End view.

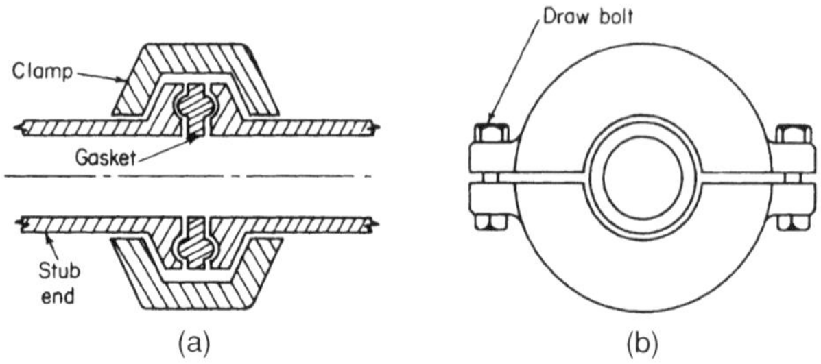

FIG. 10-143 V-clamp joint. (*a*) Section. (*b*) End view.

coupling is crimped down onto the pipe. In some designs, the seal between the pipe and coupling is metal to metal, while in others elastomer O-rings are employed. Joints of these types typically are quickly and easily made with specialized equipment, and they may be particularly attractive in maintenance applications since no welding is involved. The designer should clearly understand the limitations of the joint design and should verify the success of its long-term service in similar applications.

Grooved Joints These joints (Fig. 10-142) are divided into two classes, cut grooves and rolled grooves. Rolled grooves are preferred because, compared with cut grooves, they are easier to form and reduce the metal wall less. However, they slightly reduce the flow area. They are limited to thin walls of ductile material, while cut grooves, because of their reduction of the pipe wall, are limited to thicker walls. In the larger pipe sizes, some commonly used wall thicknesses are too thick for rolled grooves but too thin for cut grooves. The thinning of the walls impairs resistance to corrosion and erosion but not to internal pressure, because the thinned area is reinforced by the coupling.

Control of outside diameter is important. Permissible minus tolerance is limited, since it impairs the grip of the couplings. Plus tolerance makes it necessary to cut the cut grooves more deeply, increasing the thinning of the wall. Plus tolerance is not a problem with rolled grooves, since they are confined to walls thin enough so that the couplings can compress the pipe. Pipe is available from vendors already grooved and also with heavier-wall grooved ends welded on.

Grooved joints resist axial forces tending to separate the joints. Angular deflection, up to the limit specified by the manufacturer, may be used to absorb thermal expansion and to permit the piping to be laid on uneven ground. Grooved joints provide quick and easy assembly and disassembly when compared with flanges, but may require more support than welded joints.

Gaskets are self-sealing against both internal and external pressure and are available in a wide variety of elastomers. However, successful performance of an elastomer as a flange gasket does not necessarily mean equally satisfactory performance in a grooved joint, since exposure to the fluid in the latter is much greater and hardening has a greater unfavorable effect. It is advisable to select coupling material that is suitably corrosion-resistant with respect to the service; but with proper gasket style it may be permissible to use a coupling material that might otherwise be unacceptable with respect to fluid contamination.

V-Clamp Joints These joints (Fig. 10-143) are attached to the pipe by butt-weld or expanded joints. Theoretically, there is only one relative position of the parts in which the conical surfaces of the clamp are completely in contact with the conical surfaces of the stub ends. In actual practice, there is considerable flexing of the stub ends and the clamp; also complete contact is not required. This permits use of elastomeric gaskets as well as metal gaskets. Fittings are also available with integral conical shouldered ends.

Conical ends vary from machined forgings to roll-formed tubing, and clamps vary from machined forgings to bands to which several roll-formed channels are attached at their centers by spot welding. A hinge may be inserted in the band as a substitute for one of the draw bolts. Latches may also be substituted for draw bolts.

Compared with flanges, V-clamp joints use less metal, require less labor for assembly, and are less likely to leak under wide-range rapid temperature cycling. However, they are more susceptible to failure or damage from overtightening. They are widely used for high-alloy piping subject to periodic cleaning or relocation. Manufactured as forgings, they are used in carbon steel with metal gaskets for very high pressures. They resist both axial strain and bending moments. Each size of each type of joint is customarily rated by the vendor for both internal pressure and bending moment.

Seal Ring Joints These joints (Fig. 10-144) consist of hubs that are attached to pipe ends by welding. Joints of this type are proprietary, and their pressure/temperature ratings and external loading capabilities are established by the manufacturers. Variations of this design are offered by various manufacturers. Many of these designs have been widely used in critical high-pressure/temperature applications. They are particularly cost-effective in high-pressure alloy material applications.

Pressure-Seal Joints These joints (Fig. 10-145) are used for pressures of ASME Class 600 and higher. They use less metal than flanged joints but require much more machining of surfaces. There are several designs, in all of which increasing fluid pressure increases the force holding the sealing surfaces against each other. These joints are widely used as bonnet joints in carbon and alloy steel valves.

Tubing Joints **Flared-fitting joints** (see Fig. 10-146) are used for ductile tubing when the ratio of wall thickness to the diameter is small enough to permit flaring without cracking the inside surface. The tubing must have a smooth interior surface. The three-piece type avoids torsional strain on the tubing and minimizes vibration fatigue on the flared portion of the tubing. More labor is required for assembly, but the fitting is more resistant to temperature cycling than other tubing fittings and is less likely to be damaged by overtightening. Its efficiency is not impaired by repeated assembly and disassembly. Size is limited because of the large number of machined surfaces. The nut and, in the three-piece type, the sleeve need not be of the same material as the tubing. For these fittings, less control of tubing diameter is required.

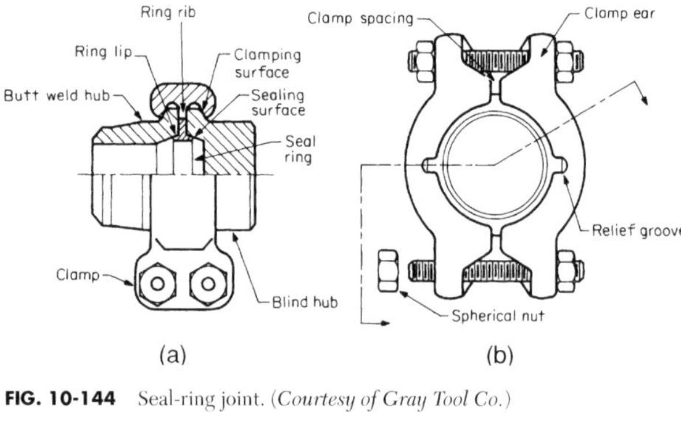

FIG. 10-144 Seal-ring joint. (*Courtesy of Gray Tool Co.*)

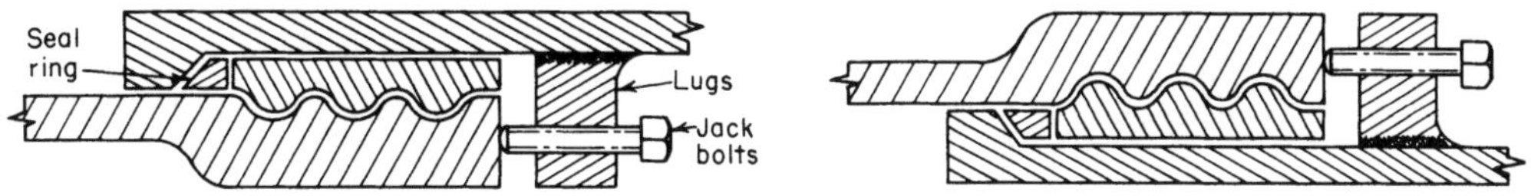

FIG. 10-145 Pressure-seal joint.

Compression-Fitting Joints These joints (Fig. 10-147) are used for ductile tubing with thin walls. The outside of the tubing must be clean and smooth. Assembly consists only of inserting the tubing and tightening the nut. These are the least costly tubing fittings but are not resistant to vibration or temperature cycling.

Bite-Type-Fitting Joints These joints (Fig. 10-148) are used when the tubing has too high a ratio of wall thickness to diameter for flaring, when the tubing lacks sufficient ductility for flaring, and for low assembly-labor cost. The outside of the tubing must be clean and smooth. Assembly consists in merely inserting the tubing and tightening the nut. The sleeve must be considerably harder than the tubing yet still ductile enough to be diametrally compressed and must be as resistant to corrosion by the fluid handled as the tubing. The fittings are resistant to vibration but not to wide-range rapid temperature cycling. Compared with flared fittings, they are less suited for repeated assembly and disassembly, require closer diametral control of the tubing, and are more susceptible to damage from overtightening. They are widely used for oil-filled hydraulic systems at all pressures.

O-ring Seal Joints These joints (Fig. 10-149) are also used for applications requiring heavy-wall tubing. The outside of the tubing must be clean and smooth. The joint may be assembled repeatedly, and as long as the tubing is not damaged, leaks can usually be corrected by replacing the O ring and the antiextrusion washer. This joint is used extensively in oil-filled hydraulic systems.

Soldered Joints These joints (Fig. 10-150) require precise control of the diameter of the pipe or tubing and of the cup or socket in the fitting in order to cause the solder to draw into the clearance between the cup and the tubing by capillary action (Fig. 10-150). Extrusion provides this diametral control, and the joints are most widely used in copper. A 50 percent lead, 50 percent tin solder is used for temperatures up to 93°C (200°F). Careful cleaning of the outside of the tubing and inside of the cup is required.

Heat for soldering is usually obtained from torches. The high conductivity of copper makes it necessary to use large flames for the larger sizes, and for this reason the location in which the joint will be made must be carefully considered. Soldered joints are most widely used in sizes 2 in and smaller for which heat requirements are less burdensome. Soldered joints should not be used in areas where plant fires are likely because exposure to fires results in rapid and complete failure of the joints. Properly made, the joints are completely impervious. The code permits the use of soldered joints only for Category D fluid service and then only if the system is not subject to severe cyclic conditions.

Silver Brazed Joints These are similar to soldered joints except that a temperature of about 600°C (1100°F) is required. A 15 percent silver, 80 percent copper, 5 percent phosphorus solder is used for copper and copper alloys, while 45 percent silver, 15 percent copper, 16 percent zinc, 24 percent cadmium solders are used for copper, copper alloys, carbon steel, and alloy steel. Silver-brazed joints are used for temperatures up to 200°C (400°F). Cast-bronze fittings and valves with preinserted rings of 15 percent silver, 80 percent copper, 5 percent phosphorus brazing alloy are available.

Silver-brazed joints are used when temperature or the combination of temperature and pressure is beyond the range of soldered joints. They are also more reliable in the event of plant fires and are more resistant to vibration. If they are used for fluids that are flammable, toxic, or damaging to human tissue, appropriate safeguarding is required by the code. There are OSHA regulations governing the use of silver brazing alloys containing cadmium and other toxic materials.

Pipe Fittings and Bends Directional changes in piping systems are typically made with bends or welded fittings. Bends are made as either hot bends or cold bends. Cold bending is done at temperatures below the material transformation temperature. Depending on the material and the amount of strain involved, annealing or stress relief may be required after bending. The bend radius that may be achieved for pipe of a given size, material, and thickness depends upon the bending machine capabilities and bending procedures used. When contemplating bending, the bending limitations should be reviewed with the pipe fabricators being considered for the project. Because bends are not generally made to radii as small as those of standard butt-weld or socket-weld fittings, the use of bends must be considered during piping layout. Wall thinning resulting from bending must also be considered when specifying the wall thickness of material that is to be bent. A detailed bending specification that addresses all aspects of bending, including requirements for bending procedure specifications, availability of bending procedure qualification records and bending operator qualification records, the range of bends covered by a single bending procedure qualification, in-process nondestructive examination requirements (including minimum wall thickness verification), dimensional tolerance requirements, etc., should be part of the bending agreement. Some bending operations and subsequent heat treatment can result in tenacious oxide formation on certain materials (such as 9Cr-1Mo-V). Removal of this oxide by conventional means such as abrasive blasting may be very difficult. Methods of avoiding this formation or of removing it should be discussed prior to bending when the application requires a high level of cleanliness, such as is the case with steam supply lines to turbines.

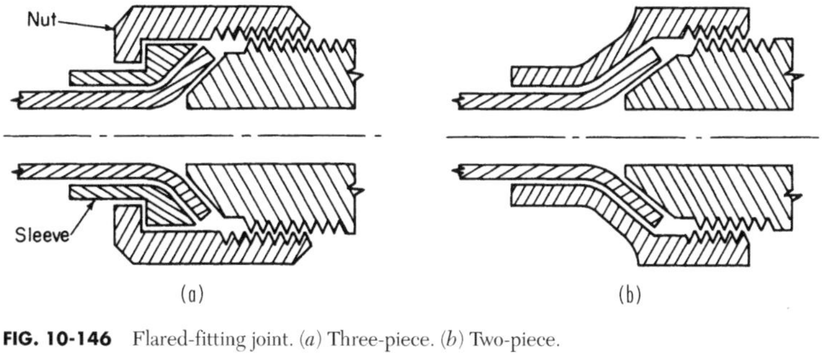

FIG. 10-146 Flared-fitting joint. (*a*) Three-piece. (*b*) Two-piece.

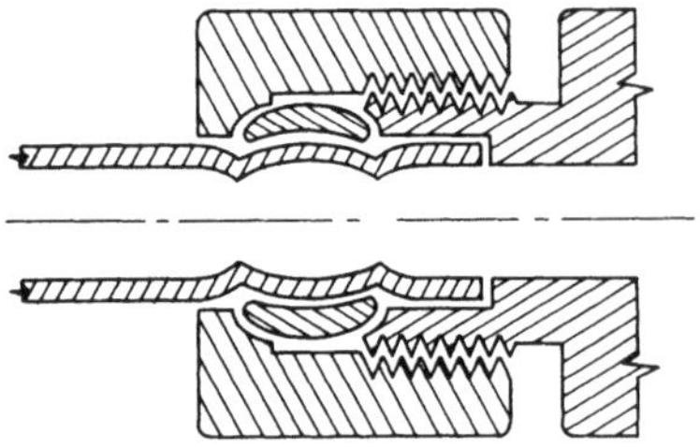

FIG. 10-147 Compression-fitting joint.

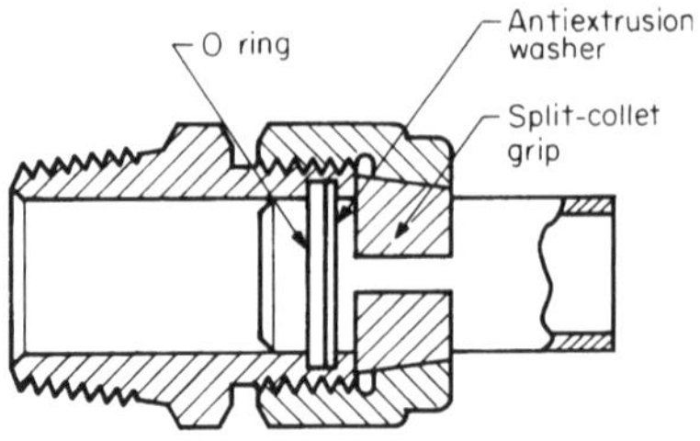

FIG. 10-149 O-ring seal joint. (*Courtesy of the Lenz Co.*)

Elbow Fittings These fittings may be cast, forged, or hot- or cold-formed from short pieces of pipe or made by welding together pieces of miter-cut pipe. The thinning of pipe during the forming of elbows is compensated for by starting with heavier walls.

Flow in bends and elbow fittings is more turbulent than in straight pipe, thus increasing corrosion and erosion. This can be countered by selecting a component with greater radius of curvature, thicker wall, or smoother interior contour, but this is seldom economical in miter elbows.

Compared with elbow fittings, bends with a centerline radius of three or five nominal pipe diameters save the cost of joints and reduce pressure drop. It is sometimes difficult to nest bends of unequal pipe size when they lie in the same plane.

Flanged fittings are used when pipe is likely to be dismantled for frequent cleaning or extensive revision, or for lined piping systems. They are also used in areas where welding is not permitted. Cast fittings are usually flanged. Table 10-30 gives dimensions for flanged fittings.

Dimensions of carbon and alloy steel **butt-welding fittings** are shown in Table 10-31. Butt-welding fittings are available in the wall thicknesses shown in Table 10-22. Larger sizes and other wall thicknesses are also available. Schedule 5 and Schedule 10 stainless-steel butt-welding fittings are available with extensions for expanding into stainless-steel hubs mechanically locked in carbon steel ASME B16.5 dimension flanges. The use of expanded joints (Fig. 10-141) is restricted by the code.

Depending on the size, forged fittings are available with socket-weld (Fig. 10-129) or screwed ends in sizes 4 in and smaller; however, 2 in is the upper size limit normally used. ASME B16.11 gives minimum dimensions for Class 3000, 6000, and 9000 socket-weld fittings, and for Class 2000, 3000, and 6000 threaded fittings. The use of socket-weld and threaded fittings is restricted by the code.

Steel forged fittings with screwed ends may be installed without pipe dope in the threads and seal-welded (Fig. 10-133) to secure bubble-tight joints.

ASME B16.3–1998 gives pressure ratings and dimensions for Class 150 and Class 300 **malleable-iron threaded fittings**. Primary usage is 2 in and smaller; however, Class 150 fittings are available in 6 in and smaller, and Class 300 fittings are available in 3 in and smaller. Malleable-iron fittings are generally less expensive than forged carbon-steel fittings, but cannot be seal-welded. Threaded ends are typically female; however, male threads or a combination of male and female is available in some fittings. Among other restrictions, the code does not permit the use of malleable iron in severe cyclic conditions, in situations subject to thermal or mechanical shock, or in any fluid service below −29°C (−20°F) or above 343°C (650°F). It also does not permit its use in flammable fluid service at temperatures above 149°C (300°F) or at gauge pressures above 2.76 MPa (400 lbf/in^2). ASME B16.3 ratings for Class 150 fittings are 2.07 MPa (300 lbf/in^2) at 66°C (150°F) and below and 1.03 MPa (150 lbf/in^2) at 186°C (366°F). ASME B16.3 ratings for Class 300 fittings are size-dependent, but at least 6.89 MPa (1000 lbf/in^2) at 66°C (150°F) and below and 2.07 MPa (300 lbf/in^2) at 288°C (550°F).

ASME B16.4–1998 gives pressure ratings and dimensions for Class 125 and Class 250 **gray-iron (cast-iron) threaded fittings**. Threaded fittings in both classes are available in sizes 12 in and smaller; however, consideration should be given to other types of end connections prior to using threaded fittings in sizes larger than 2 in. Threaded ends are typically female. Cast-iron fittings are less expensive than forged carbon steel fittings, but cannot be seal-welded. The code places significant restrictions on the use of cast iron, and its use is typically limited to low-pressure, noncritical, nonflammable services. Its brittle nature should be considered before using it for compressed gas services. The minimum permissible design temperature is 29°C (−20°F). ASME B16.4 ratings for Class 125 fittings are 1.21 MPa (175 lbf/in^2) at 66°C (150°F) and below and 0.86 MPa (125 lbf/in^2) at 178°C (353°F). ASME B16.4 ratings for Class 250 fittings are 2.76 MPa (400 lbf/in^2) at 66°C (150°F) and below and 1.72 MPa (250 lbf/in^2) at 208°C (406°F).

Tees Tees may be cast, forged, or hot- or cold-formed from plate or pipe. Tees are typically stocked with both header (run) ends of the same size. In general, run ends of different sizes are not typically stocked or specified; however, occasionally run ends of different sizes are specified in threaded or socket-welded sizes. Branch connections may be full size or reducing sizes. Branch reductions two sizes smaller than the header are routinely stocked, and it is not typically difficult to purchase reducing tees with branches as small as those listed in ASME B16.9 (i.e., approximately one-half the header size). Economics, stress intensification factors, and nondestructive examination requirements typically dictate the branch connection type.

Reducers Reducers may be cast, forged, or hot- or cold-formed from pipe or plate. End connections may be concentric or eccentric, that is, tangent to the same plane at one point on their circumference. For pipe supported by hangers, concentric reducers permit maintenance

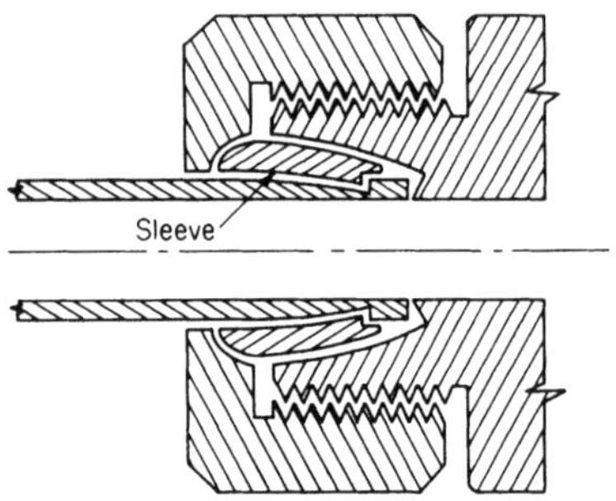

FIG. 10-148 Bite-type-fitting joint.

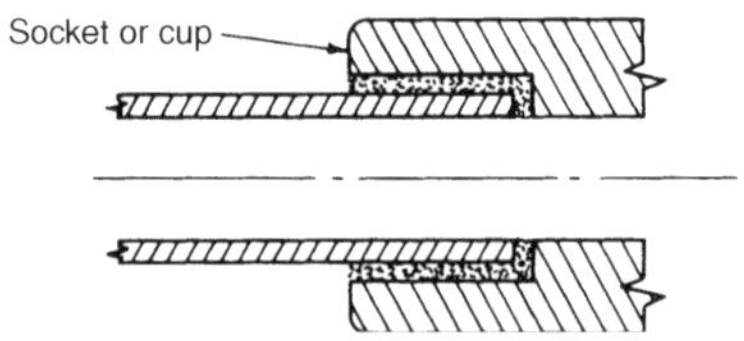

FIG. 10-150 Soldered, brazed, or cemented joint.

TABLE 10-30 Dimensions of Flanged Fittings*

All dimensions in inches

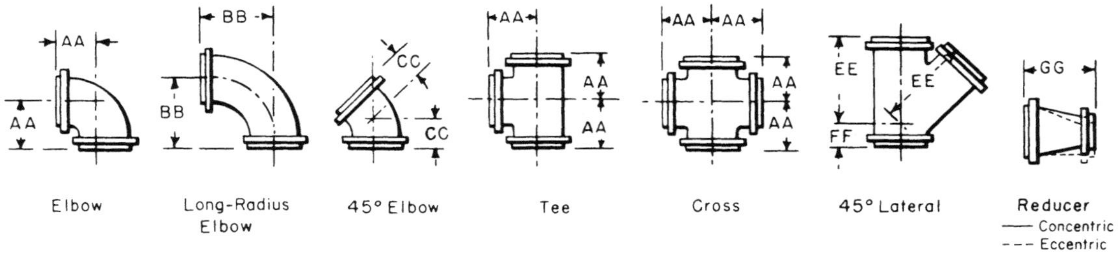

Nominal pipe size	ASME B16.5, Class 150 ASME B16.1, Class 125						ASME B16.5, Class 300 ASME B16.1, Class 250						ASME B16.5, Class 400					ASME B16.5, Class 600				
	AA	BB	CC	EE	FF	GG	AA	BB	CC	EE	FF	GG	AA	CC	EE	FF	GG	AA	CC	EE	FF	GG
½													Use Class 600 dimensions in these sizes					3.25	2.00	5.75	1.75	5.00
¾																		3.75	2.50	6.75	2.00	5.00
1	3.50	5.00	1.75	5.75	1.75	4.50	4.00	5.00	2.25	6.50	2.00	4.50						4.25	2.50	7.25	2.25	5.00
1¼	3.75	5.50	2.00	6.25	1.75	4.50	4.25	5.50	2.50	7.25	2.25	4.50						4.50	2.75	8.00	2.50	5.00
1½	4.00	6.00	2.25	7.00	2.00	4.50	4.50	6.00	2.75	8.50	2.50	4.50						4.75	3.00	9.00	2.75	5.00
2	4.50	6.50	2.50	8.00	2.50	5.00	5.00	6.50	3.00	9.00	2.50	5.00						5.75	4.25	10.25	3.50	6.00
2½	5.00	7.00	3.00	9.50	2.50	5.50	5.50	7.00	3.50	10.50	2.50	5.50						6.50	4.50	11.50	3.50	6.75
3	5.50	7.75	3.00	10.00	3.00	6.00	6.00	7.75	3.50	11.00	3.00	6.00						7.00	5.00	12.75	4.00	7.25
3½	6.00	8.50	3.50	11.50	3.00	6.50	6.50	8.50	4.00	12.50	3.00	6.50						7.50	5.50	14.00	4.50	7.75
4	6.50	9.00	4.00	12.00	3.00	7.00	7.00	9.00	4.50	13.50	3.00	7.00	8.00	5.50	16.00	4.50	8.25	8.50	6.00	16.50	4.50	8.75
5	7.50	10.25	4.50	13.50	3.50	8.00	8.00	10.25	5.00	15.00	3.50	8.00	9.00	6.00	16.75	5.00	9.25	10.00	7.00	19.50	6.00	10.25
6	8.00	11.50	5.00	14.50	3.50	9.00	8.50	11.50	5.50	17.50	4.00	9.00	9.75	6.25	18.75	5.25	10.00	11.00	7.50	21.00	6.50	11.25
8	9.00	14.00	5.50	17.50	4.50	11.00	10.00	14.00	6.00	20.50	5.00	11.00	11.75	6.75	22.25	5.75	12.00	13.00	8.50	24.50	7.00	13.25
10	11.00	16.50	6.50	20.50	5.00	12.00	11.50	16.50	7.00	24.00	5.50	12.00	13.25	7.75	25.75	6.25	13.50	15.50	9.50	29.50	8.00	15.75
12	12.00	19.00	7.50	24.50	5.50	14.00	13.00	19.00	8.00	27.50	6.00	14.00	15.00	8.75	29.75	6.50	15.25	16.50	10.00	31.50	8.50	16.75
14	14.00	21.50	7.50	27.00	6.00	16.00	15.00	21.50	8.50	31.00	6.50	16.00	16.25	9.25	32.75	7.00	16.50	17.50	10.75	34.25	9.00	17.75
16	15.00	24.00	8.00	30.00	6.50	18.00	16.50	24.00	9.50	34.50	7.50	18.00	17.75	10.25	36.25	8.00	18.50	19.50	11.75	38.50	10.00	19.75
18	16.50	26.50	8.50	32.00	7.00	19.00	18.00	26.50	10.00	37.50	8.00	19.00	19.25	10.75	39.25	8.50	19.50	21.50	12.25	42.00	10.50	21.75
20	18.00	29.00	9.50	35.00	8.00	20.00	19.50	29.00	10.50	40.50	8.50	20.00	20.75	11.25	42.75	9.00	21.00	23.50	13.00	45.50	11.00	23.75
24	22.00	34.00	11.00	40.50	9.00	24.00	22.50	34.00	12.00	47.50	10.00	24.00	24.25	12.75	50.25	10.50	24.50	27.50	14.75	53.00	13.00	27.75

TABLE 10-30 Dimensions of Flanged Fittings* *(Concluded)*

Nominal pipe size	ASME B16.5, Class 900					ASME B16.5, Class 1500					ASME B16.5, Class 2500				
	AA	CC	EE	FF	GG	AA	CC	EE	FF	GG	AA	CC	EE	FF	GG
	Use Class 1500 dimensions in these sizes														
½						4.25	3.00				5.19				
¾						4.50	3.25				5.37				
1						5.00	3.50	9.00	2.50	5.00	6.06	4.00			
1¼						5.50	4.00	10.00	3.00	5.75	6.87	4.25			
1½						6.00	4.25	11.00	3.50	6.25	7.56	4.75			
2						7.25	4.75	13.25	4.00	7.25	8.87	5.75	15.25	5.25	9.50
2½						8.25	5.25	15.25	4.50	8.25	10.00	6.25	17.25	5.75	10.50
3	7.50	5.50	14.50	4.50	7.75	9.25	5.75	17.25	5.00	9.25	11.37	7.25	19.75	6.75	11.75
4	9.00	6.50	17.50	5.50	9.25	10.75	7.25	19.25	6.00	10.75	13.25	8.50	23.00	7.75	13.50
5	11.00	7.50	21.00	6.50	11.25	13.25	8.75	23.25	7.50	13.75	15.62	10.00	27.25	9.25	15.75
6	12.00	8.00	22.50	6.50	12.25	13.88	9.38	24.88	8.12	14.50	18.00	11.50	31.25	10.50	18.00
8	14.50	9.00	27.50	7.50	14.75	16.38	10.88	29.88	9.12	17.00	20.12	12.75	35.25	11.75	20.50
10	16.50	10.00	31.50	8.50	16.75	19.50	12.00	36.00	10.25	20.25	25.00	16.00	43.25	14.75	25.50
12	19.00	11.00	34.50	9.00	17.75	22.25	13.25	40.75	12.00	23.00	28.00	17.75	49.25	16.25	29.00
14	20.25	11.50	36.50	9.50	19.00	24.75	14.25	44.00	12.50	25.75					
16	22.25	12.50	40.75	10.50	21.00	27.25	16.25	48.25	14.75	28.25					
18	24.00	13.25	45.50	12.00	24.50	30.25	17.75	53.25	16.50	31.50					
20	26.00	14.50	50.25	13.00	26.50	32.75	18.75	57.75	17.75	34.00					
24	30.50	18.00	60.00	15.50	30.50	38.25	20.75	67.25	20.50	39.75					

*Outline drawings show a ¼-in (6.5-mm) raised face machined onto the flange for ASME Class 400 and higher. ASME B16.1 Class 250 and ASME B16.5 Classes 150 and 300 have a 1/16-in. (1.5-mm) raised face. ASME B16.1 Class 125 has no raised face. See Tables 10-23 though 10-29 for flange drillings. Dimensions for Class 400 and Class 600 fittings are identical for sizes ½ through 3½ in. Dimensions for Class 900 and Class 1500 fittings are identical for sizes ½ through 2½ in. To convert inches to millimeters, multiply by 25.4. The dimensions were extracted from *Cast Iron Pipe Flanges and Flanged Fittings*, ASME B16.1-1998, and *Pipe Flanges and Flanged Fittings*, ASME B16.5-2003, with permission of the publisher, the American Society of Mechanical Engineers, New York.

TABLE 10-31 Butt-Welding Fittings*

All dimensions in inches

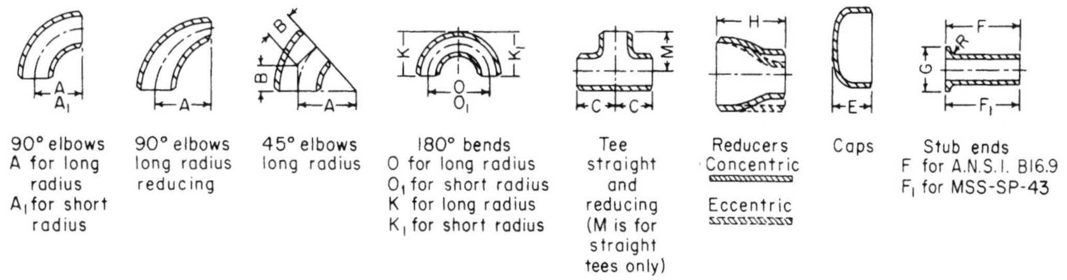

Pipe size	A	K	A1	K1	B	O	O1	M, C	H	E†	G	F	F1	R‡
½	1.50	1.88			0.62	3.00		1.00		1.00	1.38	3.00	2.00	0.12
¾	1.50	2.00			0.75	3.00		1.12	1.50	1.00	1.69	3.00	2.00	0.12
1	1.50	2.19	1.00	1.62	0.88	3.00	2.00	1.50	2.00	1.50	2.00	4.00	2.00	0.12
1¼	1.88	2.75	1.25	2.06	1.00	3.75	2.50	1.88	2.00	1.50	2.50	4.00	2.00	0.19
1½	2.25	3.25	1.50	2.44	1.12	4.50	3.00	2.25	2.50	1.50	2.88	4.00	2.00	0.25
2	3.00	4.19	2.00	3.19	1.38	6.00	4.00	2.50	3.00	1.50	3.62	6.00	2.50	0.31
2½	3.75	5.19	2.50	3.94	1.75	7.50	5.00	3.00	3.50	1.50	4.12	6.00	2.50	0.31
3	4.50	6.25	3.00	4.75	2.00	9.00	6.00	3.38	3.50	2.00	5.00	6.00	2.50	0.38
3½	5.25	7.25	3.50	5.50	2.25	10.50	7.00	3.75	4.00	2.50	5.50	6.00	3.00	0.38
4	6.00	8.25	4.00	6.25	2.50	12.00	8.00	4.12	4.00	2.50	6.19	6.00	3.00	0.44
5	7.50	10.31	5.00	7.75	3.12	15.00	10.00	4.88	5.00	3.00	7.31	8.00	3.00	0.44
6	9.00	12.31	6.00	9.31	3.75	18.00	12.00	5.62	5.50	3.50	8.50	8.00	3.50	0.50
8	12.00	16.31	8.00	12.31	5.00	24.00	16.00	7.00	6.00	4.00	10.62	8.00	4.00	0.50
10	15.00	20.38	10.00	15.38	6.25	30.00	20.00	8.50	7.00	5.00	12.75	10.00	5.00	0.50
12	18.00	24.38	12.00	18.38	7.50	36.00	24.00	10.00	8.00	6.00	15.00	10.00	6.00	0.50
14	21.00	28.00	14.00	21.00	8.75	42.00	28.00	11.00	13.00	6.50	16.25	12.00	6.00	0.50
16	24.00	32.00	16.00	24.00	10.00	48.00	32.00	12.00	14.00	7.00	18.50	12.00	6.00	0.50
18	27.00	36.00	18.00	27.00	11.25	54.00	36.00	13.50	15.00	8.00	21.00	12.00	6.00	0.50
20	30.00	40.00	20.00	30.00	12.50	60.00	40.00	15.00	20.00	9.00	23.00	12.00	6.00	0.50
24	36.00	48.00	24.00	36.00	15.00	72.00	48.00	17.00	20.00	10.50	27.25	12.00	6.00	0.50

*Extracted from *Factory-Made Wrought Buttwelding Fittings*, ASME B16.9-2003, with permission of the publisher, the American Society of Mechanical Engineers, New York. O and K dimensions of 2.25 and 1.69 in respectively may be furnished for NPS ¾ at the manufacturer's option.

†For wall thicknesses greater than extra heavy, E is greater than shown here for sizes 2 in and larger.

‡For MSS SP-43 type B stub ends, which are designed to be backed up by slip-on flanges, R = 1⁄32 in for 4 in and smaller and 1⁄16 in for 6 through 12 in. To convert inches to millimeters multiply by 25.4.

of the same hanger length; for pipe laid on structural steel, eccentric reducers permit maintaining the same elevation of top of steel. Eccentric reducers with the common tangent plane on the bottom side permit complete drainage of branched horizontal piping systems. With the common tangent plane on the top side, they permit liquid flow in horizontal lines to sweep the line free of gas or vapor.

Reducing Elbow Fittings These permit change of direction and concentric size reduction in the same fitting.

Valves Valve bodies may be cast, forged, machined from bar stock, or fabricated from welded plate. Steel valves are available with screwed or socket-weld ends in the smaller sizes. Bronze and brass screwed-end valves are widely used for low-pressure service in steel systems. Table 10-32 gives contact-surface-of-face to contact-surface-of-face dimensions for flanged ferrous valves and end-to-end dimensions for butt-welding ferrous valves. Drilling of end flanges is shown in Tables 10-23 to 10-29. Bolt holes are located so that the stem is equidistant from the centerline of two bolt holes. Even if removal for maintenance is not anticipated, flanged valves are frequently used instead of butt-welding-end valves because they permit insertion of blanks for isolating sections of a loop piping system.

Ferrous valves are also available in nodular (ductile) iron, which has tensile strength and yield point approximately equal to cast carbon steel at temperatures of 343°C (650°F) and below and only slightly less elongation.

Valves serve not only to regulate the flow of fluids but also to isolate piping or equipment for maintenance without interrupting other connected units. Valve designers attempt to minimize body distortion due to pressure, changes in temperature, and applied loads. The sealing mechanisms of certain valve designs are inherently more tolerant of these factors than are others. The selection of valve type and materials of construction should provide a valve that functions reliably and that is acceptably tight across the sealing surfaces for the lowest lifetime cost. Valve manufacturers are a valuable source of information when evaluating the suitability of specific designs. The principal types are named, described, compared, and illustrated with line diagrams in subsequent subsections. In the line diagrams, the operating stem is shown in solid black, direction of flow by arrows on a thin solid line, and motion of valve parts by arrows on a dotted line. Moving parts are drawn with solid lines in the nearly closed position and with dotted lines in the fully open position. Packing is represented by an X in a square.

Gate Valves These valves are designed in two types (Fig. 10-151). The wedge-shaped-gate, **inclined-seat** type is most commonly used. The wedge gate may be solid or flexible (partly cut into halves by a plane at right angles to the pipe) or split (completely cleft by such a plane). Flexible and split wedges minimize galling of the sealing surfaces by distorting more easily to match angularly misaligned seats. In the double-disk **parallel-seat** type, an inclined-plane device mounted between the disks converts stem force to axial force, pressing the disks against the

TABLE 10-32 Dimensions of Valves*

All dimensions in inches

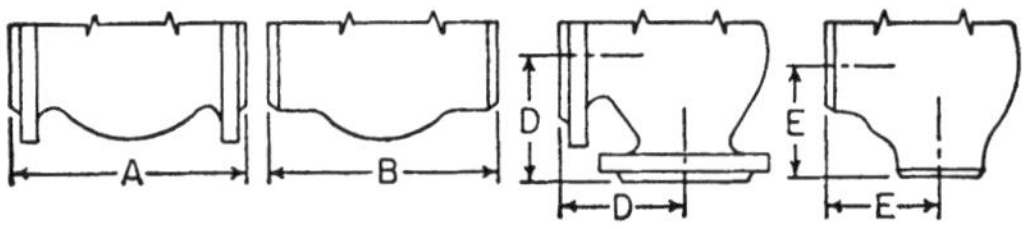

Nominal valve size	Class 125 cast iron, Flanged end, Gate, Solid wedge A	Class 125 cast iron, Flanged end, Gate, Double disk A	Class 125 cast iron, Flanged end, Globe and lift check A	Class 125 cast iron, Flanged end, Angle and lift check D	Class 125 cast iron, Flanged end, Swing check A	Class 150 steel, MSS-SP-42 through 12-in size, Flanged end, Gate, Solid wedge and double disk A	Class 150 steel, Welding end, Gate, Solid wedge and double disk B	Class 150 steel, Flanged end and welding end, Globe and lift check A and B	Class 150 steel, Flanged end and welding end, Angle and lift check D and E	Class 150 steel, Flanged end and welding end, Swing check A and B	Class 250 cast iron, Flanged end, Gate, Solid wedge and double disk A	Class 250 cast iron, Flanged end, Globe, lift check, and swing check A	Class 250 cast iron, Flanged end, Angle and lift check D
¼						4	4	4	2	4			
⅜						4	4	4	2	4			
½						4¼	4¼	4¼	2¼	4¼			
¾						4⅝	4⅝	4⅝	2½	4⅝			
1						5	5	5	2¾	5			
1¼						5½	5½	5½	3	5½			
1½						6½	6½	6½	3¼	6½			
2	7	7	8	4	8	7	8½	8	4	8	8½	10½	5¼
2½	7½	7½	8½	4¼	8½	7½	9½	8½	4¼	8½	9½	11½	5¾
3	8	8	9½	4¾	9½	8	11⅛	9½	4¾	9½	11⅛	12½	6¼
3½	8½	8½				†					12	14	7
4	9	9	11½	5¾	11½	9	12	11½	5¾	11½	15	15¾	7⅞
5	10	10	13	6½	13	10	15	14	7	13	15⅞	17½	8¾
6	10½	10½	14	7	14	10½	15⅞	16	8	14	16½	21	10½
8	11½	11½	19½	9¾	19½	11½	16½	19½	9¾	19½	18	24½	12¼
10	13	13	24½	12¼	24½	13	18	24½	12¼	24½	19¾	28	14
12	14	14	27½	13¾	27½	14	19¾	27½	13¾	27½	22½	†	
14	15	†	31	15½	31	15	22½	31	15½	31	24	†	
16	16	†	36	18	†	16	24	36	18	†	26		
18	17	†			†	17	26			†	28		
20	18	†			†	18	28			†			
24	20	†			†	20	32			†	31		

Nominal valve size	Class 300 steel, Flanged end and welding end, Gate, Solid wedge and double disk A and B	Class 300 steel, Globe and lift check A and B	Class 300 steel, Angle and lift check D and E	Class 300 steel, Swing check A and B	Class 400 steel, Flanged end and welding end, Gate, Solid wedge A and B	Class 400 steel, Gate, Double disk A and B	Class 400 steel, Globe, lift check, and swing check A and B	Class 400 steel, Angle and lift check D and E	Class 600 steel, Flanged end and welding end, Gate, Solid wedge A and B	Class 600 steel, Gate, Double disk A and B	Class 600 steel, Gate, Short pattern‡ B	Class 600 steel, Regular globe, regular lift check, swing check A and B	Class 600 steel, Short pattern‡ globe, short pattern lift check B	Class 600 steel, Angle and lift check, Regular D and E	Class 600 steel, Angle and lift check, Short pattern E
½	5½	6	3		6½		6½	3¼	6½			6½		3¼	
¾	6	7	3½		7½		7½	3¾	7½			7½		3¾	
1	6½§	8	4	8½	8½	8½	8½	4¼	8½	8½	5¼	8½	5¼	4¼	
1¼	7§	8½	4¼	9	9	9	9	4½	9	9	5¾	9	5¾	4½	
1½	7½	9	4½	9½	9½	9½	9½	4¾	9½	9½	6	9½	6	4¾	
2	8½	10½	5¼	10½	11½	11½	11½	5¾	11½	11½	7	11½	7	5¾	4¼
2½	9½	11½	5¾	11½	13	13	13	6½	13	13	8½	13	8½	6½	5
3	11⅛	12½	6¼	12½	14	14	14	7	14	14	10	14	10	7	6
4	12	14	7	14	16	16	16	8	17	17	12	17	12	8½	7
5	15	15¾	7⅞	15¾	18	18	18	9	20	20	15	20	15	10	8½
6	15⅞	17½	8¾	17½	19½	19½	19½	9¾	22	22	18	22	18	11	10
8	16½	22	11	21	23½	23½	23½	11¾	26	26	23	26	23	13	
10	18	24½	12¼	24½	26½	26½	26½	13¼	31	31	28	31	28	15½	
12	19¾	28	14	28	30	30	30	15	33	33	32	33	32	16½	
14	30			†	32½	30½	†		35	35	35	†			
16	33			†	35½	35½	†		39	39	39	†			
18	36			†	38½	38½	†		43	43	43	†			
20	39			†	41½	41½	†		47	47	47	†			
22	43			†	45	45	†		51	51		†			
24	45			†	48½	48½	†		55	55	55	†			

TABLE 10-32 Dimensions of Valves *(Concluded)*

	Class 900 steel							Class 1500 steel				
	Flanged end and welding end							Flanged end and welding end				
	Gate			Regular globe regular lift check, swing check	Short pattern‡ globe, short pattern lift check	Angle and lift check		Gate			Globe, lift check, swing check	Angle and lift check
Nominal valve size	Solid wedge *A* and *B*	Double disk *A* and *B*	Short pattern‡ *B*	*A* and *B*	*B*	Regular *D* and *E*	Short pattern *E*	Solid wedge *A* and *B*	Double disk *A* and *B*	Short pattern‡ *B*	*A* and *B*	*D* and *E*
¾				9		4½					9	4½
1	10		5½	10		5		10		5½	10	5
1¼	11		6½	11		5½		11		6½	11	5½
1½	12		7	12		6		12		7	12	6
2	14½	14½	8½	14½		7¼		14½	14½	8½	14½	7¼
2½	16½	16½	10	16½		8¼		16½	16½	10	16½	8¼
3	15	15	12	15	12	7½	6	18½	18½	12	18½	9¼
4	18	18	14	18	14	9	7	21½	21½	16	21½	10¾
5	22	22	17	22	17	11	8½	26½	26½	19	26½	13¼
6	24	24	20	24	20	12	10	27¾	27¾	22	27¾	13⅞
8	29	29	26	29	26	14½	13	32¾	32¾	28	32¾	16⅜
10	33	33	31	33	31	16½	15½	39	39	34	39	19½
12	38	38	36	38	36	19	18	44½	44½	39	44½	22¼
14	40½	40½	39	40½	39	20¼	19½	49½	49½	42	49½	24¾
16	44½	44½	43					54½	54½	47		
18	48	48	†					60½	60½	53		
20	52	52	†					65½	65½	58		
24	61	61	†					76½	76½			

	Class 2500 steel				
	Flanged end and welding end				
	Gate			Globe, lift check, swing check	Angle and lift check
Nominal valve size	Solid wedge *A* and *B*	Double disk *A* and *B*	Short pattern‡ *B*	*A* and *B*	*B*
½	10⅜			10⅜	5 3/16
¾	10¾			10¾	5⅜
1	12⅛		7 5/16	12⅛	6 1/16
1¼	13¾		9⅛	13¾	6⅞
1½	15⅛		9⅛	15⅛	7 9/16
2	17¾	17¾	11	17¾	8⅞
2½	20	20	13	20	10
3	22¾	22¾	14½	22¾	11⅜
4	26½	26½	18	26½	13¼
5	31¼	31¼	21	31¼	15⅝
6	36	36	24	36	18
8	40¼	40¼	30	40¼	20⅛
10	50	50	36	50	25
12	56	56	41	56	28
14			44		
16			49		
18			55		

NOTE: Outline drawings for flanged valves show ¼-in raised face machined onto flange, as for Class 400 cast-steel valves; Class 150 and 300 cast-steel valves and Class 250 cast-iron valves have 1/16-in raised faces; Class 125 cast-iron have no raised faces.

*Extracted from Face-to-Face and End-to-End Dimensions of Valves, ASME B16.10, with permission of the publisher, the American Society of Mechanical Engineers, New York. To convert inches to millimeters, multiply by 25.4.

†Not shown in ANSI B16.10 but commercially available.

‡These dimensions apply to pressure-seal or flangeless bonnet valves only.

§Solid wedge only.

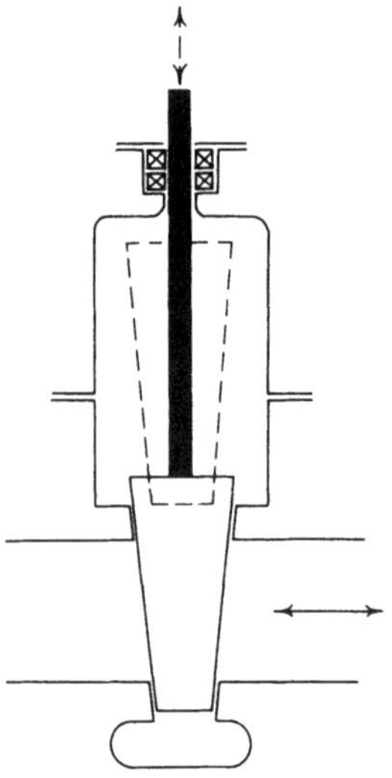

FIG. 10-151 Gate valve.

seats after the disks have been positioned for closing. This gate assembly distorts automatically to match both angular misalignment of the seats and longitudinal shrinkage of the valve body on cooling.

During opening and closing, some parallel-seat designs are more subject to vibration resulting from fluid flow than are wedge gates. Specific applications should be discussed with the manufacturer. In some applications it may be advisable to use a small bypass around the in-line valve to help lower opening and closing forces and to relieve binding between the gate and the seat due to high differential pressure or temperature. Double-disk parallel seat valves should be installed with the stem essentially vertical unless otherwise recommended by the manufacturer. All wedge gate valves are equipped with tongue-and-groove guides to keep the gate sealing surfaces from clattering on the seats and marring them during opening and closing. Depending on the velocity and density of the fluid stream being sheared, these guiding surfaces may be specified as cast, machined, or hard-surfaced and ground.

Gate valves may have nonrising stems, inside-screw rising stems, or outside-screw rising stems, listed in order of decreasing exposure of the stem threads to the fluid handled. Rising-stem valves require more space, but the position of the stem visually indicates the position of the gate. Indication is clearest on the outside-screw rising-stem valves, and on these the stem threads and thrust collars may be lubricated, reducing operating effort. The stem connection to the gate assembly prevents the stem from rotating.

Gate valves are used to minimize pressure drop in the open position and to stop the flow of fluid rather than to regulate it. The problem, when the valve is closed, of pressure buildup in the bonnet from cold liquids expanding or chemical action between fluid and bonnet should be solved by a relief valve or by notching the upstream seat ring.

Globe Valves (Fig. 10-152) These are designed as either inside-screw rising stem or outside-screw rising stem. In most designs the disk is free to rotate on the stem; this prevents galling between the disk and the seat. Various designs are used to maintain alignment between the disk and the seat, and to keep the keep fluid flow from vibrating or rotating the disk. Disks are typically guided either by the valve stem or against the valve body. Body guiding reduces side thrust loads on the stem. The suitability of each design can be determined by reviewing specific applications with valve manufacturers.

Disk shapes are commonly flat or conical. Conical designs provide either line or area contact between the seat and disk, and are generally more suitable than flat disks for high pressures and temperatures. Needle-type disks provide better flow control and are commonly available in valves 1 in and smaller.

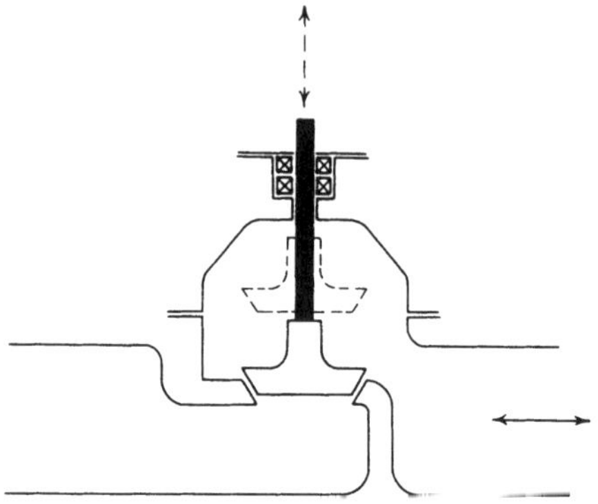

FIG. 10-152 Globe valve.

For certain valve designs, sizes, and applications, globe valves may be installed with the stem in the horizontal position; however, unless approved by the manufacturer, the stem orientation should be vertical. Globe valves are not symmetric with respect to flow. Generally globe valves are installed with pressure under the seat when the valve is in the closed position, and with the flow direction coming from under the seat. Opposite-flow direction provides pressure-assisted seating, lower seating torque requirements, and higher opening torques, and may result in blockage in dirty services. Consult the manufacturer before installing globe valves in the opposite-flow direction.

Pressure drop through globe valves is much greater than that for gate valves. In Y-type globe valves, the stem and seat are at about 45° to the pipe instead of 90°. This reduces pressure drop but presents design challenges with regard to disk alignment.

Globe valves in horizontal lines prevent complete drainage.

Angle Valves These valves are similar to globe valves; the same bonnet, stem, and disk are used for both (Fig. 10-153). They combine an elbow fitting and a globe valve into one component with a substantial saving in pressure drop.

Diaphragm Valves These valves are limited to pressures of approximately 50 lbf/in^2 (Fig. 10-154). The fabric-reinforced diaphragms may be made from natural rubber, from a synthetic rubber, or from natural or synthetic rubbers faced with Teflon* fluorocarbon resin. The simple shape of the body makes lining it economical. Elastomers have shorter lives as diaphragms than as linings because of

*Du Pont TFE fluorocarbon resin.

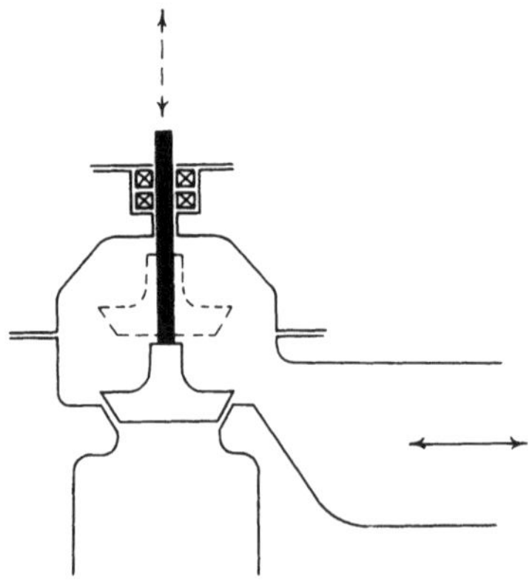

FIG. 10-153 Angle valve.

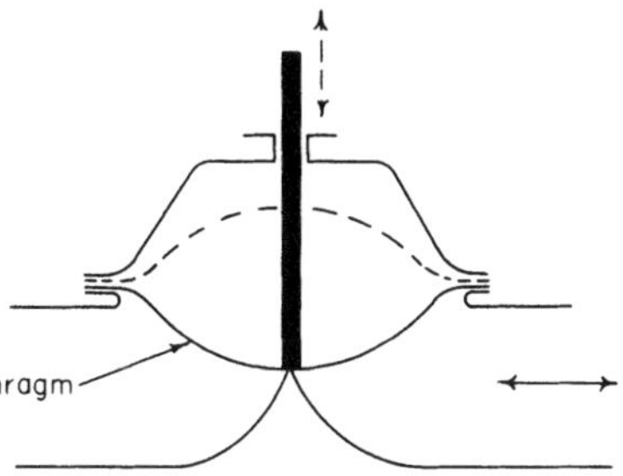

FIG. 10-154 Diaphragm valve.

flexing but still provide satisfactory service. Plastic bodies, which have low moduli of elasticity compared with metals, are practical in diaphragm valves since alignment and distortion are minor problems.

These valves are excellent for fluids containing suspended solids and can be installed in any position. Models in which the dam is very low, reducing pressure drop to a negligible quantity and permitting complete drainage in horizontal lines, are available. However, drainage can be obtained with any model simply by installing it with the stem horizontal. The only maintenance required is replacement of the diaphragm, which can be done very quickly without removing the valve from the line.

Plug Valves These valves (Fig. 10-155) are typically limited to temperatures below 260°C (500°F) since differential expansion between the plug and the body results in seizure. The size and shape of the port divide these valves into different types. In order of increasing cost they are short venturi, reduced rectangular port; long venturi, reduced rectangular port; full rectangular port; and full round port.

In lever-sealed plug valves, tapered plugs are used. The plugs are raised by turning one lever, rotated by another lever, and reseated by the first lever. **Lubricated** plug valves may use straight or tapered plugs. The tapered plugs may be raised slightly, to reduce turning effort, by injection of the lubricant, which also acts as a seal. Plastic is used in nonlubricated plug valves as a body liner, a plug coating, or port seals in the body or on the plug.

In plug valves other than lever-sealed plug valves, the contact area between plug and body is large, and gearing is usually used in sizes 6 in and larger to minimize operating effort. There are several lever-sealed plug valves incorporating mechanisms which convert the rotary motion of a handwheel into sequenced motion of the two levers.

For lubricated plug valves, the lubricant must have limited viscosity change over the range of operating temperature, must have low solubility in the fluid handled, and must be applied regularly. There must be no chemical reaction between the lubricant and the fluid which would harden or soften the lubricant or contaminate the fluid. For these reasons, lubricated plug valves are most often used when there are a large number handling the same or closely related fluids at approximately the same temperature.

Lever-sealed plug valves are used for throttling service. Because of the large contact area between plug and body, if a plug valve is operable, there is little likelihood of leakage when closed, and the handle position is a clearly visible indication of the valve position.

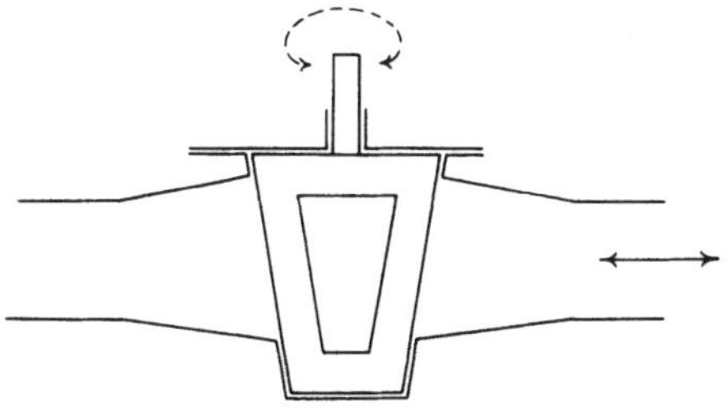

FIG. 10-155 Plug valve.

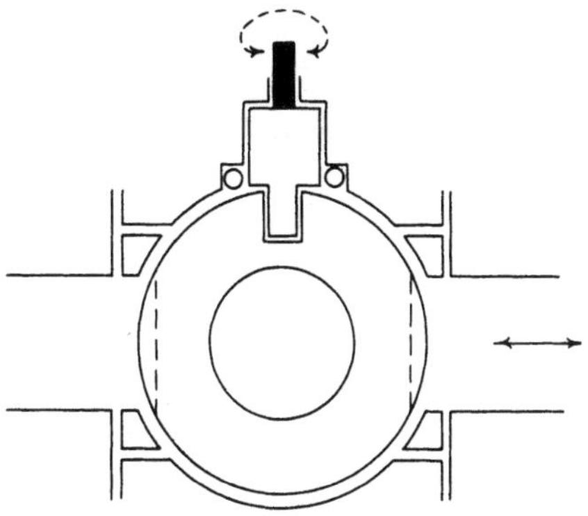

FIG. 10-156 Ball valve; floating ball.

Ball Valves Ball valves (Figs. 10-156 and 10-157) are of two primary designs, floating ball and trunnion-mounted ball. In floating ball designs, the ball is supported by the downstream seat. In trunnion ball designs, the ball is supported by the trunnion and the seat loads are less than those in floating ball valves. Because of operating torque and shutoff pressure ratings, trunnion ball valves are available in larger sizes and higher pressure ratings than floating ball valves.

Both floating and trunnion-mounted designs are available with other design variations that include metal seated valves, soft seated valves, top-entry valves, end-entry valves, and split body valves. Valves in all these design variations are available as either full port or reduced port. *Port* refers to the round-bore fluid flow area through the ball. Full port valves have a bore that is approximately equal to the inside diameter of the mating pipe. Reduced port valves have a bore that is approximately equal to the inside diameter of pipe one size smaller than full bore.

A variety of soft seat materials are available, including PTFE and nylon. Since the shutoff pressure capability of ball valves is limited by the load capabilities of the seat material, the upper temperature limit of soft seated valves is limited by the seat material selection. The shutoff pressure rating of soft seated valves typically declines rapidly with increasing temperature, and the shutoff rating is often less than the body pressure rating. Metal seated valves do not share this characteristic.

For equal port size, ball valves share the low pressure drop characteristics of gate valves. Also as is the case with gate valves, consideration must be given to venting the valve when the expansion of fluid trapped within the body cavity could overpressurize the valve body. Some seat designs are inherently self-venting to either the upstream or the downstream side of the valve. In floating ball valves, venting may result in a unidirectional valve that seats against flow in only one direction.

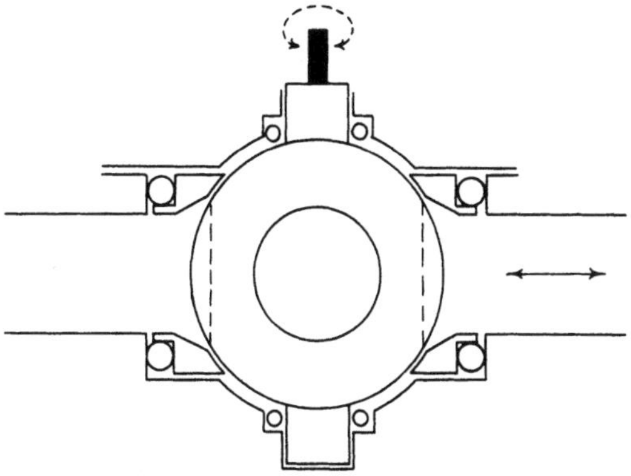

FIG. 10-157 Ball valve; trunnion-mounted ball.

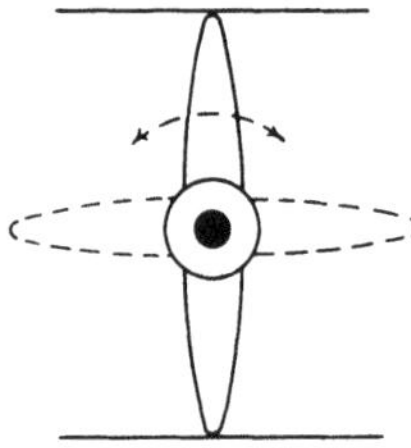

FIG. 10-158 Butterfly valve.

With split body designs, internals must be removed by separating the valve body in the axial direction of the mating pipe. Top-entry design permits removal of the internals through the top of the valve. When valves are butt-welded, top entry may be specified to permit repair without removing the valve from the piping. Top-entry valves are significantly more expensive than split body and end-entry valves, and full port valves are more expensive than reduced port in any body design. Metal seated valves are significantly more expensive than soft seated valves and are typically used only when other types of valves are unsuitable for the application.

Butterfly Valves These valves (Fig. 10-158) occupy less space, are much lighter than other types of block valves, and are available in body styles that include wafer, lugged (drilled-through or tapped), and flanged. They are available in ASME Class 900 and lower pressure ratings. The maximum size available varies with pressure rating. Valves in Class 150 are available in sizes exceeding 60-in diameter. Within limits, they may be used for throttling. Their relatively high pressure drop must be considered during design.

Like ball valves, butterfly valves are fully opened in one-quarter turn and are therefore well suited to automation. *Butterfly Valves: Double Flange, Lug- and Wafer-Type*, API 609, is one of the standards commonly used to specify butterfly valves. API 609 defines two major categories of butterfly valves: Category A and Category B. Category A valves are typically soft seated valves with shell pressure ratings that may be less than the flange rating of the valve. They are typically used for utility services and are commonly referred to as *utility* valves. Category B may be soft seated or metal seated, and must have shell pressure ratings equal to the full pressure rating of the valve flange, and seat ratings that essentially meet the shell rating within the temperature capability of the seat material. Within Category B, valves may be further divided into concentric shaft, double-offset shaft, and triple-offset shaft designs. *Offset* refers to the position of the shaft with respect to seat area. With minor exception, double- and triple-offset valve designs are metal-to-metal seated. They are distinguished from other designs by their exceptional seat tightness (often "zero" leakage) that is maintained throughout the life of the valve. Their tightness exceeds the seat tightness capability and reliability of wedge-type gate valves. Double-offset valves minimize rubbing between the disk and the seat, and triple-offset valves virtually eliminate rubbing. Although double- and triple-offset valves are more expensive than other butterfly valve designs, because of their weight they are often more economical than gate valves for some combinations of pressure class, size, and materials.

Check Valves These valves are used to prevent reversal of flow. They must be located where flow turbulence or instability does not result in chatter (high-frequency opening and closing of the valve) and in systems designed to prevent sudden high-velocity flow reversal which results in slamming upon closure. Many valve manufacturers can provide application advice.

Swing Check Valves These valves (Fig. 10-159) are normally designed for use in horizontal lines or in vertical lines with normally upward flow. Since their seating force is primarily due to pipeline pressure, they may not seal as tightly at low pressures as at higher pressures. When suitable, nonmetallic seats may be used to minimize this problem.

Lift Check Valves These valves (Figs. 10-160 through 10-162) are made in three styles. Vertical lift check valves are for installation in vertical lines with flow normally upward. Globe (or piston) valves with a 90° bonnet (Fig. 10-161) are for installation in horizontal lines, although inclined bonnet versions (approximately 45°) with spring assist may be used in vertical lines with normally upward flow. Globe and angle check valves often incorporate mechanisms to control the opening or closing rate of the piston, or to promote full opening under low-flow conditions. In some designs, spring-assisted closure is available, but this increases pressure drop. Lift check valves should not be used when the fluid contains suspended solids. Ball check valves having designs similar to those in Figs. 10-160 and 10-161 are available in sizes 2 in and smaller. They promote even wear of the seat area and are more suitable for viscous services, or services with limited solids.

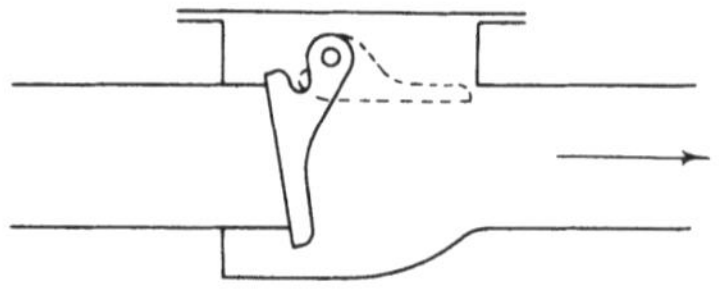

FIG. 10-159 Swing check valve.

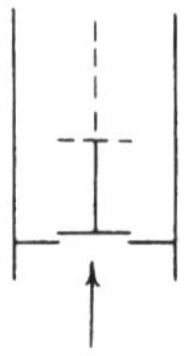

FIG. 10-160 Lift check valve, vertical.

Tilting-Disk Check Valves These valves (Fig. 10-163) may be installed in a horizontal line or in lines in which the flow is vertically upward. The pivot point is located so that the distribution of pressure in the fluid handled speeds the closing but arrests slamming. Compared with swing check valves of the same size, pressure drop is less at low velocities but greater at high velocities.

Closure at the instant of flow reversal is most nearly attained with tilting-disk, dual-plate, and specialty axial-flow check valves. However, quick closure is not the solution to all noise, shock, and water hammer problems. External dashpots are available when a controlled rate of closure is desired. Nonmetallic seats are also available.

Dual-Plate Check Valves These valves (Fig. 10-163) occupy less space, are much lighter than other types of check valves, and are available in body styles that include wafer, lugged (drilled-through or tapped), and flanged. They are available in all ASME pressure classes. The maximum size available varies with pressure rating. Valves in Class 150 are available in sizes 60 in or larger. They are available with either metallic or nonmetallic seats. Pressure drop is greater than that in a fully open swing check valve. Plate closure is spring-assisted, and the rate of closure can be controlled with proper spring selection. High-performance valves with fast closure rates are available to address water hammer problems. They typically weigh as little as 15 to 30 percent as much as swing check valves. Because of their weight they are often more economical than other types of check valves.

Valve Trim Various alloys are available for valve parts such as seats, disks, and stems which must retain smooth finish for successful operation. The problem in seat materials is fivefold: (1) resistance to corrosion by the fluid handled and to oxidation at high temperatures, (2) resistance to erosion by suspended solids in the fluid, (3) prevention of galling (seizure at point of contact) by differences in material or hardness or both, (4) maintenance of high strength at high temperature, and (5) avoidance of distortion.

Standard valve trims are defined by standards such as API 600 and API 602. Elastomer or plastic inserts may be specified to achieve bubble-tight shutoff. Valve manufacturers may be consulted for recommended trims.

CAST IRON, DUCTILE IRON, AND HIGH-SILICON IRON PIPING SYSTEMS

Cast Iron and Ductile Iron Cast iron and ductile iron provide more metal for less cost than steel in piping systems and are widely used in low-pressure services in which internal and external corrosion

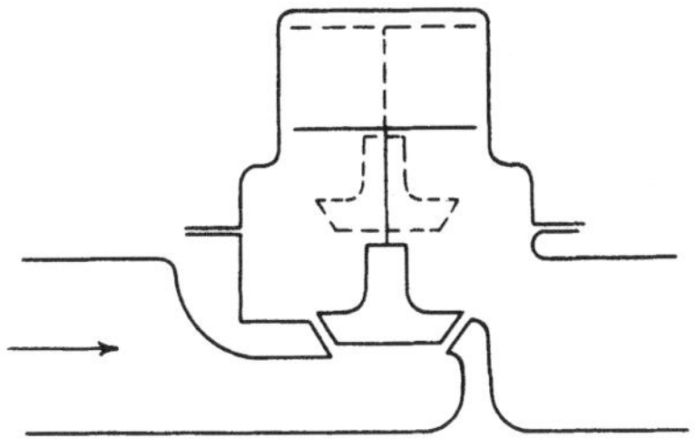

FIG. 10-161 Lift check valve, globe.

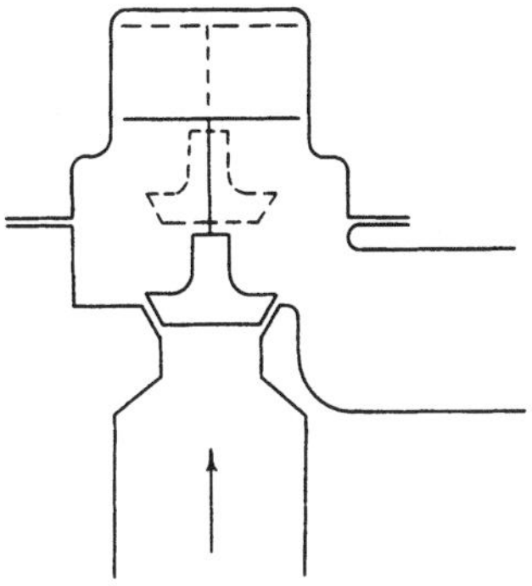

FIG. 10-162 Lift check valve, angle.

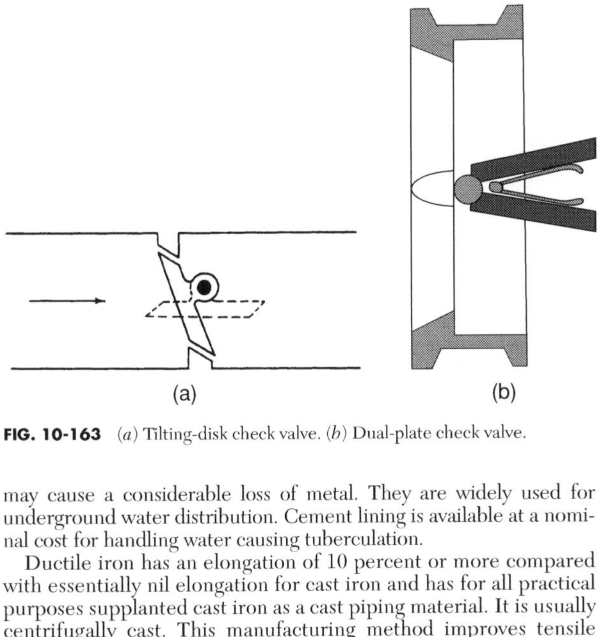

(a) (b)

FIG. 10-163 (*a*) Tilting-disk check valve. (*b*) Dual-plate check valve.

may cause a considerable loss of metal. They are widely used for underground water distribution. Cement lining is available at a nominal cost for handling water causing tuberculation.

Ductile iron has an elongation of 10 percent or more compared with essentially nil elongation for cast iron and has for all practical purposes supplanted cast iron as a cast piping material. It is usually centrifugally cast. This manufacturing method improves tensile strength and reduces porosity. Ductile-iron pipe is manufactured to AWWA C151/A21.51-2002 and is available in nominal sizes from 3 through 64 in. Wall thicknesses are specified by seven standard thickness classes. Table 10-33 gives the outside diameter and standard thickness for various rated water working pressures for centrifugally cast ductile-iron pipe. The required wall thickness for underground installations increases with internal pressure, depth of laying, and weight of vehicles operating over the pipe. It is reduced by the degree to which the soil surrounding the pipe provides uniform support along the pipe and around the lower 180°. Tables are provided in AWWA C151/A21.51 for determining wall-thickness-class recommendations for various installation conditions. The poured joint (Fig. 10-139) has been almost entirely superseded by the mechanical joint (Fig. 10-138) and the push-on joint (Fig. 10-140), which are better suited to wet trenches, bad weather, and unskilled labor. Such joints also minimize strain on the pipe from ground settlement. Lengths vary between 5 and 6 m (between 18 and 20 ft), depending on the supplier. Stock fittings are designed for 1.72-MPa (250-lbf/in^2) cast iron or 2.41-MPa (350-lbf/in^2) ductile iron in sizes through 12 in and for 1.0- and 1.72-MPa (150- and 250-lbf/in^2) cast iron or 2.41-MPa (350-lbf/in^2) ductile iron in sizes 14 in and larger. Stock fittings include 22½° and 11¼° bends. Ductile-iron pipe is also supplied with flanges that match the dimensions of Class 125 flanges shown in ASME B16.1 (see Table 10-23). These flanges are assembled to the pipe barrel by threaded joints.

High-Silicon Iron Pipe and fittings are cast products of material typically conforming to ASTM A518. Nominal silicon content is 14.5 percent, and nominal carbon content is approximately 0.85 percent. This material is corrosion-resistant to most chemicals, highly abrasion-resistant, and suitable for applications to 260°C (500°F). Applications are primarily gravity drain. Pipe and fittings are available under the trade name Duriron®.

Pipe and fitting sizes typically available are shown in Table 10-34. Pipe and fitting dimensions commonly conform to ASTM A861. Bell and spigot connections are sealed with chemical-resistant rope packing and molten lead. Mechanical joint connections are made with TFE gaskets and stainless steel clamps.

The coefficient of linear expansion of this alloy in the temperature range of 21 to 100°C (70 to 212°F) is 12.2×10^{-6}/°C (6.8×10^{-6}/°F), which is slightly above that of cast iron (National Bureau of Standards). Since this material has practically no elasticity, the need for expansion joints should be considered. Connections for flanged pipe, fittings, valves, and pumps are made to ASME B16.1, Class 125.

The use of high-silicon iron in flammable-fluid service or in Category M fluid service is prohibited by the code.

NONFERROUS METAL PIPING SYSTEMS

Aluminum Seamless aluminum pipe and tube are produced by extrusion in essentially pure aluminum and in several alloys; 6-, 9-, and 12-m (20-, 30-, and 40-ft) lengths are available. Alloying and mill treatment improve physical properties, but welding reduces them. Essentially pure aluminum has an ultimate tensile strength of 58.6 MPa (8500 lbf/in^2) subject to a slight increase by mill treatment which is lost during welding. Alloy 6061, which contains 0.25 percent copper, 0.6 percent silicon, 1 percent magnesium, and 0.25 percent chromium, has an ultimate tensile strength of 124 MPa (18,000 lbf/in^2) in the annealed condition, 262 MPa (38,000 lbf/in^2), mill-treated as 6061-T6, and 165 MPa (24,000 lbf/in^2) at welded joints. Extensive use is made of alloy 1060, which is 99.6 percent pure aluminum, for hydrogen peroxide; of alloy 3003, which contains 1.2 percent manganese, for high-purity chemicals; and of alloys 6063 and 6061 for many other services. Alloy 6063 is the same as 6061 minus the chromium and has slightly lower mechanical properties.

Aluminum is not embrittled by low temperatures and is not subject to external corrosion when exposed to normal atmospheres. At 200°C (400°F) its strength is less than half that at room temperature. It is attacked by alkalies, by traces of copper, nickel, mercury, and other heavy-metal ions, and by prolonged contact with wet insulation. It suffers from galvanic corrosion when coupled to copper, nickel, or lead-base alloys but not when coupled to galvanized iron.

Aluminum pipe schedules conform to those in Table 10-22. Consult suppliers for available sizes and schedules.

Threaded aluminum fittings are seldom recommended for process piping. Wrought fittings with welding ends (see Table 10-31 for dimensions) and with grooved joint ends are available. Wrought 6061-T6 flanges with dimensions per Table 10-23 are also available. Consult suppliers on availability of cast flanges and fittings. Castings manufactured in accordance with ASTM B26 are available in several grades. Refer to Table 10-30 for dimensions. The low strength and

TABLE 10-33 Dimensions of Ductile-Iron Pipe*

Standard thickness for internal pressure†

		Rated water working pressure, lbf/in²‡									
		150		200		250		300		350	
Pipe size, in	Outside diameter, in	Thickness, in	Thickness class	Thickness, in	Thickness class	Thickness, in	Thickness class	Thickness, in	Thickness class	Thickness, in	Thickness class
3	3.96	0.25	51	0.25	51	0.25	51	0.25	51	0.25	51
4	4.80	0.26	51	0.26	51	0.26	51	0.26	51	0.26	51
6	6.90	0.25	50	0.25	50	0.25	50	0.25	50	0.25	50
8	9.05	0.27	50	0.27	50	0.27	50	0.27	50	0.27	50
10	11.10	0.29	50	0.29	50	0.29	50	0.29	50	0.29	50
12	13.20	0.31	50	0.31	50	0.31	50	0.31	50	0.31	50
14	15.30	0.33	50	0.33	50	0.33	50	0.33	50	0.33	50
16	17.40	0.34	50	0.34	50	0.34	50	0.34	50	0.34	50
18	19.50	0.35	50	0.35	50	0.35	50	0.35	50	0.35	50
20	21.60	0.36	50	0.36	50	0.36	50	0.36	50	0.39	51
24	25.80	0.38	50	0.38	50	0.38	50	0.41	51	0.44	52
30	32.00	0.39	50	0.39	50	0.43	51	0.47	52	0.51	53
36	38.30	0.43	50	0.43	50	0.48	51	0.53	52	0.58	53
42	44.50	0.47	50	0.47	50	0.53	51	0.59	52	0.65	53
48	50.80	0.51	50	0.51	50	0.58	51	0.65	52	0.72	53
54	57.56	0.57	50	0.57	50	0.65	51	0.73	52	0.81	53

*Extracted from *Ductile Iron Pipe, Centrifugally Cast,* AWWA C151/A21.51-2002, with permission of the publisher, the American Water Works Association, Denver, Colo.

†To convert from inches to millimeters, multiply by 25.4; to convert pounds-force per square inch to megapascals, multiply by 0.006895.

‡These pipe walls are adequate for the rated working pressure plus a surge allowance of 100 lbf/in². For the effect of laying conditions and depth of bury, see ANSI A21.51.

modulus of elasticity of aluminum must be considered when using flanged connections.

Aluminum-body diaphragm and ball valves are used extensively.

Copper and Copper Alloys Seamless pipe and tubing manufactured from copper, bronze, brass, and copper-nickel alloys are produced by extrusion. The availability of pipe or tubing depends on the metallurgy and size. Copper tubing is widely used for plumbing, steam tracing, compressed air, instrument air, and inert gas applications. Copper tubing specifications are generally segregated into three types: water and general service, refrigeration service (characterized by cleanliness requirements), and drain/waste/vent (DWV) service. Tubing is available in the annealed or hard-drawn condition. Hard-drawn products are available only as straight lengths. Annealed products are often available in coils or as straight lengths. Suppliers should be consulted regarding wall thickness availability and standard lengths; however, many products are available in 3.0-m (10-ft) and 6.1-m (20-ft) straight lengths. Coil lengths are generally 12.2 m (40 ft) to 30 m (100 ft).

ASTM copper tubing specifications for water and general-purpose applications include B75 and B88 (Table 10-35). Tubing, fittings, and solders specified for potable water services must always be approved by the appropriate authority, such as the National Sanitation Foundation (NSF). ASTM copper tubing specifications for refrigeration services include B68 and B280 (Table 10–36). ASTM specifications for DWV services include B306 (1¼ through 8-in OD). Pipe conforming to the diameters shown in Table 10-37 is available as ASTM B42 and B302. ASTM B302 has diameters matching ASTM B42, but is available with thinner walls (Table 10-38). Red brass pipe is available as ASTM B43.

Joints are typically soldered, silver-brazed, or mechanical. When using flare, compression, or other mechanical joint fittings, consideration must be given to the fitting manufacturer's recommendations regarding hardness, minimum and maximum wall thickness, finish requirements, and diameter tolerances. Flanges and flanged fittings are seldom used since soldered and brazed joints can be easily disassembled. Brass and bronze valves are available with female ends for soldering.

70 percent copper, 30 percent nickel and 90 percent copper, 10 percent nickel ASTM B466 are available as seamless pipe (ASTM B466) or welded pipe (ASTM B467) and welding fittings for handling brackish water in Schedule 10 and regular copper pipe thicknesses.

Nickel and Nickel Alloys A wide range of ferrous and nonferrous nickel and nickel-bearing alloys are available. They are usually selected because of their improved resistance to chemical attack or their superior resistance to the effects of high temperature. In general terms their cost and corrosion resistance are somewhat a function of their nickel content. The 300 Series stainless steels are the most generally used. Some other frequently used alloys are listed in Table 10-39 together with their nominal compositions. For metallurgical and corrosion resistance data, see Sec. 25.

TABLE 10-34 High-Silicon Iron Pipe*

	No hub (MJ) ends				Bell-and-spigot ends			
Size, inside diam., in	Outside diam., in	Wall thickness, in	Standard† length, ft	Est. weight per piece, lb	Outside diam., in	Wall thickness, in	Standard† length, ft	Est. weight per piece, lb
1½	2 3/16	5/16	7	46				
2	2 11/16	5/16	7	56	2 11/16	5/16	7	62
3	3 49/64	5/16	7	79	3 25/32	5/16	7	90
4	4 49/64	5/16	7	100	4 25/32	5/16	7	114
6					6 11/16	13/32	7	169
8					9	13/32	7	234
10					11¾	5/8	7	340
12					13¼	5/8	5	470
15					16¼	7/8	5	800

*Extracted from *Standard Specification for High-Silicon Iron Pipe and Fittings,* ASTM A861-2004, with permission of the American Society for Testing and Materials, West Conshocken, Pa.

TABLE 10-35 Dimensions, Weights, and Tolerances in Diameter and Wall Thickness for Nominal or Standard Copper Water Tube Sizes (ASTM B88-2003)*

All tolerances are plus and minus except as otherwise noted.

Nominal standard size, in	Outside diameter, in	Average outside diameter[A] tolerance, in		Wall thickness and tolerances, in						Theoretical weight, lb/ft		
				Type K		Type L		Type M				
		Annealed	Drawn	Wall thickness	Tolerance[B]	Wall thickness	Tolerance[B]	Wall thickness	Tolerance[B]	Type K	Type L	Type M
¼	0.375	0.002	0.001	0.035	0.0035	0.030	0.003	[C]	[C]	0.145	0.126	[C]
⅜	0.500	0.0025	0.001	0.049	0.005	0.035	0.004	0.025	0.002	0.269	0.198	0.145
½	0.625	0.0025	0.001	0.049	0.005	0.040	0.004	0.028	0.003	0.344	0.285	0.204
⅝	0.750	0.0025	0.001	0.049	0.005	0.042	0.004	[C]	[C]	0.418	0.362	[C]
¾	0.875	0.003	0.001	0.065	0.006	0.045	0.004	0.032	0.003	0.641	0.455	0.328
1	1.125	0.0035	0.0015	0.065	0.006	0.050	0.005	0.035	0.004	0.839	0.655	0.465
1¼	1.375	0.004	0.0015	0.065	0.006	0.055	0.006	0.042	0.004	1.04	0.884	0.682
1½	1.625	0.0045	0.002	0.072	0.007	0.060	0.006	0.049	0.005	1.36	1.14	0.940
2	2.125	0.005	0.002	0.083	0.008	0.070	0.007	0.058	0.006	2.06	1.75	1.46
2½	2.625	0.005	0.002	0.095	0.010	0.080	0.008	0.065	0.006	2.93	2.48	2.03
3	3.125	0.005	0.002	0.109	0.011	0.090	0.009	0.072	0.007	4.00	3.33	2.68
3½	3.625	0.005	0.002	0.120	0.012	0.100	0.010	0.083	0.008	5.12	4.29	3.58
4	4.125	0.005	0.002	0.134	0.013	0.110	0.011	0.095	0.010	6.51	5.38	4.66
5	5.125	0.005	0.002	0.160	0.016	0.125	0.012	0.109	0.011	9.67	7.61	6.66
6	6.125	0.005	0.002	0.192	0.019	0.140	0.014	0.122	0.012	13.9	10.2	8.92
8	8.125	0.006	+0.002 −0.004	0.271	0.027	0.200	0.020	0.170	0.017	25.9	19.3	16.5
10	10.125	0.008	+0.002 −0.006	0.338	0.034	0.250	0.025	0.212	0.021	40.3	30.1	25.6
12	12.125	0.008	+0.002 −0.006	0.405	0.040	0.280	0.028	0.254	0.025	57.8	40.4	36.7

*Extracted from ASTM B88-2003 with permission of the publisher, the American Society for Testing and Materials, West Conshohocken, Pa.

[A] The average outside diameter of a tube is the average of the maximum and minimum outside diameter, as determined at any one cross section of the tube.

[B] Maximum deviation at any one point.

[C] Indicates that the material is not generally available or that no tolerance has been established.

Titanium Seamless pipe is available as ASTM B861, and welded pipe is available as ASTM B862. Both standards offer numerous grades of unalloyed and alloyed materials. While the alloys often have higher tensile strengths, corrosion resistance may be sacrificed. Forged or wrought fittings and forged or cast valves are available. For many applications, elastomer-lined valves having carbon-steel or ductile-iron bodies and titanium trim offer an economical alternative to valves with titanium bodies. Titanium pipe is available with wall thicknesses conforming to many of those listed in Table 10-22, including Schedule 10S and Standard Weight. Properly selected and specified, titanium can be a good choice for seawater systems such as offshore fire water systems. Seamless and welded tubing is manufactured to ASTM B338; however, availability may be limited.

Flexible Metal Hose Flexible hoses provide flexible connections for conveying gases or liquids, wherever rigid pipes are impractical. There are two basic types of flexible hose: corrugated hoses and interlocked hoses. These flexible hoses can absorb vibrations and noise. They can also provide some flexibility for misaligned rigid piping or equipment during construction. Corrugated or interlocked thin brass, bronze, Monel, aluminum, and steel tubes are covered with flexible braided-wire jackets to form flexible metal hose. Both tube and braid are brazed or welded to pipe-thread, union, or flanged ends. Failures are often the

TABLE 10-36a Standard Dimensions and Weights, and Tolerances in Diameter and Wall Thickness for Coil Lengths (ASTM B280-2003*)

Standard size, in	Outside diameter, in (mm)	Wall thickness, in (mm)	Weight, lb/ft (kg/m)	Tolerances: Average[A] outside diameter, plus and minus, in (mm)	Tolerances: Wall[B] thickness, plus and minus, in (mm)
⅛	0.125 (3.18)	0.030 (0.762)	0.0347 (0.0516)	0.002 (0.051)	0.003 (0.08)
3⁄16	0.187 (4.75)	0.030 (0.762)	0.0575 (0.0856)	0.002 (0.051)	0.003 (0.08)
¼	0.250 (6.35)	0.030 (0.762)	0.0804 (0.120)	0.002 (0.051)	0.0025 (0.08)
5⁄16	0.312 (7.92)	0.032 (0.813)	0.109 (0.162)	0.002 (0.051)	0.003 (0.08)
⅜	0.375 (9.52)	0.032 (0.813)	0.134 (0.199)	0.002 (0.051)	0.003 (0.08)
½	0.500 (12.7)	0.032 (0.813)	0.182 (0.271)	0.002 (0.051)	0.003 (0.08)
⅝	0.625 (15.9)	0.035 (0.889)	0.251 (0.373)	0.002 (0.051)	0.004 (0.011)
¾	0.750 (19.1)	0.042 (0.889)	0.305 (0.454)	0.0025 (0.064)	0.004 (0.011)
¾	0.750 (19.1)	0.042 (1.07)	0.362 (0.539)	0.0025 (0.064)	0.004 (0.011)
⅞	0.875 (22.3)	0.045 (1.14)	0.455 (0.677)	0.003 (0.076)	0.004 (0.11)
1⅛	1.125 (28.6)	0.050 (1.27)	0.665 (0.975)	0.0035 (0.089)	0.005 (0.13)
1⅜	1.375 (34.9)	0.055 (1.40)	0.884 (1.32)	0.004 (0.10)	0.006 (0.15)
1⅝	1.625 (41.3)	0.060 (1.52)	1.14 (1.70)	0.0045 (0.11)	0.006 (0.15)

*Extracted from ASTM B280-2003 with permission of the publisher, the American Society for Testing and Materials, West Conshohocken, Pa.

[A] The average outside diameter of a tube is the average of the maximum and minimum outside diameters as determined at any one cross section of the tube.

[B] The tolerances listed represent the maximum deviation at any point.

TABLE 10-36*b* Standard Dimensions and Weights, and Tolerances in Diameter and Wall Thickness for Straight Lengths (ASTM B280-2003*)

Note 1—Applicable to drawn temper tube only.

Standard size, in	Outside diameter, in (mm)	Wall thickness, in (mm)	Weight, lb/ft (kg/m)	Tolerances: Average[A] outside diameter, plus and minus, in (mm)	Tolerances: Wall[B] thickness, plus and minus, in (mm)
3/8	0.375 (9.52)	0.030 (0.762)	0.126 (0.187)	0.001 (0.025)	0.0035 (0.08)
1/2	0.500 (12.7)	0.035 (0.889)	0.198 (0.295)	0.001 (0.025)	0.004 (0.09)
5/8	0.625 (15.9)	0.040 (1.02)	0.285 (0.424)	0.001 (0.025)	0.004 (0.10)
3/4	0.750 (19.1)	0.042 (1.07)	0.362 (0.539)	0.001 (0.025)	0.004 (0.11)
7/8	0.875 (22.3)	0.045 (1.14)	0.455 (0.677)	0.001 (0.025)	0.004 (0.11)
1 1/8	1.125 (28.6)	0.050 (1.27)	0.655 (0.975)	0.0015 (0.038)	0.004 (0.13)
1 3/8	1.375 (34.9)	0.055 (1.40)	0.884 (1.32)	0.0015 (0.038)	0.006 (0.11)
1 5/8	1.625 (41.3)	0.060 (1.52)	1.14 (1.70)	0.002 (0.051)	0.006 (0.11)
2 1/8	2.125 (54.0)	0.070 (1.78)	1.75 (2.60)	0.002 (0.051)	0.007 (0.15)
2 5/8	2.625 (66.7)	0.080 (2.03)	2.48 (3.69)	0.002 (0.051)	0.008 (0.15)
3 1/8	3.125 (79.4)	0.090 (2.29)	3.33 (4.96)	0.002 (0.051)	0.009 (0.23)
3 5/8	3.625 (92.1)	0.100 (2.54)	4.29 (6.38)	0.002 (0.051)	0.010 (0.25)
4 1/8	4.125 (105)	0.110 (2.79)	5.38 (8.01)	0.002 (0.051)	0.011 (0.28)

*Extracted from ASTM B280-2003, with permission of the publisher, the American Society for Testing and Materials, West Conshohocken, Pa.

[A] The average outside diameter of a tube is the average of the maximum and minimum outside diameters as determined at any one cross section of the tube.

[B] The tolerances listed represent the maximum deviation at any point.

TABLE 10-37 Copper and Red-Brass Pipe (ASTM B42-2002 and B43-2004)*: Standard Dimensions, Weights, and Tolerances

Standard pipe size, in	Nominal outside diameter, in (mm)	Average outside diameter tolerances, in (mm), all minus†	Nominal wall thickness, in (mm)	Tolerance, in (mm)‡	Theoretical weight, lb/ft (kg/m)	Theoretical weight, lb/ft (kg/m)
			Regular pipe		Red brass	Copper
1/8	0.405 (10.3)	0.004 (0.10)	0.062 (1.57)	0.004 (0.10)	0.253 (0.376)	0.259 (0.385)
1/4	0.540 (13.7)	0.004 (0.10)	0.082 (2.08)	0.005 (0.13)	0.447 (0.665)	0.457 (0.680)
3/8	0.675 (17.1)	0.005 (0.13)	0.090 (2.29)	0.005 (0.13)	0.627 (0.933)	0.641 (0.954)
1/2	0.840 (21.3)	0.005 (0.13)	0.107 (2.72)	0.006 (0.15)	0.934 (1.39)	0.955 (1.42)
3/4	1.050 (26.7)	0.006 (0.15)	0.114 (2.90)	0.006 (0.15)	1.27 (1.89)	1.30 (1.93)
1	1.315 (33.4)	0.006 (0.15)	0.126 (3.20)	0.007 (0.18)	1.78 (2.65)	1.82 (2.71)
1 1/4	1.660 (42.2)	0.006 (0.15)	0.146 (3.71)	0.008 (0.20)	2.63 (3.91)	2.69 (4.00)
1 1/2	1.900 (48.3)	0.006 (0.15)	0.150 (3.81)	0.008 (0.20)	3.13 (4.66)	3.20 (4.76)
2	2.375 (60.3)	0.008 (0.20)	0.156 (3.96)	0.009 (0.23)	4.12 (6.13)	4.22 (6.28)
2 1/2	2.875 (73.0)	0.008 (0.20)	0.187 (4.75)	0.010 (0.25)	5.99 (8.91)	6.12 (9.11)
3	3.500 (88.9)	0.010 (0.25)	0.219 (5.56)	0.012 (0.30)	8.56 (12.7)	8.76 (13.0)
3 1/2	4.000 (102)	0.010 (0.25)	0.250 (6.35)	0.013 (0.33)	11.2 (16.7)	11.4 (17.0)
4	4.500 (114)	0.012 (0.30)	0.250 (6.35)	0.014 (0.36)	12.7 (18.9)	12.9 (19.2)
5	5.562 (141)	0.014 (0.36)	0.250 (6.35)	0.014 (0.36)	15.8 (23.5)	16.2 (24.1)
6	6.625 (168)	0.016 (0.41)	0.250 (6.35)	0.014 (0.36)	19.0 (28.3)	19.4 (28.9)
8	8.625 (219)	0.020 (0.51)	0.312 (7.92)	0.022 (0.56)	30.9 (46.0)	31.6 (47.0)
10	10.750 (273)	0.022 (0.56)	0.365 (9.27)	0.030 (0.76)	45.2 (67.3)	46.2 (68.7)
12	12.750 (324)	0.024 (0.61)	0.375 (9.52)	0.030 (0.76)	55.3 (82.3)	56.5 (84.1)
			Extrastrong pipe		Red brass	Copper
1/8	0.405 (10.3)	0.004 (0.10)	0.100 (2.54)	0.006 (0.15)	0.363 (0.540)	0.371 (0.552)
1/4	0.540 (13.7)	0.004 (0.10)	0.123 (3.12)	0.007 (0.18)	0.611 (0.909)	0.625 (0.930)
3/8	0.675 (17.1)	0.005 (0.13)	0.127 (3.23)	0.007 (0.18)	0.829 (1.23)	0.847 (1.26)
1/2	0.840 (21.3)	0.005 (0.13)	0.149 (3.78)	0.008 (0.20)	1.23 (1.83)	1.25 (1.86)
3/4	1.050 (26.7)	0.006 (0.15)	0.157 (3.99)	0.009 (0.23)	1.67 (2.48)	1.71 (2.54)
1	1.315 (33.4)	0.006 (0.15)	0.182 (4.62)	0.010 (0.25)	2.46 (3.66)	2.51 (3.73)
1 1/4	1.660 (42.2)	0.006 (0.15)	0.194 (4.93)	0.010 (0.25)	3.39 (5.04)	3.46 (5.15)
1 1/2	1.900 (48.3)	0.006 (0.15)	0.203 (5.16)	0.011 (0.28)	4.10 (6.10)	4.19 (6.23)
2	2.375 (60.3)	0.008 (0.20)	0.221 (5.61)	0.012 (0.30)	5.67 (8.44)	5.80 (8.63)
2 1/2	2.875 (73.0)	0.008 (0.20)	0.280 (7.11)	0.015 (0.38)	8.66 (12.9)	8.85 (13.2)
3	3.500 (88.9)	0.010 (0.25)	0.304 (7.72)	0.016 (0.41)	11.6 (17.3)	11.8 (17.6)
3 1/2	4.000 (102)	0.010 (0.25)	0.321 (8.15)	0.017 (0.43)	14.1 (21.0)	14.4 (21.4)
4	4.500 (114)	0.012 (0.30)	0.341 (8.66)	0.018 (0.46)	16.9 (25.1)	17.3 (25.7)
5	5.562 (141)	0.014 (0.36)	0.375 (9.52)	0.019 (0.48)	23.2 (34.5)	23.7 (35.3)
6	6.625 (168)	0.016 (0.41)	0.437 (11.1)	0.027 (0.69)	32.2 (47.9)	32.9 (49.0)
8	8.625 (219)	0.020 (0.51)	0.500 (12.7)	0.035 (0.89)	48.4 (72.0)	49.5 (73.7)
10	10.750 (273)	0.022 (0.56)	0.500 (12.7)	0.040 (1.0)	61.1 (90.9)	62.4 (92.9)

*Copyright American Society for Testing and Materials, West Conshohocken, PA; reprinted/adapted with permission. All tolerances are plus and minus except as otherwise indicated.

†The average outside diameter of a tube is the average of the maximum and minimum outside diameters as determined at any one cross section of the tube.

‡Maximum deviation at any one point.

TABLE 10-38 Hard-Drawn Copper Threadless Pipe (ASTM B302)*

Standard pipe size, in	Nominal dimensions, in (mm) Outside diameter	Inside diameter	Wall thickness	Cross-sectional area of bore, in^2 (cm^2)	Nominal weight, lb/ft (kg/m)	Tolerances, in (mm) Average outside diameter, all minus†	Wall thickness, plus and minus
¼	0.540 (13.7)	0.410 (10.4)	0.065 (1.65)	0.132 (0.852)	0.376 (0.559)	0.004 (0.10)	0.0035 (0.089)
⅜	0.675 (17.1)	0.545 (13.8)	0.065 (1.65)	0.233 (1.50)	0.483 (0.719)	0.004 (0.10)	0.004 (0.10)
½	0.840 (21.3)	0.710 (18.0)	0.065 (1.65)	0.396 (2.55)	0.613 (0.912)	0.005 (0.13)	0.004 (0.10)
¾	1.050 (26.7)	0.920 (23.4)	0.065 (1.65)	0.665 (4.29)	0.780 (1.16)	0.005 (0.13)	0.004 (0.10)
1	1.315 (33.4)	1.185 (30.1)	0.065 (1.65)	1.10 (7.10)	0.989 (1.47)	0.005 (0.13)	0.004 (0.10)
1¼	1.660 (42.2)	1.530 (38.9)	0.065 (1.65)	1.84 (11.9)	1.26 (1.87)	0.006 (0.15)	0.004 (0.10)
1½	1.900 (48.3)	1.770 (45.0)	0.065 (1.65)	2.46 (15.9)	1.45 (2.16)	0.006 (0.15)	0.004 (0.10)
2	2.375 (60.3)	2.245 (57.0)	0.065 (1.65)	3.96 (25.5)	1.83 (272)	0.007 (0.18)	0.006 (0.15)
2½	2.875 (73.0)	2.745 (69.7)	0.065 (1.65)	5.92 (38.2)	2.22 (3.30)	0.007 (0.18)	0.006 (0.15)
3	3.500 (88.9)	3.334 (84.7)	0.083 (2.11)	8.73 (56.3)	3.45 (5.13)	0.008 (0.20)	0.007 (0.18)
3½	4.000 (102)	3.810 (96.8)	0.095 (2.41)	11.4 (73.5)	4.52 (6.73)	0.008 (0.20)	0.007 (0.18)
4	4.500 (114)	4.286 (109)	0.107 (2.72)	14.4 (92.9)	5.72 (8.51)	0.010 (0.25)	0.009 (0.23)
5	5.562 (141)	5.298 (135)	0.132 (3.40)	22.0 (142)	8.73 (13.0)	0.012 (0.30)	0.010 (0.25)
6	6.625 (168)	6.309 (160)	0.158 (4.01)	31.3 (202)	12.4 (18.5)	0.014 (0.36)	0.010 (0.25)
8	8.625 (219)	8.215 (209)	0.205 (5.21)	53.0 (342)	21.0 (31.2)	0.018 (0.46)	0.014 (0.36)
10	10.750 (273)	10.238 (260)	0.256 (6.50)	82.3 (531)	32.7 (48.7)	0.018 (0.46)	0.016 (0.41)
12	12.750 (324)	12.124 (308)	0.313 (7.95)	115 (742)	47.4 (70.5)	0.018 (0.46)	0.020 (0.51)

*Copyright American Society for Testing and Materials, West Conshohocken, PA; reprinted/adapted with permission.
†The average outside diameter of a tube is the average of the maximum and minimum outside diameters, as determined at any one cross section of the tube.

result of corrosion of the braided-wire jacket or of a poor jacket-to-fitting weld. Maximum recommended temperature for bronze hose is approximately 230°C (450°F). Metal thickness is much less than for straight tube for the same pressure-temperature conditions; so accurate data on corrosion and erosion are required to make proper selection.

NONMETALLIC PIPE AND METALLIC PIPING SYSTEMS WITH NONMETALLIC LININGS

Cement-Lined Carbon-Steel Pipe Cement-lined carbon-steel pipe is made by lining steel pipe with special cement. The cement lining prevents pickup of iron by the fluid handled, corrosion of the metal by brackish or saline water, and growth of tuberculation. Various grades of cement are available, and the proper grade should be selected to match the application.

Cement-lined pipe in sizes smaller than 1½ in is not generally recommended. Cement-lined carbon steel pipe can be supplied with butt-weld or flanged ends. Butt-welded construction involves the use of special joint grouts at the weld joint and controlled welding procedures. See Table 10-40.

American Water Works Association Standard C205 addresses shop-applied cement mortar lining of pipe sizes 4 in and larger. Fittings are

TABLE 10-39 Common Nickel and Nickel-Bearing Alloys

Common trade name or registered trademark	Code designation	Alloy no.	ASTM specification (pipe)	Nominal composition, % Ni	Cr	Mo	Fe	C[a]	Si[a]	Mn	Cu	Cb	Co	W
Type 304 stainless steel		S30400	A312	8	18		BAL	0.08	1.00	2.0				
Type 304L stainless steel		S30403	A312	8	18		BAL	0.03	1.00	2.0				
Type 316 stainless steel		S31600	A312	12	16	2	BAL	0.08	1.00	2.0				
Type 316L stainless steel		S31603	A312	12	16	2	BAL	0.03	1.00	2.0				
Carpenter 20cb[b]	Ni-Cr-Fe-Mo-Cu-Cb stabilized	N08020	B464	33	20	2.5	38.5	0.06		2.0	3	1		
Incoloy 800[c]	Ni-Fe-Cr	N08800	B407	32.5	21		46	0.05	0.5	0.8	0.4			
Incoloy 825[c]	Ni-Fe-Cr-Mo-Cu	N08825	B423	42	21.5	3	30	0.03	0.2	0.5	2.2			
Hastelloy C-276[d]	Ni-Mo-Cr low carbon	N10276	B575[e]	54	15	16	5	0.02	0.08	1			2.5	4
Hastelloy B-2[d]	Ni-Mo	N10001	B333[e]	64	1	28	2	0.02	0.1	1				
Inconel 625[c]	Ni-Cr-Mo-Cb	N06625	B444	61	21.5	9	2.5	0.05	0.2	0.2		4		
Inconel 600[c]	Ni-Cr-Fe	N06600	B167	76	15.5		8	0.08	0.2	0.5	0.2			
Monel 400[c]	Ni-Cu	N04400	B165	66			1.2	0.20	0.2	1	31.5			
Nickel 200[c]	Ni	N02200	B161	99+			0.2	0.08						
Hastelloy G[d]	Ni-Cr-Fe-Mo-Cu	N06007	B622	42	22.2	6.5	19.5	0.05	1	1.5	2	2.2[f]	2.5[a]	1[a]

[a] Maximum.
[b] Registered trademark, Carpenter Technology Corp.
[c] Registered trademark, Huntington Alloys, Inc.
[d] Registered trademark, Cabot Corp.
[e] Plate.
[f] Cb + Ta.

TABLE 10-40 Cement-Lined Carbon-Steel Pipe*

Standard pipe size, in	Inside diam. after lining, in	Typical thickness of lining, in	Weight, per ft, lb	Standard pipe size, in	Inside diam. after lining, in	Typical thickness of lining, in	Weight per ft, lb
				3	2.70	0.13	8.3
				4	3.60	.16	12.0
				6	5.40	.25	24.0
1½	1.40	.09	3.0	8	7.40	.25	32.0
2	1.80	.10	4.1	10	9.40	.30	43.0
2½	2.20	.10	6.6	12	11.40	.30	55.0

*To convert inches to millimeters, multiply by 25.4; to convert pounds per foot to kilograms per meter, multiply by 1.49.

available as cement mortar lined butt-weld or flanged carbon steel, flanged cast iron, or flanged ductile iron. AWWA C602 addresses in-place (i.e., in situ or field) application of cement mortar lining for pipe sizes 4 in and larger.

Concrete Pipe Concrete piping and nonmetallic piping such as PVC, RTR, and HDPE are commonly used for buried gravity drain and pressurized applications. Common applications for both include construction culverts and forced water mains, sewage, industrial waste, and storm water systems. Some of the factors to be considered when deciding whether to use concrete or nonmetallic piping include local code requirements, pipe size, soil and commodity corrosivity, commodity temperature and pressure, resistance to tuberculin growth, traffic and burial loads, soil conditions and bedding requirements, groundwater level and buoyancy issues, suitability of available joining methods, ability of joints to resist internal pressure thrust without the use of thrust blocks, availability of pressure-rated and nonpressure-rated fittings, shipping weight, load capacity of available construction equipment, requirements for special equipment such as fusion bonding machines, contractor's experience and labor skill level requirements, and final installed cost.

Nonreinforced concrete culvert pipe for gravity drain applications is manufactured to ASTM C14 in strength Classes 1, 2, and 3. It is available with internal diameters 4 through 36 in. Reinforced concrete culvert for gravity drain applications is manufactured to ASTM C76 with internal diameters of 12 through 144 in. Metric sizes are manufactured to ASTM C14M and C76M. Joints are typically bell and spigot (or a similar variation) with rubber gaskets.

Concrete pressure pipe is typically custom-designed to three different specifications. Each design provides a cement mortar lining or concrete interior. It is advisable to consult manufacturers regarding the most appropriate specification for a given application and the availability of fittings. The names of some manufacturers can be obtained through the American Concrete Pressure Pipe Association. American Water Works Association standard AWWA C300 addresses steel cylinder reinforced concrete pressure pipe in inside-diameter sizes 30 through 144 in. AWWA C301 addresses prestressed reinforced pipe with a steel cylinder wrapped with steel wire. Inside-diameter sizes are 16 through 144 in. AWWA C302 addresses circumferentially reinforced pipe without a steel cylinder or prestress. Inside diameter sizes are 12 through 144 in, with continuous pressure ratings to 0.38 MPa (55 lbf/in^2) and total pressure (including surge) to 0.45 MPa (65 lbf/in^2). AWWA C303 addresses reinforced pipe with a steel cylinder helically wrapped with steel bar. Inside-diameter sizes are 10 through 60 in, with pressure ratings to 2.7-MPa (400 lbf/in^2) working pressure. Joints are typically bell and spigot (or a similar variation) with rubber gaskets. In addition to the gasket, grouting is used on the exterior and interior of the joint to seal otherwise exposed steel.

Glass Pipe and Fittings These are made from heat- and chemical-resistant borosilicate glass in accordance with ASTM E-438 Type 1 Class A. This glass is resistant to chemical attack by almost all products, which makes its resistance much more comprehensive than that of other well-known materials. It is highly resistant to water, saline solutions, organic substances, halogens such as chlorine and bromine, and many acids. There are only a few chemicals that can cause noticeable corrosion of the glass surface, such as hydrofluoric acid, concentrated phosphoric acid, and strong caustic solutions at elevated temperatures. Some important physical properties are as follows:

Mean linear expansion coefficient between 20 and 300°C: $[(3.3 \pm 0.1) \times 10^{-6}]$/K
Mean thermal conductivity between 20 and 200°C: 1.2 W/(m K)
Mean specific heat capacity between 20 and 100°C: 0.8 kJ/kg(K)
Mean specific heat capacity between 101 and 200°C: 0.9 kJ/kg(K)
Density at 20°C: 2.23 kg/dm^3

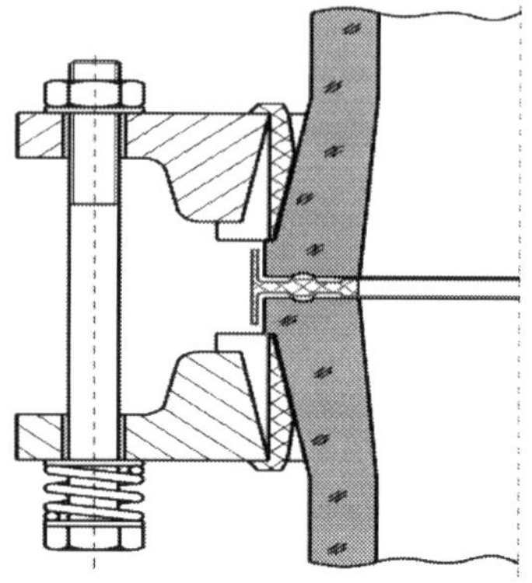

FIG. 10-164 Flanged joint with conical ends. (*Adapted with permission of De Dietrich Process Systems, Mountainside, N.J.*)

Flanged glass pipe with conical ends (Fig. 10-164) should be used in applications requiring a pressure rating or that are expected to see thermal cycling. Flanged ends are normally required to mate with other glass components (vessels, coil heat exchangers, etc). The flanges are specially designed plastic backing flanges that are cushioned from the glass by either molded plastic or fiber inserts. The liquid seal is provided by means of a gasket that is gripped between the grooved pipe ends.

Glass pipe is made in the sizes shown in Table 10-41. Depending on the nominal diameter, lengths range from 0.075 to 3 m. Design pressure ranges from −0.10 MPa (−14.5 lbf/in^2) vacuum to 0.40 MPa (58 lbf/in^2) for nominal diameters of 15 through 50 mm, 3 MPa (43 lbf/in^2) for nominal diameter of 80 mm, 0.20 MPa (29 lbf/in^2) for nominal diameters of 100 and 150 mm, and 0.10 MPa (14.5 lbf/in^2) for nominal diameters of 200 and 300 mm. Maximum permissible thermal shock as a general guide is 120 K. Maximum operating temperature is 200°C (248°F). A complete line of fittings is available, and special parts can be made to order. Thermal expansion stresses should be completely relieved by tied PTFE corrugated expansion joints and offsets. Temperature rating may be limited by joint design and materials.

Beaded pipe is used for process waste lines and vent lines. Some applications have also been made in low-pressure and vacuum lines. Beaded end pipe is available in nominal diameters of 1½ through 6 in. For operating conditions manufacturers should be consulted. In this system the ends of the pipe and fittings are formed into a bead, as

TABLE 10-41 Dimensions for Glass Pipe and Flanged Joints (see Fig. 10-165)*

Nominal pipe size, mm	D1	D2	D3	D4	Type
15	16.8	28.6	23	15.5–17.5	A
25	26.5	42.2	34	25–27	A
40	38.5	57.4	48	36.5–39.75	A
50	50.5	70	60.5	48–52	A
80	76	99.2	88	72–78	A
100	104.5	132.6	120.5	97.6–110	A
150	154	185	172	150–156	A
200	203	235	220	197–205	B
300	300	340	321	299–303	B

*Adapted with permission of De Dietrich Process Systems, Mountainside, N.J.

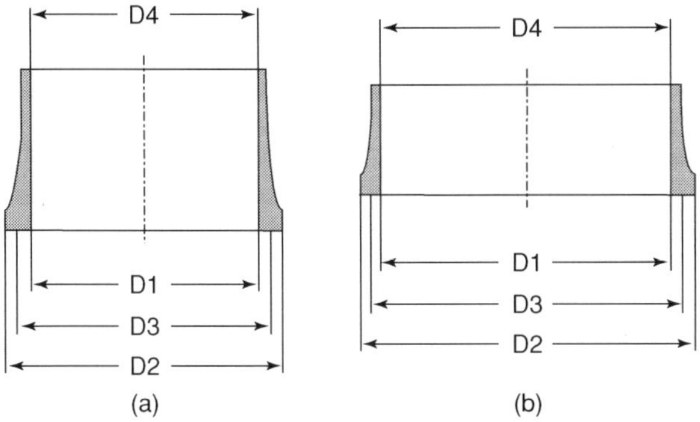

FIG. 10-165 Flanged pipe ends. (*Adapted with permission of De Dietrich Process Systems, Mountainside, N.J.*)

shown in Fig. 10-166. The coupling consists of a stainless steel outer shell, a rubber collar, and a TFE liner-gasket. When the single coupling nut is tightened, the thick rubber sleeve forces the TFE gasket against the glass to make the seal.

Glass-Lined Steel Pipe and Fittings This pipe is fully resistant to all acids except hydrofluoric and concentrated phosphoric acids at temperatures up to 120°C (248°F). It is also resistant to alkaline solutions at moderate temperatures. Glass-lined steel pipe can be used at temperatures up to 220°C (428°F) under some exposure conditions provided there are no excessive temperature changes. The operating pressure rating of commonly available systems is 1 MPa (145 lbf/in^2). The glass lining is approximately 1.6 mm (1⁄16 in) thick. It is made by lining Schedule 40 steel pipe. Fittings are available in glass-lined cast steel. Standard nominal pipe sizes available are 1½ through 8 in. Larger-diameter pipe up to 48 in is available on a custom-order basis. A range of lengths are generally available. See Table 10-42 for dimensional data. Steel split flanges drilled to ANSI B16.5 Class 150 dimensions along with PTFE envelope gaskets are used for the assembly of the system.

Fused Silica or Fused Quartz Containing 99.8 percent silicon dioxide, fused silica or fused quartz can be obtained as opaque or transparent pipe and tubing. The melting point is 1710°C (3100°F). Tensile strength is approximately 48 MPa (7000 lbf/in^2); specific gravity is about 2.2. The pipe and tubing can be used continuously at temperatures up to 1000°C (1830°F) and intermittently up to 1500°C (2730°F). The material's chief assets are noncontamination of most chemicals in high-temperature service, thermal-shock resistance, and high-temperature electrical insulating characteristics.

Transparent tubing is available in inside diameters from 1 to 125 mm in a range of wall thicknesses. Satin-surface tubing is available in inside diameters from 1⁄16 to 2 in, and sand-surface pipe and tubing are available in ½- to 24-in inside diameters and lengths up to 6 m (20 ft). Sand-surface pipe and tubing are obtainable in wall thicknesses varying from ⅛ to 1 in. Pipe and tubing sections in both opaque and transparent fused silica or fused quartz can be readily machine-ground to special tolerances for pressure joints or other purposes. Also, fused-silica piping and tubing can be reprocessed to meet special-design requirements. Manufacturers should be consulted for specific details.

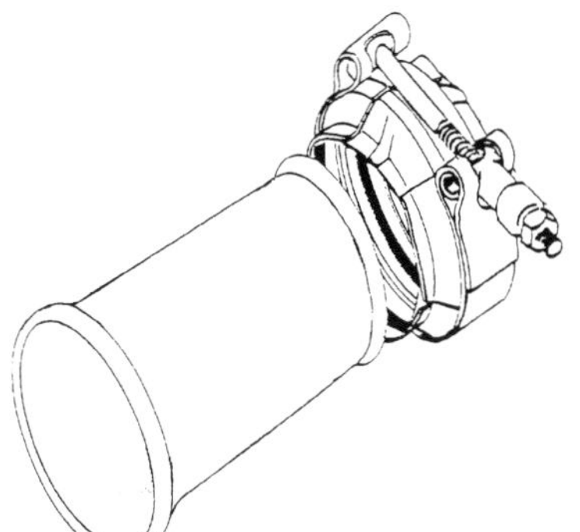

FIG. 10-166 Joint with beaded pipe ends. (*Adapted with permission of De Dietrich Process Systems, Mountainside, N.J.*)

TABLE 10-42 Glass-Lined Steel Pipe*

Size, in.	Outside diameter, in	Range of standard lengths, in	
		Minimum	Maximum
1½	1.900	6	60
2	2.375	6	84
3	3.500	6	120
4	4.500	6	120
6	6.625	6	120
8	8.625	6	120

*Adapted with permission of De Dietrich Process Systems, Mountainside, N.J. Other manufacturers may offer different standard lengths.

Plastic-Lined Steel Pipe Use of a variety of polymeric materials as liners for steel pipe rather than as piping systems solves problems which the relatively low tensile strength of the polymer at elevated temperature and high thermal expansion, compared with steel, would produce. The steel outer shell permits much wider spacing of supports, reliable flanged joints, and higher pressure and temperature in the piping. The size range is 1 through 12 in. The systems are flanged with 125-lb cast-iron, 150-lb ductile-iron, and 150- and 300-lb steel flanges. The linings are factory-installed in both pipe and fittings. Lengths are available up to 6 m (20 ft). Lined ball, diaphragm, and check valves and plug cocks are available.

One method of manufacture consists of inserting the liner into an oversize, approximately Schedule 40 steel tube and swaging the assembly to produce iron-pipe-size outside diameter, firmly engaging the liner which projects from both ends of the pipe. Flanges are then screwed onto the pipe, and the projecting liner is hot-flared over the flange faces nearly to the bolt holes. In another method, the liner is pushed into steel pipe having cold-flared laps backed up by flanges at the ends and then hot-flared over the faces of the laps. Pipe lengths made by either method may be shortened in the field and reflared with special procedures and tools. Square and tapered spacers are furnished to adjust for small discrepancies in assembly.

Liner types available are suitable for a wide variety of chemical services, including acids, alkalies, and various solvents. All liners are permeable to some degree, and manufacturers use various methods to vent gas out of the interspace between the liner and casing. All plastics are subject to environmental stress cracking (ESC). ESC can occur even when the liner is chemically resistant to the service. Lined pipe manufacturers should always be consulted regarding liner selections and service applications. Also consult manufacturers regarding vacuum service limits.

Polyvinylidene Chloride Liners Polyvinylidene chloride liners have excellent resistance to hydrochloric acid. Maximum temperature is 80°C (175°F). Polyvinylidene chloride is also known as Saran, a product of the Dow Chemical Co.

Polypropylene Liners Polypropylene liners are used in sulfuric acid service. At 10 to 30 percent concentration the upper temperature limit is 93°C (200°F). Polypropylene is also suitable for higher concentrations at lower temperatures.

Kynar Liners Kynar (Pennwalt Chemicals Corp.) polyvinylidene fluoride liners are used for many chemicals, including bromine and 50 percent hydrochloric acid.

PTFE and PFA Lined Steel Pipe These are available in sizes from 1 through 12 in and in lengths through 6 m (20 ft). Experience has determined that practical upper temperature limits are approximately 204°C (400°F) for PTFE (polytetrafluoroethylene) and PFA (perfluoroalkoxy) and 149°C (300°F) for FEP (fluoroethylene polymer); Class 150 and 300 ductile-iron or steel flanged lined fittings and valves are used. The nonadhesive properties of the liner make it ideal for handling sticky or viscous substances. Thickness of the lining varies from 1.5 to 3.8 mm (60 to 150 mil), depending on pipe size. Only flanged joints are used.

Rubber-Lined Steel Pipe This pipe is made in lengths up to 6 m (20 ft) with seamless, straight seam-welded and some types of spiral-welded pipe using various types of natural and synthetic adhering rubber. The type of rubber is selected to provide the most suitable lining for the specific service. In general, soft rubber is used for abrasion resistance, semihard for general service, and hard for the more severe service conditions. Multiple-ply lining and combinations of hard and soft rubber are available. The thickness of lining ranges from 3.2 to 6.4 mm (1/8 to 1/4 in) depending on the service, the type of rubber, and the method of lining. Cast-steel, ductile-iron, and cast-iron flanged fittings are available rubber-lined. The fittings are usually purchased by the vendor since absence of porosity on the inner surface is essential. Pipe is flanged before rubber lining, and welding elbows and tees may be incorporated at one end of the length of pipe, subject to the conditions that the size of the pipe and the location of the fittings are such that the operator doing the lining can place a hand on any point on the interior surface of the fitting. Welds must be ground smooth on the inside, and a radius is required at the inner edge of the flange face.

The rubber lining is extended out over the face of flanges. With hard-rubber lining, a gasket is required. With soft-rubber lining, a release coating or a polyethylene sheet is required in place of a gasket to avoid bonding of the lining of one flange to the lining on the other and to permit disassembly of the flanged joint. Also, for pressures over 0.86 MPa (125 lbf/in^2), the tendency of soft-rubber linings to extrude out between the flanges may be prevented by terminating the lining inside the bolt holes and filling the balance of the space between the flange faces with a spacer of the proper thickness. Hard-rubber-lined gate, diaphragm, and swing check valves are available. In gate valves, the stem, wedge assembly, and seat rings, and in the check valves, the hinge pin, flapper arm, disk, and seat ring must be made of metal resistant to the solution handled.

Plastic Pipe In contrast to other piping materials, plastic pipe is free from internal and external corrosion, is easily cut and joined, and does not cause galvanic corrosion when coupled to other materials. Allowable stresses and upper temperature limits are low. Normal operation is in the creep range. Fluids for which a plastic is not suited penetrate and soften it rather than dissolve surface layers. Coefficients of thermal expansion are high.

Plastic pipe or tubing may be used for a wide variety of services. As with all nonmetallic materials, code restrictions limit the applications in which their use is permitted. In general, their use in flammable or toxic service is limited. Plastic tubing of various types may be used for instrument air-signal connections; however, as is the case with all nonmetallic applications, the need for fire resistance must be considered. When used in specialized applications such as potable water or underground fire water, care should be taken to ensure that the specified products are certified by appropriate agencies such as the National Sanitation Foundation and Factory Mutual.

Support spacing must be much closer than for carbon steel. As temperature increases, the allowable stress for many plastic pipes decreases very rapidly, and heat from sunlight or adjacent hot uninsulated equipment has a marked effect. Many plastics deteriorate with exposure to ultraviolet light if not provided with a UV-resistant coating or other surface barrier. Successful economical underground use of plastic pipe does not necessarily indicate similar economies outdoors aboveground.

Methods of joining include threaded joints with IPS dimensions, solvent-welded joints, heat-fused joints, and insert fittings. Schedules 40 and 80 (see Table 10-22) have been used as a source for standardized dimensions at joints. Some plastics are available in several grades with allowable stresses varying by a factor of 2 to 1. For the same plastic, 1/2-in Schedule 40 pipe of the strongest grade may have 4 times the allowable internal pressure of the weakest grade of a 2-in Schedule 40 pipe. For this reason, the plastic-pipe industry is shifting to standard dimension ratios (approximately the same ratio of diameter to wall thickness over a wide range of pipe sizes).

ASTM and the Plastics Pipe Institute have established identifications for plastic pipe in which the first group of letters identifies the plastic, the two following numbers identify the grade of that plastic, and the last two numbers represent the design stress in the nearest lower 0.7-MPa (100-lbf/in^2) unit at 23°C (73.4°F).

Polyethylene Plastics Pipe Institute (www.plasticpipe.org) is an excellent source of information regarding specification, design, fabrication, and testing of polyethylene piping. Polyethylene (PE) pipe and tubing are available in sizes 48 in and smaller. They have excellent resistance at room temperature to salts, sodium and ammonium hydroxides, and sulfuric, nitric, and hydrochloric acids. High-density polyethylene (HDPE) is often used for underground fire water. Pipe and tubing are produced by extrusion from resins whose density varies with the manufacturing process. Physical properties and therefore wall thickness depend on the particular resin used. In some products, about 3 percent carbon black is added to provide resistance to ultraviolet light. Use of higher-density resin reduces splitting and pinholing in service and increases the strength of the material and the maximum service temperature.

ASTM D2104 covers PE pipe in sizes 1/2 through 6 in with IPS Schedule 40 outside and inside diameters for insert-fitting joints. ASTM D2239 covers six standard dimension ratios of pipe diameter to wall thickness in sizes 1/2 through 6 in. ASTM D2447 covers sizes 1/2 through 12 in with IPS Schedule 40 and 80 outside and inside diameters for use with heat-fusion socket-type and butt-type fittings. ASTM D3035 covers 10 standard dimension ratios of pipe sizes from 1/2 through 24 in with IPS outside diameters. All these specifications cover various PE materials. Hydrostatic design stresses within the recommended temperature limits are given in Appendix B, Table B-1, of the code. The hydrostatic design stress is the maximum tensile hoop stress due to internal hydrostatic water pressure that can be applied continuously with a high degree of certainty that failure of the pipe will not occur. Both manufacturers and the Plastic Pipe Institute publish literature describing design calculations required to determine the required wall thickness.

Polyethylene water piping is not damaged by freezing. Pipe and tubing 2 in and smaller are shipped in coils several hundred feet in length.

Clamped-insert joints (Fig. 10-167) are used for flexible plastic pipe up through the 2-in size. Friction between the pipe and the spud is developed both by forcing the spud into the pipe and by tightening the clamp. For the larger sizes, which have thicker walls, these methods cannot develop adequate friction. Insert joints also have high pressure drop. Stainless-steel bands are available. Inserts are available in nylon, polypropylene, and a variety of metals.

Joints of all sizes may be made with heat fusion techniques. Fused joints may be made with either electrofusion or conventional heat fusion. Electrofusion joints are made with fittings which have embedded heating wires. Conventional fusion joints are made with special machines which trim the pipe ends, apply heat, and then force them together to form a bond. Consult the manufacturer regarding the sizes for which electrofusion fittings are available.

A significant use for PE and PP pipe is the technique of rehabilitating deteriorated pipe lines by lining them with plastic pipe. Lining an existing pipe with plastic pipe has a large cost advantage over replacing the line, particularly if replacement of the old line would require excavation.

Polyvinyl Chloride Polyvinyl chloride (PVC) pipe and tubing are available with socket fittings for solvent-cemented joints in sizes 24 in and smaller. PVC with gasketed bell and spigot joints is available in sizes 4 through 48 in. Chlorinated polyvinyl chloride (CPVC) pipe and tubing are available with socket fittings for solvent-cemented joints in sizes 4 in and smaller. PVC and CPVC are suitable

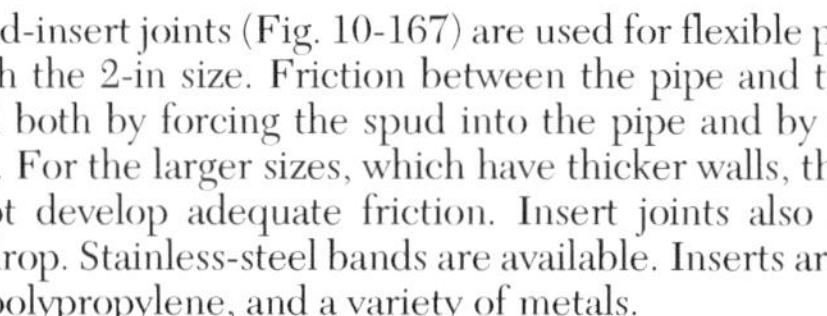

FIG. 10-167 Clamped-insert joint.

for a variety of chemical services and are commonly used for potable water. Consult manufacturers or the Plastics Pipe Institute for chemical resistance data. Hydrostatic design stresses within the temperature limits are given in Appendix B, Table B-1, of the code.

ASTM D1785 covers sizes from ⅛ through 12 in of PVC pipe in IPS Schedules 40, 80, and 120, except that Schedule 120 starts at ½ in and is not IPS for sizes from ½ through 3 in. ASTM D2241 covers sizes ⅛ through 36 in but with IPS outside diameter and seven standard dimension ratios: 13.5, 17, 21, 26, 32.5, 41, and 64.

ASTM D2513 specifies requirements for thermoplastic materials for buried fuel gas and relining applications. Materials addressed include PE, PVC, and crosslinked polyethylene (PEX). Tubing sizes covered are $\frac{1}{4}$ through 1¾ in. IPS sizes covered are $\frac{1}{2}$ through 12 in. Size availability depends upon the material.

Solvent-cemented joints (Fig. 10-150) are standard, but screwed joints are sometimes used with Schedule 80 pipe. Cemented joints must not be disturbed for 5 min and achieve full strength in 1 day. A wide variety of valve types are available in PVC and CPVC.

Polypropylene Polypropylene (PP) pipe and fittings have excellent resistance to most common organic and mineral acids and their salts, strong and weak alkalies, and many organic chemicals. They are available in sizes ½ through 6 in, in Schedules 40 and 80, but are not covered as such by ASTM specifications.

Reinforced-Thermosetting-Resin (RTR) Pipe Glass-reinforced epoxy resin has good resistance to nonoxidizing acids, alkalies, salt water, and corrosive gases. The glass reinforcement is many times stronger at room temperature than plastics, does not lose strength with increasing temperature, and reinforces the resin effectively up to 149°C (300°F). (See Table 10-21 for temperature limits.) The glass reinforcement is located near the outside wall, protected from the contents by a thick wall of resin and protected from the atmosphere by a thin wall of resin. Stock sizes are 2 through 12 in.

Pipe is supplied in 6- and 12-m (20- and 40-ft) lengths. It is more economical for long, straight runs than for systems containing numerous fittings. When the pipe is sawed to nonfactory lengths, it must be sawed very carefully to avoid cracking the interior plastic zone. A two-component cement may be used to bond lengths into socket couplings or flanges or cemented-joint fittings. Curing of the cement is temperature-sensitive; it sets to full strength in 45 min at 93°C (200°F), in 12 h at 38°C (100°F), and in 24 h at 10°C (50°F). Extensive use is made of shop-fabricated flanged preassemblies. Only flanged joints are used to connect to metallic piping systems. Compared with that of other plastics, the ratio of fitting cost to pipe cost is high. Cemented-joint fittings and flanged fittings are available. Internally lined flanged metallic valves are used.

RTR is more flexible than metallic pipe and consequently requires closer support spacing. While the recommended spacing varies among manufacturers and with the type of product, Table 10-43 gives typical hanger-spacing ranges. The pipe fabricator should be consulted for recommended hanger spacing on the specific pipe-wall construction being used.

Epoxy resin has a higher strength at elevated temperatures than polyester resins but is not as resistant to attack by some fluids. Some glass-reinforced epoxy-resin pipe is made with a polyester-resin liner. The coefficient of thermal expansion of glass-reinforced resin pipe is higher than that for carbon steel but much less than that for plastics.

Glass-reinforced polyester is the most widely used reinforced-resin system. A wide choice of polyester resins is available. The bisphenol resins resist strong acids as well as alkaline solutions. The size range is 2 through at least 36 in; the temperature range is shown in Table 10-21. Diameters are not standardized. Adhesive-cemented socket joints and hand-lay-up reinforced butt joints are used. For the latter, reinforcement consists of layers of glass cloth saturated with adhesive cement.

TABLE 10-43 Typical Hanger-Spacing Ranges Recommended for Reinforced-Thermosetting-Resin Pipe

Nominal pipe size, in	2	3	4	6	8	10	12
Hanger-spacing range, ft	5–8	6–9	6–10	8–11	9–13	10–14	11–15

NOTE: Consult pipe manufacturer for recommended hanger spacing for the specific RTR pipe being used. Tabulated values are based on a specific gravity of 1.25 for the contents of the pipe. To convert feet to meters, multiply by 0.3048.

DESIGN OF PIPING-SYSTEMS

Safeguarding Safeguarding may be defined as the provision of protective measures as required to ensure the safe operation of a proposed piping system. General considerations to be evaluated should include (1) the hazardous properties of the fluid, (2) the quantity of fluid which could be released by a piping failure, (3) the effect of a failure (such as possible loss of cooling water) on overall plant safety, (4) evaluation of effects on a reaction with the environment (i.e., possibility of a nearby source of ignition), (5) the probable extent of exposure of operating or maintenance personnel, and (6) the relative inherent safety of the piping by virtue of materials of construction, methods of joining, and history of service reliability.

Evaluation of safeguarding requirements might include engineered protection against possible failures such as thermal insulation, armor, guards, barricades, and damping for protection against severe vibration, water hammer, or cyclic operating conditions. Simple means to protect people and property such as shields for valve bonnets, flanged joints, and sight glasses should not be overlooked. The necessity for means to shut off or control flow in the event of a piping failure such as block valves or excess-flow valves should be examined.

Classification of Fluid Services The code applies to piping systems as illustrated in Fig. 10-127, but two categories of fluid services are segregated for special consideration as follows:

Category D Category D fluid service is defined as "a fluid service to which all the following apply: (1) the fluid handled is nonflammable and nontoxic; (2) the design gage pressure does not exceed 150 psi (1.0 MPa); and (3) the design temperature is between −20°F (−29°C) and 366°F (186°C)."

Category M Category M fluid service is defined as "a fluid service in which a single exposure to a very small quantity of a toxic fluid, caused by leakage, can produce serious irreversible harm to persons on breathing or bodily contact, even when prompt restorative measures are taken."

The code assigns to the owner the responsibility for identifying those fluid services which are in Categories D and M. The design and fabrication requirements for Class M toxic-service piping are beyond the scope of this *Handbook*. See ASME B31.3—2004, chap. VIII.

Design Conditions Definitions of the temperatures, pressures, and various forces applicable to the design of piping systems are as follows:

Design Pressure The design pressure of a piping system shall not be less than the pressure at the most severe condition of coincident internal or external pressure and temperature resulting in the greatest required component thickness or rating.

Design Temperature The design temperature is the material temperature representing the most severe condition of coincident pressure and temperature. For uninsulated metallic pipe with fluid below 65°C (150°F), the metal temperature is taken as the fluid temperature.

With fluid at or above 65°C (150°F) and without external insulation, the metal temperature is taken as a percentage of the fluid temperature unless a lower temperature is determined by test or calculation. For pipe, threaded and welding-end valves, fittings, and other components with a wall thickness comparable with that of the pipe, the percentage is 95 percent; for flanges and flanged valves and fittings, 90 percent; for lap-joint flanges, 85 percent; and for bolting, 80 percent.

With external insulation, the metal temperature is taken as the fluid temperature unless service data, tests, or calculations justify lower values. For internally insulated pipe, the design metal temperature shall be calculated or obtained from tests.

Ambient Influences If cooling results in a vacuum, the design must provide for external pressure or a vacuum breaker installed; also provision must be made for thermal expansion of contents trapped between or in closed valves. Nonmetallic or nonmetallic-lined pipe may require protection when ambient temperature exceeds design temperature.

Occasional variations of pressure or temperature, or both, above operating levels are characteristic of certain services. If the following criteria are met, such variations need not be considered in determining pressure-temperature design conditions. Otherwise, the most severe conditions of coincident pressure and temperature during the variation shall be used to determine design conditions. (Application of pressures exceeding pressure-temperature ratings of valves may under certain conditions cause loss of seat tightness or difficulty of operation. Such an application is the owner's responsibility.)

All the following criteria must be met:

1. The piping system shall have no pressure-containing components of cast iron or other nonductile metal.
2. Nominal pressure stresses shall not exceed the yield strength at temperature (see S_y data in BPV Code, Sec. II, Part D).
3. Combined longitudinal stresses S_L shall not exceed the limits established in the code (see pressure design of piping components for S_L limitations).
4. The total number of pressure-temperature variations above the design condition shall not exceed 1000 during the life of the piping system.
5. Occasional variations above design conditions shall remain within one of the following limits for pressure design:

- When the variation lasts no more than 10 h at any one time and no more than 100 h per year, it is permissible to exceed the pressure rating or the allowable stress for pressure design at the temperature of the increased condition by not more than 33 percent.
- When the variation lasts no more than 50 h at any one time and not more than 500 h per year, it is permissible to exceed the pressure rating or the allowable stress for pressure design at the temperature of the increased condition by not more than 20 percent.

Dynamic Effects Design must provide for impact (hydraulic shock, etc.), wind (exposed piping), earthquake, discharge reactions, and vibrations (of piping arrangement and support).

Weight Effects Weight considerations include (1) live loads (contents, ice, and snow), (2) dead loads (pipe, valves, insulation, etc.), and (3) test loads (test fluid).

Thermal Expansion and Contraction Effects Thermal-expansion and -contraction loads occur when a piping system is prevented from free thermal expansion or contraction as a result of anchors and restraints or undergoes large, rapid temperature changes or unequal temperature distribution because of an injection of cold liquid striking the wall of a pipe carrying hot gas.

Effects of Support, Anchor, and Terminal Movements The effects of movements of piping supports, anchors, and connected equipment shall be taken into account in the design of piping. These movements may result from the flexibility and/or thermal expansion of equipment, supports, or anchors; and from settlement, tidal movements, or wind sway.

Reduced Ductility The harmful effects of reduced ductility shall be taken into account in the design of piping. The effects may, e.g., result from welding, heat treatment, forming, bending, or low operating temperatures, including the chilling effect of sudden loss of pressure on highly volatile fluids. Low ambient temperatures expected during operation shall be considered.

Cyclic Effects Fatigue due to pressure cycling, thermal cycling, and other cyclic loading shall be considered in the design.

Air Condensation Effects At operating temperatures below −191°C (−312°F) in ambient air, condensation and oxygen enrichment occur. These shall be considered in selecting materials, including insulation, and adequate shielding and/or disposal shall be provided.

Design Criteria: Metallic Pipe The code uses three different approaches to design, as follows:

1. It provides for the use of dimensionally standardized components at their published pressure-temperature ratings.
2. It provides design formulas and maximum stresses.
3. It prohibits the use of materials, components, or assembly methods in certain conditions.

Components Having Specific Ratings These are listed in ASME, API, and industry standards. These ratings are acceptable for design pressures and temperatures unless limited in the code. A list of component standards is given in Appendix E of the ASME B31.3 code. Table 10-44 lists pressure–temperature ratings for

TABLE 10-44*a* Pressure–Temperature Ratings for Group 1.1 Materials (Carbon Steel)*

Nominal designation	Forgings	Castings	Plates
C–Si	A 105 (1)	A 216 Gr. WCB (1)	A 515 Gr. 70 (1)
C–Mn–Si	A 350 Gr. LF2 (1)		A 516 Gr. 70 (1), (2)
C–Mn–Si–V	A 350 Gr. LF6 Cl. 1 (4)		A 537 Cl. 1 (3)
3½ NI	A 350 Gr. LF 3		

Class temp., °F	Working pressures by classes, psig: 150	300	400	600	900	1500	2500
−20 to 100	285	740	985	1480	2220	3705	6170
200	260	680	905	1360	2035	3395	5655
300	230	655	870	1310	1965	3270	5450
400	200	635	845	1265	1900	3170	5280
500	170	605	805	1205	1810	3015	5025
600	140	570	755	1135	1705	2840	4730
650	125	550	730	1100	1650	2745	4575
700	110	530	710	1060	1590	2655	4425
750	95	505	675	1015	1520	2535	4230
800	80	410	550	825	1235	2055	3430
850	65	320	425	640	955	1595	2655
900	50	230	305	460	690	1150	1915
950	35	135	185	275	410	685	1145
1000	20	85	115	170	255	430	715

NOTES:
(1) Upon prolonged exposure to temperatures above 800°F, the carbide phase of steel may be converted to graphite. Permissible, but not recommended for prolonged use above 800°F.
(2) Not to be used over 850°F.
(3) Not to be used over 700°F.
(4) Not to be used over 500°F.

TABLE 10-44*b* Pressure–Temperature Ratings for Group 1.5 Materials (Carbon, ½Mo Steel)

Nominal designation	Forgings	Castings	Plates
C–½Mo	A 182 Gr. F1 (1)		A 204 Gr. A (1) A 204 Gr. B (1)

Working pressures by classes, psig							
Class temp., °F	150	300	400	600	900	1500	2500
−20 to 100	265	695	930	1395	2090	3480	5805
200	260	695	930	1395	2090	3480	5805
300	230	685	915	1375	2060	3435	5725
400	200	660	885	1325	1985	3310	5520
500	170	640	855	1285	1925	3210	5350
600	140	605	805	1210	1815	3025	5040
650	125	590	785	1175	1765	2940	4905
700	110	570	755	1135	1705	2840	4730
750	95	530	710	1065	1595	2660	4430
800	80	510	675	1015	1525	2540	4230
850	65	485	650	975	1460	2435	4060
900	50	450	600	900	1350	2245	3745
950	35	280	375	560	845	1405	2345
1000	20	165	220	330	495	825	1370

NOTE:

(1) Upon prolonged exposure to temperatures above 875°F, the carbide phase of carbon-molybdenum steel may be converted to graphite. Permissible, but not recommended for prolonged use above 875°F.

TABLE 10-44*c* Pressure–Temperature Ratings for Group 2.1 Materials (Type 304 Stainless Steel)

Nominal designation	Forgings	Castings	Plates
18Cr–8Ni	A 182 Gr. F304 (1) A 182 Gr. F304H	A 351 Gr. CF3 (2) A 351 Gr. CF8 (1)	A 240 Gr. 304 (1) A 240 Gr. 304H

Working pressures by classes, psig							
Class temp., °F	150	300	400	600	900	1500	2500
−20 to 100	275	720	960	1440	2160	3600	6000
200	230	600	800	1200	1800	3000	5000
300	205	540	715	1075	1615	2690	4480
400	190	495	660	995	1490	2485	4140
500	170	465	620	930	1395	2330	3880
600	140	440	590	885	1325	2210	3680
650	125	430	575	865	1295	2160	3600
700	110	420	565	845	1265	2110	3520
750	95	415	550	825	1240	2065	3440
800	80	405	540	810	1215	2030	3380
850	65	395	530	790	1190	1980	3300
900	50	390	520	780	1165	1945	3240
950	35	380	510	765	1145	1910	3180
1000	20	355	470	710	1065	1770	2950
1050	. . .	325	435	650	975	1630	2715
1100	. . .	255	345	515	770	1285	2145
1150	. . .	205	275	410	615	1030	1715
1200	. . .	165	220	330	495	825	1370
1250	. . .	135	180	265	400	670	1115
1300	. . .	115	150	225	340	565	945
1350	. . .	95	125	185	280	465	770
1400	. . .	75	100	150	225	380	630
1450	. . .	60	80	115	175	290	485
1500	. . .	40	55	85	125	205	345

NOTES:

(1) At temperatures over 1000°F, use only when the carbon content is 0.04% or higher.

(2) Not to be used over 800°F.

TABLE 10-44*d* Pressure–Temperature Ratings for Group 2.2 Materials (Type 316 Stainless Steel)

Nominal designation	Forgings	Castings	Plates
16Cr–12Ni–2Mo	A 182 Gr. F316 (1)	A 351 Gr. CF3M (2)	A 240 Gr. 316 (1)
	A 182 Gr. F316H	A 351 Gr. CF8M (1)	A 240 Gr. 316H
18Cr–13Ni–3Mo	A 182 Gr. F317 (1)		A 240 Gr. 317 (1)
19Cr–10Ni–3Mo		A 351 Gr. CG8M (3)	

Class temp., °F	Working pressures by classes, psig 150	300	400	600	900	1500	2500
–20 to 100	275	720	960	1440	2160	3600	6000
200	235	620	825	1240	1860	3095	5160
300	215	560	745	1120	1680	2795	4660
400	195	515	685	1025	1540	2570	4280
500	170	480	635	955	1435	2390	3980
600	140	450	600	900	1355	2255	3760
650	125	440	590	885	1325	2210	3680
700	110	435	580	870	1305	2170	3620
750	95	425	570	855	1280	2135	3560
800	80	420	565	845	1265	2110	3520
850	65	420	555	835	1255	2090	3480
900	50	415	555	830	1245	2075	3460
950	35	385	515	775	1160	1930	3220
1000	20	365	485	725	1090	1820	3030
1050	. . .	360	480	720	1080	1800	3000
1100	. . .	305	405	610	915	1525	2545
1150	. . .	235	315	475	710	1185	1970
1200	. . .	185	245	370	555	925	1545
1250	. . .	145	195	295	440	735	1230
1300	. . .	115	155	235	350	585	970
1350	. . .	95	130	190	290	480	800
1400	. . .	75	100	150	225	380	630
1450	. . .	60	80	115	175	290	485
1500	. . .	40	55	85	125	205	345

NOTES:
(1) At temperatures over 1000°F, use only when the carbon content is 0.04% or higher.
(2) Not to be used over 850°F.
(3) Not to be used over 1000°F.

TABLE 10-44*e* Pressure–Temperature Ratings for Group 2.2 Materials (Type 304L and 316L Stainless Steel)

Nominal designation	Forgings	Castings	Plates
16Cr–12Ni–2Mo	A 182 Gr. F316L		A 240 Gr. 316L
18Cr–8Ni	A 182 Gr. F304L (1)		A 240 Gr. 304L (1)

Class temp., °F	Working pressures by classes, psig 150	300	400	600	900	1500	2500
–20 to 100	230	600	800	1200	1800	3000	5000
200	195	510	680	1020	1535	2555	4260
300	175	455	610	910	1370	2280	3800
400	160	420	560	840	1260	2100	3500
500	150	395	525	785	1180	1970	3280
600	140	370	495	745	1115	1860	3100
650	125	365	485	730	1095	1825	3040
700	110	360	480	720	1080	1800	3000
750	95	355	470	705	1060	1765	2940
800	80	345	460	690	1035	1730	2880
850	65	340	450	675	1015	1690	2820

NOTE:
(1) Not to be used over 800°F.

flanges, flanged fittings, and flanged valves, and has been extracted from ASME B16.5 with permission of the publisher, the American Society of Mechanical Engineers, New York. Only a few of the more common materials of construction of piping are reproduced here. See ASME B16.5 for other materials. Flanged joints, flanged valves in the open position, and flanged fittings may be subjected to system hydrostatic tests at a pressure not to exceed the hydrostatic-shell test pressure. Flanged valves in the closed position may be subjected to a system hydrostatic test at a pressure not to exceed 110 percent of the 100°F rating of the valve unless otherwise limited by the manufacturer.

Pressure-temperature ratings for soldered tubing joints are given in ASME B16.22—2001.

Components without Specific Ratings Components such as pipe and butt-welding fittings are generally furnished in nominal thicknesses. Fittings are rated for the same allowable pressures as pipe of the same nominal thickness and, along with pipe, are rated by the rules for pressure design and other provisions of the code.

Limits of Calculated Stresses due to Sustained Loads and Displacement Strains

1. *Internal pressure stresses.* Stresses due to internal pressure shall be considered safe when the wall thickness of the piping component, including any reinforcement, meets the requirements of the pressure design of components defined by the ASME B31.3 code.

2. *External pressure stresses.* Stresses due to external pressure shall be considered safe when the wall thickness of the piping component, and its means of stiffening meet the requirements of the pressure design of components defined by the ASME B31.3 code.

3. *Longitudinal stresses* S_L. The sum of longitudinal stresses S_L in any component in a piping system, due to sustained loads such as pressure and weight, shall not exceed the product $S_h W$, where S_h is the basic allowable stress at maximum metal temperature expected during the displacement cycle under analysis, and W is the weld joint strength reduction factor.

4. *Allowable displacement stress range* S_A. The computed displacement stress range S_E in a piping system shall not exceed the allowable displacement stress range S_A.

$$S_A = f(1.25S_c + 0.25S_h) \tag{10-93}$$

When S_h is greater than S_L, the difference between them may be added to the term $0.25S_h$ in Eq. (10-93). In that case, the allowable stress range is calculated by

$$S_A = f[1.25(S_c + S_h) - S_L] \tag{10-94}$$

where S_A = allowable displacement stress range
f = stress range factor (see Fig. 10-168)
S_c = basic allowable stress at minimum metal temperature expected during displacement cycle under analysis
S_h = basic allowable stress at maximum metal temperature expected during displacement cycle under analysis
S_L = longitudinal stresses, including pressure and weight

5. *Weld joint strength reduction factor W.* It is very important to include the weld joint strength reduction factor W in the design consideration. Especially, at elevated temperatures, the long-term strength of weld joints may be lower than the long-term strength of the base material. The weld joint strength reduction factor only applies at weld locations.

Pressure Design of Metallic Components External-pressure stress evaluation of piping is the same as for pressure vessels. But an important difference exists when one is establishing design pressure and wall thickness for internal pressure as a result of the ASME Boiler and Pressure Vessel Code's requirement that the relief-valve setting be not higher than the design pressure. For vessels this means

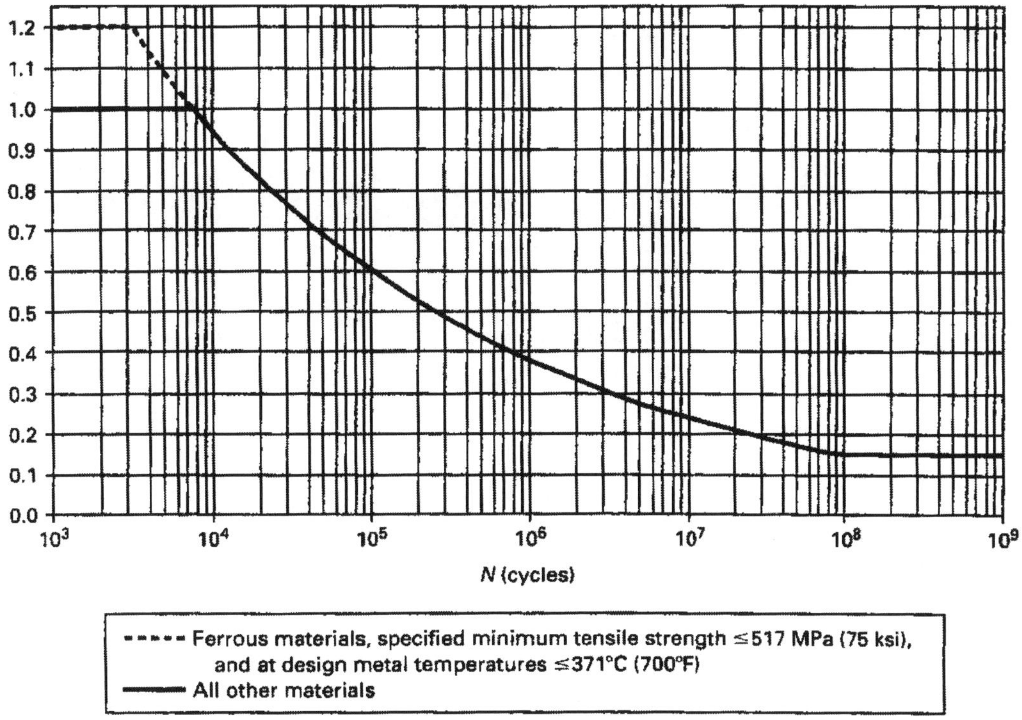

FIG. 10-168 Stress range reduction factor, *f*. (*Reproduced from ASME B31.3-2004 with permission of the publisher, the American Society of Mechanical Engineers, New York.*)

that the design is for a pressure 10 percent more or less above the intended maximum operating pressure to avoid popping or leakage from the valve during normal operation. However, on piping the design pressure and temperature are taken as the maximum intended operating pressure and coincident temperature combination which results in the maximum thickness. The temporary increased operating conditions listed under "Design Criteria" cover temporary operation at pressures that cause relief valves to leak or open fully. Allowable stresses for nearly 1000 materials are contained in the code. For convenience, the allowable stresses for commonly used carbon and stainless steels have been extracted from the code and listed in Table 10-44.

For **straight metal pipe under internal pressure** the formula for minimum required wall thickness t_m is applicable for:

1. $t < D/6$. The internal pressure design thickness for straight pipe shall be not less than that calculated in accordance with the equation below. The more conservative Barlow and Lamé equations may also be used. Equation (10-95) includes a factor Y varying with material and temperature to account for the redistribution of circumferential stress which occurs under steady-state creep at high temperature and permits slightly lesser thickness at this range.

$$t_m = \frac{PD_o}{2(SE + PY)} \div C \qquad (10\text{-}95)$$

where (in consistent units)

P = design pressure
D_o = outside diameter of pipe
C = sum of allowances for corrosion, erosion, and any thread or groove depth. For threaded components the depth is h of ASME B1.20.1, and for grooved components the depth is the depth removed (0.02 in when no tolerance is specified).
SE = allowable stress (see Table 10-44)
S = basic allowable stress for materials, excluding casting, joint, or structural-grade quality factors
E = quality factor. The quality factor E is one or the product of more than one of the following quality factors: casting quality factor E_c, joint quality factor E_j (see Fig. 10-169).

No.	Type of Joint		Type of Seam	Examination	Factor E_j
1	Furnace but weld, continuous weld		Straight	As required by listed specification	0.60 [Note (1)]
2	Electric resistance weld		Straight or spiral	As required by listed specification	0.85 [Note (1)]
3	Electric fusion weld				
	(a) Single butt weld		Straight or spiral	As required by listed specification or this code	0.80
	(with or without filler metal)			Additionally spot radiographed per para. 341.5.1, ASME B31.3	0.90
				Additionally 100% radiographed per para. 344.5.1 and Table 341.3.2, ASME B31.3	1.00
	(b) Double butt weld		Straight or spiral [except as provided in 4(a) below]	As required by listed specification or this code	0.85
	(with or without filler metal)	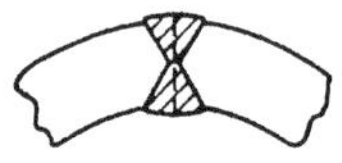		Additionally spot radiographed per para. 341.5.1, ASME B31.3	0.90
				Additionally 100% radiographed per para. 344.5.1 and Table 341.3.2, ASME B31.3	1.00
4	Per specific specification				
	(a) API 5L	Submerged arc weld (SAW) Gas metal arc weld (GMAW) Combined GMAW, SAW 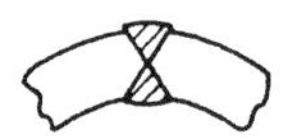	Straight with one or two seams Spiral	As required by specification	0.95

FIG. 10-169 Longitudinal weld joint quality factor, E_j. [NOTE (1): It is not permitted to increase the joint quality factor by additional examination for joint 1 or 2.] (*Reproduced from ASME B31.3 with permission of the publisher, the American Society of Mechanical Engineers, New York.*)

Y = coefficient having value in Table 10-45 valid for $t < D/6$ and for materials shown
t_m = minimum required thickness, in, including mechanical, corrosion, and erosion allowances

2. $t \geq D/6$ or $P/SE > 0.385$. Calculation of pressure design thickness for straight pipe requires special consideration of factors such as theory of failure, effects of fatigue, and thermal stress.

For flanges of nonstandard dimensions or for sizes beyond the scope of the approved standards, design shall be in accordance with the requirements of the ASME Boiler and Pressure Vessel Code, Sec. VIII, except that requirements for fabrication, assembly, inspection testing, and the pressure and temperature limits for materials of the Piping Code are to prevail. Countermoment flanges of flat face or otherwise providing a reaction outside the bolt circle are permitted if designed or tested in accordance with code requirements under pressure-containing components "not covered by standards and for which design formulas or procedures are not given."

Test Conditions The shell pressure test for flanged fittings shall be at a pressure no less than 1.5 times the 38°C (100°F) pressure rating rounded off to the next higher 1-bar (psi) increment.

In accordance with listed standards, **blind flanges** may be used at their pressure-temperature ratings. The minimum thickness of non-standard blind flanges shall be the same as for a bolted flat cover, in accordance with the rules of the ASME Boiler and Pressure Vessel Code, Sec. VIII.

Operational blanks shall be of the same thickness as blind flanges or may be calculated by the following formula (use consistent units):

$$t = d\sqrt{3P/16SE} + C \qquad (10\text{-}96)$$

where d = inside diameter of gasket for raised- or flat (plain)-face flanges, or the gasket pitch diameter for retained gasketed flanges
P = internal design pressure or external design pressure
S = applicable allowable stress
E = quality factor
C = sum of the mechanical allowances

Valves must comply with the applicable standards listed in Appendix E of the code and with the allowable pressure-temperature limits established thereby but not beyond the code-established service or materials limitations. Special valves must meet the same requirements as for countermoment flanges.

The code contains no specific rules for the design of **fittings** other than as branch openings. Ratings established by recognized standards are acceptable, however. ASME Standard B16.5 for steel-flanged fittings incorporates a 1.5 shape factor and thus requires the entire fitting to be 50 percent heavier than a simple cylinder in order to provide reinforcement for openings and/or general shape. ASME B16.9 for butt-welding fittings, on the other hand, requires only that the fittings be able to withstand the calculated bursting strength of the straight pipe with which they are to be used.

The thickness of **pipe bends** shall be determined as for straight pipe, provided the bending operation does not result in a difference between maximum and minimum diameters greater than 8 and 3 percent of the nominal outside diameter of the pipe for internal and external pressure, respectively.

TABLE 10-45 Values of Coefficient Y for $t < D/6$

	Temperature, °C (°F)					
Materials	≤482 (900 and lower)	510 (950)	538 (1000)	566 (1050)	593 (1100)	≥621 (1150 and up)
Ferritic steels	0.4	0.5	0.7	0.7	0.7	0.7
Austenitic steels	0.4	0.4	0.4	0.4	0.5	0.7
Other ductile metals	0.4	0.4	0.4	0.4	0.4	0.4
Cast iron	0.0	. . .	. . .	. . .	. . .	. . .

The maximum allowable internal pressure for multiple miter bends shall be the lesser value calculated from Eqs. (10-97) and (10-98). These equations are not applicable when θ exceeds 22.5°.

$$P_m = \frac{SEW(T-C)}{r_2} = \left[\frac{T-C}{(T-C) + 0.643\tan\theta\sqrt{r_2(T-C)}}\right] \qquad (10\text{-}97)$$

$$P_m = \frac{SEW(T-C)}{r_2} = \left(\frac{R_1 - r_2}{R_1 - 0.5r_2}\right) \qquad (10\text{-}98)$$

where nomenclature is the same as for straight pipe except as follows (see Fig. 10-170):

S = stress value for material
C = sum of mechanical allowances
r_2 = mean radius of pipe
R_1 = effective radius of miter bend, defined as the shortest distance from the pipe centerline to the intersection of the planes of adjacent miter joints
θ = angle of miter cut,°
α = angle of change in direction at miter joint $= 2\theta$,°
T = pipe wall thickness
W = weld joint strength reduction factor
E = quality factor

For compliance with the code, the value of R_1 shall not be less than that given by Eq. (10-99):

$$R_1 = A/\tan\theta + D/2 \qquad (10\text{-}99)$$

where A has the following empirical values:

t, in	A
≤ 0.5	1.0
$0.5 < t < 0.88$	$2(T - C)$
≥ 0.88	$[2(T - C)/3] + 1.17$

Piping branch connections involve the same considerations as pressure-vessel nozzles. However, outlet size in proportion to piping header size is unavoidably much greater for piping. The current Piping Code rules for calculation of branch-connection reinforcement are similar to those of the ASME Boiler and Pressure Vessel Code, Sec. VIII, Division I—2004 for a branch with axis at right angles to the header axis. If the branch connection makes an angle β with the header axis from 45 to 90°, the Piping Code requires that the area to be replaced be increased by dividing it by sin β. In such cases the half width of the reinforcing zone measured along the header axis is similarly increased, except that it may not exceed the outside diameter of the header. Some details of commonly used reinforced branch connections are given in Fig. 10-171.

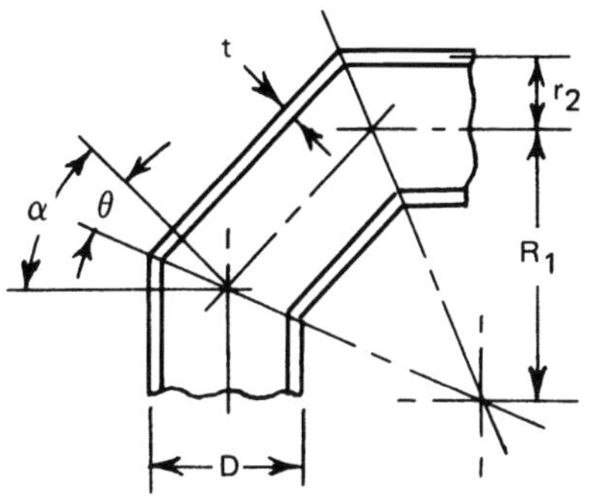

FIG. 10-170 Nomenclature for miter bends. (*Extracted from the Process Piping Code, ASME B31.3—2004, with permission of the publisher, the American Society of Mechanical Engineers, New York.*)

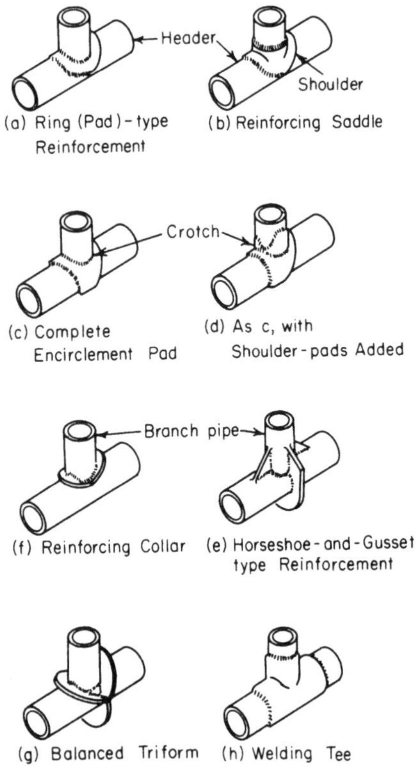

FIG. 10-171 Types of reinforcement for branch connections. (*From Kellogg,* Design of Piping Systems, *Wiley, New York, 1965.*)

The rules provide that a branch connection has adequate strength for pressure if a fitting (tee, lateral, or cross) is in accordance with an approved standard and is used within the pressure-temperature limitations or if the connection is made by welding a coupling or half coupling (wall thickness not less than the branch anywhere in reinforcement zone or less than extra heavy or 3000 lb) to the run and provided the ratio of branch to run diameters is not greater than one-fourth and that the branch is not greater than 2 in nominal diameter.

Dimensions of extra-heavy couplings are given in the *Steel Products Manual* published by the American Iron and Steel Institute. In ASME B16.11—2001, 2000-lb couplings were superseded by 3000-lb couplings.

ASME B31.3 states that the reinforcement area for resistance to external pressure is to be at least one-half of that required to resist internal pressure.

The code provides no guidance for analysis but requires that external and internal **attachments** be designed to avoid flattening of the pipe, excessive localized bending stresses, or harmful thermal gradients, with further emphasis on minimizing stress concentrations in cyclic service.

The code provides design requirements for **closures** which are flat, ellipsoidal, spherically dished, hemispherical, conical (without transition knuckles), conical convex to pressure, toriconical concave to pressure, and toriconical convex to pressure.

Openings in closures over 50 percent in diameter are designed as flanges in flat closures and as reducers in other closures. Openings of not over one-half of the diameter are to be reinforced as branch connections.

Thermal Expansion and Flexibility: Metallic Piping ASME B31.3 requires that piping systems have sufficient flexibility to prevent thermal expansion or contraction or the movement of piping supports or terminals from causing (1) failure of piping supports from overstress or fatigue; (2) leakage at joints; or (3) detrimental stresses or distortions in piping or in connected equipment (pumps, turbines, or valves, for example), resulting from excessive thrusts or movements in the piping.

To assure that a system meets these requirements, the computed displacement–stress range S_E shall not exceed the allowable stress range S_A [Eqs. (10-93) and (10-94)], the reaction forces R_m [Eq. (10-106)] shall not be detrimental to supports or connected equipment, and movement of the piping shall be within any prescribed limits.

Displacement Strains Strains result from piping being displaced from its unrestrained position:

1. *Thermal displacements.* A piping system will undergo dimensional changes with any change in temperature. If it is constrained from free movement by terminals, guides, and anchors, it will be displaced from its unrestrained position.
2. *Reaction displacements.* If the restraints are not considered rigid and there is a predictable movement of the restraint under load, this may be treated as a compensating displacement.
3. *Externally imposed displacements.* Externally caused movement of restraints will impose displacements on the piping in addition to those related to thermal effects. Such movements may result from causes such as wind sway or temperature changes in connected equipment.

Total Displacement Strains Thermal displacements, reaction displacements, and externally imposed displacements all have equivalent effects on the piping system and must be considered together in determining total displacement strains in a piping system.

Expansion strains may be taken up in three ways: by bending, by torsion, or by axial compression. In the first two cases maximum stress occurs at the extreme fibers of the cross section at the critical location. In the third case the entire cross-sectional area over the entire length is for practical purposes equally stressed.

Bending or torsional flexibility may be provided by bends, loops, or offsets; by corrugated pipe or expansion joints of the bellows type; or by other devices permitting rotational movement. These devices must be anchored or otherwise suitably connected to resist end forces from fluid pressure, frictional resistance to pipe movement, and other causes.

Axial flexibility may be provided by expansion joints of the slipjoint or bellows types, suitably anchored and guided to resist end forces from fluid pressure, frictional resistance to movement, and other causes.

Displacement Stresses Stresses may be considered proportional to the total displacement strain only if the strains are well distributed and not excessive at any point. The methods outlined here and in the code are applicable only to such a system. Poor distribution of strains (unbalanced systems) may result from:

1. Highly stressed small-size pipe runs in series with large and relatively stiff pipe runs
2. Local reduction in size or wall thickness or local use of a material having reduced yield strength (for example, girth welds of substantially lower strength than the base metal)
3. A line configuration in a system of uniform size in which expansion or contraction must be absorbed largely in a short offset from the major portion of the run

If unbalanced layouts cannot be avoided, appropriate analytical methods must be applied to assure adequate flexibility. If the designer determines that a piping system does not have adequate inherent flexibility, additional flexibility may be provided by adding bends, loops, offsets, swivel joints, corrugated pipe, expansion joints of the bellows or slip-joint type, or other devices. Suitable anchoring must be provided.

As contrasted with stress from sustained loads such as internal pressure or weight, displacement stresses may be permitted to cause limited overstrain in various portions of a piping system. When the system is operated initially at its greatest displacement condition, any yielding reduces stress. When the system is returned to its original condition, there occurs a redistribution of stresses which is referred to as self-springing. It is similar to cold springing in its effects.

While stresses resulting from thermal strain tend to diminish with time, the algebraic difference in displacement condition and in either the original (as-installed) condition or any anticipated condition with a greater opposite effect than the extreme displacement condition remains substantially constant during any one cycle of operation. This difference is defined as the displacement-stress range, and it is a determining factor in the design of piping for flexibility. See Eqs. (10-93) and (10-94) for the allowable stress range S_A and Eq. (10-101) for the computed stress range S_E.

Cold Spring Cold spring is the intentional deformation of piping during assembly to produce a desired initial displacement and stress. For pipe operating at a temperature higher than that at which it was installed, cold spring is accomplished by fabricating it slightly shorter than design length. Cold spring is beneficial in that it serves to balance the magnitude of stress under initial and extreme displacement conditions. When cold spring is properly applied, there is less likelihood of overstrain during initial operation; hence, it is recommended especially for piping materials of limited ductility. There is also less deviation from as-installed dimensions during initial operation, so that hangers will not be displaced as far from their original settings.

Inasmuch as the service life of a system is affected more by the range of stress variation than by the magnitude of stress at a given time, no credit for cold spring is permitted in stress-range calculations. However, in calculating the thrusts and moments when actual reactions as well as their range of variations are significant, credit is given for cold spring.

Values of thermal-expansion coefficients to be used in determining total displacement strains for computing the stress range are determined from Table 10-46 as the algebraic difference between the value at design maximum temperature and that at the design minimum temperature for the thermal cycle under analysis.

Values for Reactions Values of thermal displacements to be used in determining total displacement strains for the computation of reactions on supports and connected equipment shall be determined as the algebraic difference between the value at design maximum (or minimum) temperature for the thermal cycle under analysis and the value at the temperature expected during installation.

The as-installed and maximum or minimum moduli of elasticity, E_a and E_m respectively, shall be taken as the values shown in Table 10-47.

Poisson's ratio may be taken as 0.3 at all temperatures for all metals.

The allowable stress range for displacement stresses S_A and permissible additive stresses shall be as specified in Eqs. (10-93) and (10-94) for systems primarily stressed in bending and/or torsion. For pipe or piping components containing longitudinal welds the basic allowable stress S may be used to determine S_A.

Nominal thicknesses and outside diameters of pipe and fittings shall be used in flexibility calculations.

In the absence of more directly applicable data, the flexibility factor k and stress-intensification factor i shown in Table 10-48 may be used in flexibility calculations in Eq. (10-102). For piping components or attachments (such as valves, strainers, anchor rings, and bands) not covered in the table, suitable stress-intensification factors may be assumed by comparison of their significant geometry with that of the components shown.

Requirements for Analysis No formal analysis of adequate flexibility is required in systems which (1) are duplicates of successfully operating installations or replacements without significant change of systems with a satisfactory service record; (2) can readily be judged adequate by comparison with previously analyzed systems; or (3) are of uniform size, have no more than two points of fixation, have no intermediate restraints, and fall within the limitations of empirical Eq. (10-100):*

$$\frac{Dy}{(L-U)^2} \leq K_1 \qquad (10\text{-}100)$$

where D = ouside diameter of pipe, in (mm)
y = resultant of total displacement strains, in (mm), to be absorbed by the piping system
L = developed length of piping between anchors, ft (m)
U = anchor distance, straight line between anchors, ft (m)
K_1 = 0.03 for U.S. customary units listed
= 208.3 for SI units listed in parentheses

*WARNING: No general proof can be offered that this equation will yield accurate or consistently conservative results. It is not applicable to systems used under severe cyclic conditions. It should be used with caution in configurations such as unequal leg U bends ($L/U > 2.5$) or near-straight sawtooth runs, or for large thin-wall pipe ($i \geq 5$), or when extraneous displacements (not in the direction connecting anchor points) constitute a large part of the total displacement. There is no assurance that terminal reactions will be acceptably low even if a piping system falls within the limitations of Eq. (10-100).

1. All systems not meeting these criteria shall be analyzed by simplified, approximate, or comprehensive methods of analysis appropriate for the specific case.
2. Approximate or simplified methods may be applied only if they are used in the range of configurations for which their adequacy has been demonstrated.
3. Acceptable comprehensive methods of analysis include analytical and chart methods which provide an evaluation of the forces, moments, and stresses caused by displacement strains.
4. Comprehensive analysis shall take into account stress-intensification factors for any component other than straight pipe. Credit may be taken for the extra flexibility of such a component.

In calculating the flexibility of a piping system between anchor points, the system shall be treated as a whole. The significance of all parts of the line and of all restraints introduced for the purpose of reducing moments and forces on equipment or small branch lines and also the restraint introduced by support friction shall be recognized. Consider all displacements over the temperature range defined by operating and shutdown conditions.

Flexibility Stresses Bending and torsional stresses shall be computed using the as-installed modulus of elasticity E_a and then combined in accordance with Eq. (10-101) to determine the computed displacement stress range S_E, which shall not exceed the allowable stress range S_A [Eqs. (10-93) and (10-94)]:

$$S_E = \sqrt{S_b^2 + 4S_t^2} \qquad (10\text{-}101)$$

where S_b = resultant bending stress, lbf/in² (MPa)
$S_t = M_t/2Z$ = torsional stress, lbf/in² (MPa)
M_t = torsional moment, in·lbf (N·mm)
Z = section modulus of pipe, in³ (mm³)

The resultant bending stresses S_b to be used in Eq. (10-101) for elbows and miter bends shall be calculated in accordance with Eq. (10-102), with moments as shown in Fig. (10-172):

$$S_b = \frac{\sqrt{(i_iM_i)^2 + (i_oM_o)^2}}{Z} \qquad (10\text{-}102)$$

where S_b = resultant bending stress, lbf/in² (MPa)
i_i = in-plane stress-intensification factor from Table 10-48
i_o = out-plane stress-intensification factor from Table 10-48
M_i = in-plane bending moment, in·lbf (N·mm)
M_o = out-plane bending moment, in·lbf (N·mm)
Z = section modulus of pipe, in³ (mm³)

The resultant bending stresses S_b to be used in Eq. (10-101) for branch connections shall be calculated in accordance with Eqs. (10-103) and (10-104), with moments as shown in Fig. 10-173.

For header (legs 1 and 2):

$$S_b = \frac{\sqrt{(i_iM_i)^2 + (i_oM_o)^2}}{Z} \qquad (10\text{-}103)$$

For branch (leg 3):

$$S_b = \frac{\sqrt{(i_iM_i)^2 + (i_oM_o)^2}}{Z_e} \qquad (10\text{-}104)$$

where S_b = resultant bending stress, lbf/in² (MPa)
Z_e = effective section modulus for branch, in³ (mm³)

$$Z_e = \pi r_2^2 T_S \qquad (10\text{-}105)$$

r_2 = mean branch cross-sectional radius, in (mm)
T_s = effective branch wall thickness, in (mm) [lesser of $\overline{T}_h$ and $(i_o)(\overline{T}_b)$]
$\overline{T}_h$ = thickness of pipe matching run of tee or header exclusive of reinforcing elements, in (mm)

TABLE 10-46 Thermal Coefficients, U.S. Units, for Metals

Mean coefficient of linear thermal expansion between 70°F and indicated temperature, μin/(in·°F)

	Material															
Temp., °F	Carbon steel carbon–moly–low–chrome (through 3Cr–Mo)	5Cr–Mo through 9Cr–Mo	Austenitic stainless steels 18Cr–8Ni	12Cr, 17Cr, 27Cr	25Cr–20Ni	UNS N04400 Monel 67Ni–30Cu	3½Ni	Copper and copper alloys	Aluminum	Gray cast iron	Bronze	Brass	70Cu–30Ni	UNS N08XXX series Ni–Fe–Cr	UNS N06XXX series Ni–Cr–Fe	Ductile iron
−450	. . .	. . .	. . .	. . .	. . .	. . .	. . .	6.30	. . .	. . .	. . .	. . .	. . .	. . .	. . .	. . .
−425	. . .	. . .	. . .	. . .	. . .	. . .	. . .	6.61	. . .	. . .	. . .	. . .	. . .	. . .	. . .	. . .
−400	. . .	. . .	. . .	. . .	. . .	. . .	. . .	6.93	. . .	. . .	. . .	. . .	. . .	. . .	. . .	. . .
−375	. . .	. . .	. . .	. . .	. . .	. . .	. . .	7.24	. . .	. . .	. . .	. . .	. . .	. . .	. . .	. . .
−350	. . .	. . .	. . .	. . .	. . .	. . .	. . .	7.51	. . .	. . .	. . .	. . .	. . .	. . .	. . .	. . .
−325	5.00	4.70	8.15	4.30	. . .	5.55	4.76	7.74	9.90	. . .	8.40	8.20	6.65	. . .	. . .	. . .
−300	5.07	4.77	8.21	4.36	. . .	5.72	4.90	7.94	10.04	. . .	8.45	8.24	6.76	. . .	. . .	. . .
−275	5.14	4.84	8.28	4.41	. . .	5.89	5.01	8.11	10.18	. . .	8.50	8.29	6.86	. . .	. . .	. . .
−250	5.21	4.91	8.34	4.47	. . .	6.06	5.15	8.26	10.33	. . .	8.55	8.33	6.97	. . .	. . .	. . .
−225	5.28	4.98	8.41	4.53	. . .	6.23	5.30	8.40	10.47	. . .	8.60	8.37	7.08	. . .	. . .	. . .
−200	5.35	5.05	8.47	4.59	. . .	6.40	5.45	8.51	10.61	. . .	8.65	8.41	7.19	. . .	. . .	4.65
−175	5.42	5.12	8.54	4.64	. . .	6.57	5.52	8.62	10.76	. . .	8.70	8.46	7.29	. . .	. . .	4.76
−150	5.50	5.20	8.60	4.70	. . .	6.75	5.59	8.72	10.90	. . .	8.75	8.50	7.40	. . .	. . .	4.87
−125	5.57	5.26	8.66	4.78	. . .	6.85	5.67	8.81	11.08	. . .	8.85	8.61	7.50	. . .	. . .	4.98
−100	5.65	5.32	8.75	4.85	. . .	6.95	5.78	8.89	11.25	. . .	8.95	8.73	7.60	. . .	. . .	5.10
−75	5.72	5.38	8.83	4.93	. . .	7.05	5.83	8.97	11.43	. . .	9.05	8.84	7.70	. . .	. . .	5.20
−50	5.80	5.45	8.90	5.00	. . .	7.15	5.88	9.04	11.60	. . .	9.15	8.95	7.80	. . .	. . .	5.30
−25	5.85	5.51	8.94	5.05	. . .	7.22	5.94	9.11	11.73	. . .	9.23	9.03	7.87	. . .	. . .	5.40
0	5.90	5.56	8.98	5.10	. . .	7.28	6.00	9.17	11.86	. . .	9.32	9.11	7.94	. . .	. . .	5.50
25	5.96	5.62	9.03	5.14	. . .	7.35	6.08	9.23	11.99	. . .	9.40	9.18	8.02	. . .	. . .	5.58
50	6.01	5.67	9.07	5.19	. . .	7.41	6.16	9.28	12.12	. . .	9.49	9.26	8.09	. . .	. . .	5.66
70	6.07	5.73	9.11	5.24	. . .	7.48	6.25	9.32	12.25	. . .	9.57	9.34	8.16	. . .	7.13	5.74
100	6.13	5.79	9.16	5.29	. . .	7.55	6.33	9.39	12.39	. . .	9.66	9.42	8.24	. . .	7.20	5.82
125	6.19	5.85	9.20	5.34	. . .	7.62	6.36	9.43	12.53	. . .	9.75	9.51	8.31	. . .	7.25	5.87
150	6.25	5.92	9.25	5.40	. . .	7.70	6.39	9.48	12.67	. . .	9.85	9.59	8.39	. . .	7.30	5.92
175	6.31	5.98	9.29	5.45	. . .	7.77	6.42	9.52	12.81	. . .	9.93	9.68	8.46	. . .	7.35	5.97
200	6.38	6.04	9.34	5.50	8.79	7.84	6.45	9.56	12.95	5.75	10.03	9.76	8.54	7.90	7.40	6.02
225	6.43	6.08	9.37	5.54	8.81	7.89	6.50	9.60	13.03	5.80	10.05	9.82	8.58	8.01	7.44	6.08
250	6.49	6.12	9.41	5.58	8.83	7.93	6.55	9.64	13.12	5.84	10.08	9.88	8.63	8.12	7.48	6.14
275	6.54	6.15	9.44	5.62	8.85	7.98	6.60	9.68	13.20	5.89	10.10	9.94	8.67	8.24	7.52	6.20
300	6.60	6.19	9.47	5.66	8.87	8.02	6.65	9.71	13.28	5.93	10.12	10.00	8.71	8.35	7.56	6.25
325	6.65	6.23	9.50	5.70	8.89	8.07	6.69	9.74	13.36	5.97	10.15	10.06	8.76	8.46	7.60	6.31
350	6.71	6.27	9.53	5.74	8.90	8.11	6.73	9.78	13.44	6.02	10.18	10.11	8.81	8.57	7.63	6.37
375	6.76	6.30	9.56	5.77	8.91	8.16	6.77	9.81	13.52	6.06	10.20	10.17	8.85	8.69	7.67	6.43
400	6.82	6.34	9.59	5.81	8.92	8.20	6.80	9.84	13.60	6.10	10.23	10.23	8.90	8.80	7.70	6.48
425	6.87	6.38	9.62	5.85	8.92	8.25	6.83	9.86	13.68	6.15	10.25	10.29	. . .	8.82	7.72	6.57
450	6.92	6.42	9.65	5.89	8.92	8.30	6.86	9.89	13.75	6.19	10.28	10.35	. . .	8.85	7.75	6.66
475	6.97	6.46	9.67	5.92	8.92	8.35	6.89	9.92	13.83	6.24	10.30	10.41	. . .	8.87	7.77	6.75
500	7.02	6.50	9.70	5.96	8.93	8.40	6.93	9.94	13.90	6.28	10.32	10.47	. . .	8.90	7.80	6.85
525	7.07	6.54	9.73	6.00	8.93	8.45	6.97	9.97	13.98	6.33	10.35	10.53	. . .	8.92	7.82	6.88
550	7.12	6.58	9.76	6.05	8.93	8.49	7.01	9.99	14.05	6.38	10.38	10.58	. . .	8.95	7.85	6.92
575	7.17	6.62	9.79	6.09	8.93	8.54	7.04	10.1	14.13	6.42	10.41	10.64	. . .	8.97	7.88	6.95
600	7.23	6.66	9.82	6.13	8.94	8.58	7.08	10.04	14.20	6.47	10.44	10.69	. . .	9.00	7.90	6.98
625	7.28	6.70	9.85	6.17	8.94	8.63	7.12	. . .	. . .	6.52	10.46	10.75	. . .	9.02	7.92	7.02
650	7.33	6.73	9.87	6.20	8.95	8.68	7.16	. . .	. . .	6.56	10.48	10.81	. . .	9.05	7.95	7.04
675	7.38	6.77	9.90	6.23	8.95	8.73	7.19	. . .	. . .	6.61	10.50	10.86	. . .	9.07	7.98	7.08
700	7.44	6.80	9.92	6.26	8.96	8.78	7.22	. . .	. . .	6.65	10.52	10.92	. . .	9.10	8.00	7.11
725	7.49	6.84	9.95	6.29	8.96	8.83	7.25	. . .	. . .	6.70	10.55	10.98	. . .	9.12	8.02	7.14
750	7.54	6.88	9.99	6.33	8.96	8.87	7.29	. . .	. . .	6.74	10.57	11.04	. . .	9.15	8.05	7.18
775	7.59	6.92	10.02	6.36	8.96	8.92	7.31	. . .	. . .	6.79	10.60	11.10	. . .	9.17	8.08	7.22
800	7.65	6.96	10.05	6.39	8.97	8.96	7.34	. . .	. . .	6.83	10.62	11.16	. . .	9.20	8.10	7.25
825	7.70	7.00	10.08	6.42	8.97	9.01	7.37	. . .	. . .	6.87	10.65	11.22	. . .	9.22	. . .	7.27

850	7.75	7.03	10.11	6.46	8.98	9.06	7.40	...	...	6.92	10.67	11.28	...	9.25	...	7.31
875	7.79	7.07	10.13	6.49	8.99	9.11	7.43	...	...	6.96	10.70	11.34	...	9.27	...	7.34
900	7.84	7.10	10.16	6.52	9.00	9.16	7.45	...	...	7.00	10.72	11.40	...	9.30	...	7.37
925	7.87	7.13	10.19	6.55	9.05	9.21	7.47	...	...	7.05	10.74	11.46	...	9.32	...	7.41
950	7.91	7.16	10.23	6.58	9.10	9.25	7.49	...	...	7.10	10.76	11.52	...	9.35	...	7.44
975	7.94	7.19	10.26	6.60	9.15	9.30	7.52	...	...	7.14	10.78	11.57	...	9.37	...	7.47
1000	7.97	7.22	10.29	6.63	9.18	9.34	7.55	...	...	7.19	10.80	11.63	...	9.40	...	7.50
1025	8.01	7.25	10.32	6.65	9.20	9.39	...	...	...	...	10.83	11.69	...	9.42	...	...
1050	8.05	7.27	10.34	6.68	9.22	9.43	...	...	...	...	10.85	11.74	...	9.45	...	...
1075	8.08	7.30	10.37	6.70	9.24	9.48	...	...	...	...	10.88	11.80	...	9.47	...	...
1100	8.12	7.32	10.39	6.72	9.25	9.52	...	...	...	...	10.90	11.85	...	9.50	...	...
1125	8.14	7.34	10.41	6.74	9.29	9.57	...	...	...	...	10.93	11.91	...	9.52	...	...
1150	8.16	7.37	10.44	6.75	9.33	9.61	...	...	...	...	10.95	11.97	...	9.55	...	...
1175	8.17	7.39	10.46	6.77	9.36	9.66	...	...	...	...	10.98	12.03	...	9.57	...	...
1200	8.19	7.41	10.48	6.78	9.39	9.70	...	...	...	...	11.00	12.09	...	9.60	...	...
1225	8.21	7.43	10.50	6.80	9.43	9.75	...	...	...	...	...	...	...	9.64	...	...
1250	8.24	7.45	10.51	6.82	9.47	9.79	...	...	...	...	...	...	...	9.68	...	...
1275	8.26	7.47	10.53	6.83	9.50	9.84	...	...	...	...	...	...	...	9.71	...	...
1300	8.28	7.49	10.54	6.85	9.53	9.88	...	...	...	...	...	...	...	9.75	...	...
1325	8.30	7.51	10.56	6.86	9.53	9.92	...	...	...	...	...	...	...	9.79	...	...
1350	8.32	7.52	10.57	6.88	9.54	9.96	...	...	...	...	...	...	...	9.83	...	...
1375	8.34	7.54	10.59	6.89	9.55	10.00	...	...	...	...	...	...	...	9.86	...	...
1400	8.36	7.55	10.60	6.90	9.56	10.04	...	...	...	...	...	...	...	9.90	...	...
1425	...	...	10.64	...	...	...	...	...	...	...	...	...	...	9.94	...	...
1450	...	...	10.68	...	...	...	...	...	...	...	...	...	...	9.98	...	...
1475	...	...	10.72	...	...	...	...	...	...	...	...	...	...	10.01	...	...
1500	...	...	10.77	...	...	...	...	...	...	...	...	...	...	10.05	...	...

GENERAL NOTE: For Code references to this table, see para. 319.3.1, ASME B31.3. These data are for use in the absence of more applicable data. It is the designer's responsibility to verify that materials are suitable for the intended service at the temperatures shown.

TABLE 10-47 Modulus of Elasticity, U.S. Units, for Metals

Material	E = modulus of elasticity, Msi (millions of psi), at temperature, °F																				
	−425	−400	−350	−325	−200	−100	70	200	300	400	500	600	700	800	900	1000	1100	1200	1300	1400	1500
Ferrous Metals																					
Gray cast iron	. . .	. . .	. . .	. . .	. . .	. . .	13.4	13.2	12.9	12.6	12.2	11.7	11.0	10.2	. . .	. . .	. . .	. . .	. . .	. . .	. . .
Carbon steels, $C \leq 0.3\%$	31.9	. . .	. . .	31.4	30.8	30.2	29.5	28.8	28.3	27.7	27.3	26.7	25.5	24.2	22.4	20.4	18.0	. . .	. . .	. . .	. . .
Carbon steels, $C \leq 0.3\%$	31.7	. . .	. . .	31.2	30.6	30.0	29.3	28.6	28.1	27.5	27.1	26.5	25.3	24.0	22.2	20.2	17.9	15.4	. . .	. . .	. . .
Carbon-moly steels	31.7	. . .	. . .	31.1	30.5	29.9	29.2	28.5	28.0	27.4	27.0	26.4	25.3	23.9	22.2	20.1	17.8	15.3	. . .	. . .	. . .
Nickel steels, Ni 2%–9%	30.1	. . .	. . .	29.6	29.1	28.5	27.8	27.1	26.7	26.1	25.7	25.2	24.6	23.0	. . .	. . .	. . .	. . .	. . .	. . .	. . .
Cr-Mo steels, Cr ½%–2%	32.1	. . .	. . .	31.6	31.0	30.4	29.7	29.0	28.5	27.9	27.5	26.9	26.3	25.5	24.8	23.9	23.0	21.8	20.5	18.9	. . .
Cr-Mo steels, Cr 2¼%–3%	33.1	. . .	. . .	32.6	32.0	31.4	30.6	29.8	29.4	28.8	28.3	27.7	27.1	26.3	25.6	24.6	23.7	22.5	21.1	19.4	. . .
Cr-Mo steels, Cr 5%–9%	33.4	. . .	. . .	32.9	32.3	31.7	30.9	30.1	29.7	29.0	28.6	28.0	27.3	26.1	24.7	22.7	20.4	18.2	15.5	12.7	. . .
Chromium steels, Cr 12%, 17%, 27%	31.8	. . .	. . .	31.2	30.7	30.1	29.2	28.5	27.9	27.3	26.7	26.1	25.6	24.7	22.2	21.5	19.1	16.6	. . .	. . .	. . .
Austenitic steels (TP304, 310, 316, 321, 347)	30.8	. . .	. . .	30.3	29.7	29.0	28.3	27.6	27.0	26.5	25.8	25.3	24.8	24.1	23.5	22.8	22.1	21.2	20.2	19.2	18.1
Copper and Copper Alloys (UNS Nos.)																					
Comp. and leaded Sn-bronze (C83600, C92200)	. . .	. . .	. . .	14.8	14.6	14.4	14.0	13.7	13.4	13.2											
Naval brass, Si- & Al-bronze (C46400, C65500, C95200, C95400)	. . .	. . .	. . .	15.9	15.6	15.4	15.0	14.6	14.4	14.1	12.9	12.5	12.0	. . .	. . .	. . .	. . .	. . .	. . .	. . .	. . .
Copper (C11000)	. . .	. . .	. . .	16.9	16.6	16.5	16.0	15.6	15.4	15.0	13.8	13.4	12.8	. . .	. . .	. . .	. . .	. . .	. . .	. . .	. . .
Copper, red brass, Al-bronze (C10200, C12000, C12200, C12500, C14200, C23000, C61400)	. . .	. . .	. . .	18.0	17.7	17.5	17.0	16.6	16.3	16.0	14.7	14.2	13.7	. . .	. . .	. . .	. . .	. . .	. . .	. . .	. . .
90Cu 10Ni (C70600)	. . .	. . .	. . .	19.0	18.7	18.5	18.0	17.6	17.3	16.9	15.6	15.1	14.5	. . .	. . .	. . .	. . .	. . .	. . .	. . .	. . .
Leaded Ni-bronze	. . .	. . .	. . .	20.1	19.8	19.6	19.0	18.5	18.2	17.9	16.6	16.0	15.4	. . .	. . .	. . .	. . .	. . .	. . .	. . .	. . .
80Cu-20Ni (C71000)	. . .	. . .	. . .	21.2	20.8	20.6	20.0	19.5	19.2	18.8	17.5	16.9	16.2	. . .	. . .	. . .	. . .	. . .	. . .	. . .	. . .
70Cu-30Ni (C71500)	. . .	. . .	. . .	23.3	22.9	22.7	22.0	21.5	21.1	20.7	18.4	17.8	17.1	. . .	. . .	. . .	. . .	. . .	. . .	. . .	. . .
											20.2	19.6	18.8	. . .	. . .	. . .	. . .	. . .	. . .	. . .	. . .
Nickel and Nickel Alloys (UNS Nos.)																					
Monel 400 N04400	28.3	. . .	. . .	27.8	27.3	26.8	26.0	25.4	25.0	24.7	24.3	24.1	23.7	23.1	22.6	22.1	21.7	21.2	. . .	. . .	. . .
Alloys N06007, N08320	30.3	. . .	. . .	29.5	29.2	28.6	27.8	27.1	26.7	26.4	26.0	25.7	25.3	24.7	24.2	23.6	23.2	22.7	. . .	. . .	. . .
Alloys N08800, N08810, N06002	31.1	. . .	. . .	30.5	29.9	29.4	28.5	27.8	27.4	27.1	26.6	26.4	25.9	25.4	24.8	24.2	23.8	23.2	. . .	. . .	. . .
Alloys N06455, N10276	32.5	. . .	. . .	31.6	31.3	30.6	29.8	29.1	28.6	28.3	27.9	27.6	27.1	26.5	25.9	25.3	24.9	24.3	. . .	. . .	. . .
Alloys N02200, N02201, N06625	32.7	. . .	. . .	32.1	31.5	30.9	30.0	29.3	28.8	28.5	28.1	27.8	27.3	26.7	26.1	25.5	25.1	24.5	. . .	. . .	. . .
Alloy N06600	33.8	. . .	. . .	33.2	32.6	31.9	31.0	30.2	29.9	29.5	29.0	28.7	28.2	27.6	27.0	26.4	25.9	25.3	. . .	. . .	. . .
Alloy N10001	33.9	. . .	. . .	33.3	32.7	32.0	31.1	30.3	29.9	29.5	29.1	28.8	28.3	27.7	27.1	26.4	26.0	25.3	. . .	. . .	. . .
Alloy N10665	34.2	. . .	. . .	33.3	33.0	32.3	31.4	30.6	30.1	29.8	29.4	29.0	28.6	27.9	27.3	26.7	26.2	25.6	. . .	. . .	. . .
Unalloyed Titanium																					
Grades 1, 2, 3, and 7	. . .	. . .	. . .	. . .	. . .	. . .	15.5	15.0	14.6	14.0	13.3	12.6	11.9	11.2	. . .	. . .	. . .	. . .	. . .	. . .	. . .
Aluminum and Aluminum Alloys (UNS Nos.)																					
Grades 443, 1060, 1100, 3003, 3004, 6061, 6063 (A24430, A91060, A91100, A93003, A93004, A96061, A96063)	11.4	. . .	. . .	11.1	10.8	10.5	10.0	9.6	9.2	8.7											
Grades 5052, 5154, 5454, 5652 (A95052, A95154, A95454, A95652)	11.6	. . .	. . .	11.3	11.0	10.7	10.2	9.7	9.4	8.9											
Grades 356, 5083, 5086, 5456 (A03560, A95083, A95086, A95456)	11.7	. . .	. . .	11.4	11.1	10.8	10.3	9.8	9.5	9.0											

TABLE 10-48 Flexibility Factor *k* and Stress Intensification Factor *i*

Description	Flexibility factor k	Stress intensification factor [Notes (2), (3)] Out-of-plane, i_o	In-plane, i_i	Flexibility characteristic h	Sketch
Welding elbow or pipe bend [Notes (2)], (4)-(7)]	$\frac{1.65}{h}$	$\frac{0.75}{h^{2/3}}$	$\frac{0.9}{h^{2/3}}$	$\frac{\bar{T}R_1}{r_2^2}$	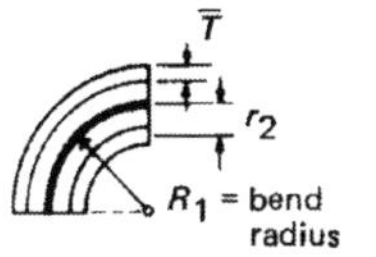
Closely spaced miter bend $s < r_2 (1 + \tan\theta)$ [Notes (2), (4), (5), (7)]	$\frac{1.52}{h^{5/6}}$	$\frac{0.9}{h^{2/3}}$	$\frac{0.9}{h^{2/3}}$	$\frac{\cot\theta}{2}\left(\frac{s\bar{T}}{r_2^2}\right)$	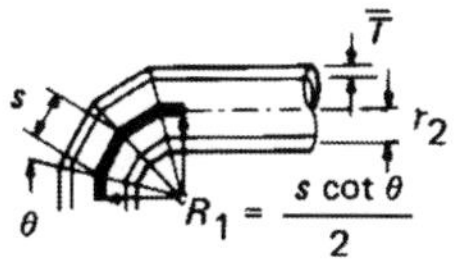
Single miter bend or widely spaced miter bend $s \geq r_2 (1 + \tan\theta)$ [Notes (2), (4), (7)]	$\frac{1.52}{h^{5/6}}$	$\frac{0.9}{h^{2/3}}$	$\frac{0.9}{h^{2/3}}$	$\frac{1+\cot\theta}{2}\left(\frac{\bar{T}}{r_2}\right)$	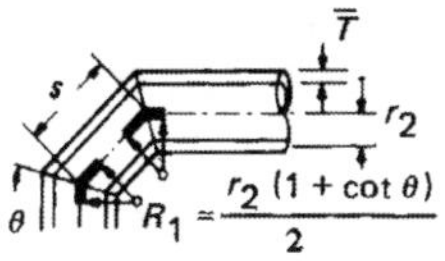
Welding tee per ASME B16.9 [Notes (2), (4), (6), (11), (13)]	1	$\frac{0.9}{h^{2/3}}$	$\frac{3}{4} i_o + \frac{1}{4}$	$3.1 \frac{\bar{T}}{r_2}$	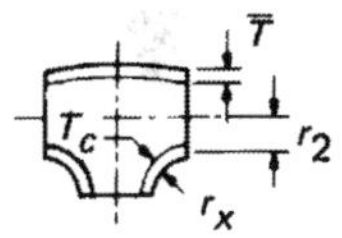
Reinforced fabricated tee with pad or saddle [Notes (2), (4), (8), (12), (13)]	1	$\frac{0.9}{h^{2/3}}$	$\frac{3}{4} i_o + \frac{1}{4}$	$\frac{(\bar{T} + \frac{1}{2}\bar{T}_r)^{2.5}}{\bar{T}^{1.5} r_2}$	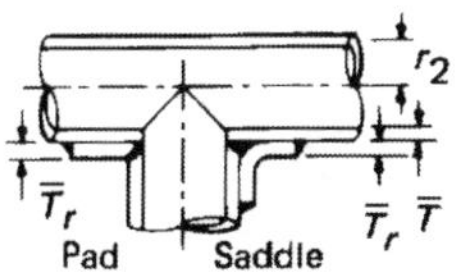
Unreinforced fabricated tee [Notes (2), (4), (12), (13)]	1	$\frac{0.9}{h^{2/3}}$	$\frac{3}{4} i_o + \frac{1}{4}$	$\frac{\bar{T}}{r_2}$	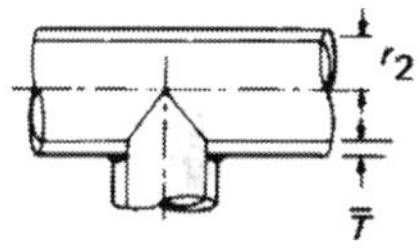
Extruded welding tee with $r_x \geq 0.05 D_b$ $T_c < 1.5 \bar{T}$ [Notes (2), (4), (13)]	1	$\frac{0.9}{h^{2/3}}$	$\frac{3}{4} i_c + \frac{1}{4}$	$\left(1 + \frac{r_x}{r_2}\right)\frac{\bar{T}}{r_2}$	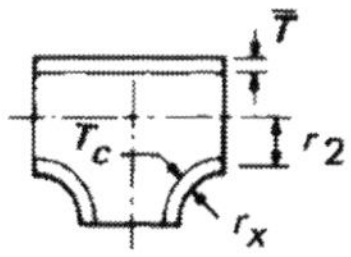
Welded-in contour insert [Notes (2), (4), (11), (13)]	1	$\frac{0.9}{h^{2/3}}$	$\frac{3}{4} i_c + \frac{1}{4}$	$3.1 \frac{\bar{T}}{r_2}$	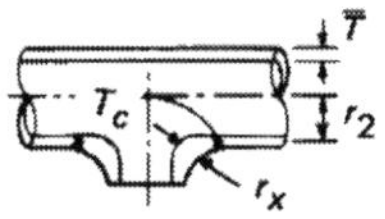
Branch welded-on fitting (integrally reinforced) [Notes (2), (4), (9), (12)]	1	$\frac{0.9}{h^{2/3}}$	$\frac{0.9}{h^{2/3}}$	$3.3 \frac{\bar{T}}{r_2}$	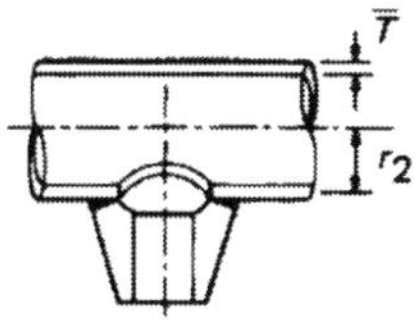

TABLE 10-48 Flexibility Factor *k* and Stress-Intensification Factor *i* *(Concluded)*

Description	Flexibility factor k	Stress intensification factor i [Note (1)]
Butt-welded joint, reducer, or weld-neck flange	1	1.0
Double-welded slip-on flange	1	1.2
Fillet welded joint, or socket weld flange or fitting	1	Note (14)
Lap joint flange (with ASME B16.9 lap joint stub)	1	1.6
Threaded pipe joint or threaded flange	1	2.3
Corrugated straight pipe, or corrugated or creased bend [Note (10)]	5	2.5

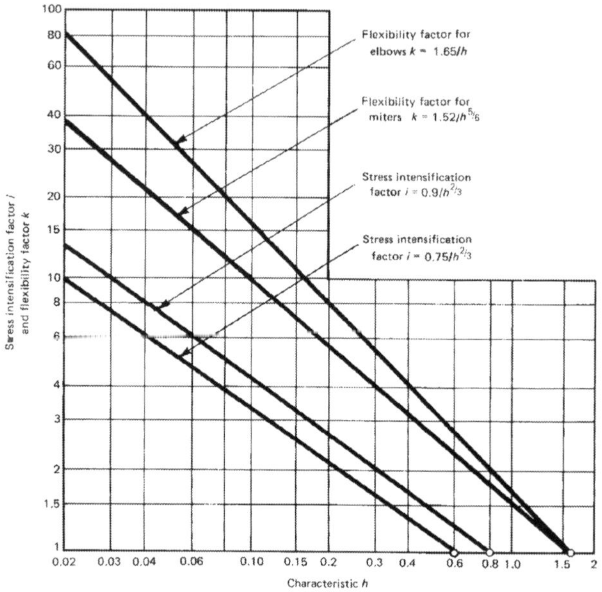

Chart A

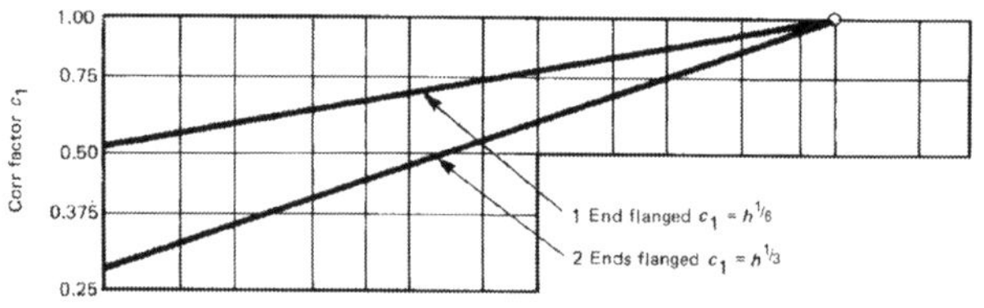

Chart B

NOTES:

(1) Stress intensification and flexibility factor data in this table are for use in the absence of more directly applicable data (see para. 319.3.6). Their validity has been demonstrated for $D/\overline{T} \leq 100$.

(2) The flexibility factor, k, in this table applies to bending in any plane. The flexibility factors, k, and stress intensification factors, i, shall not be less than unity; factors for torsion equal unity. Both factors apply over the effective arc length (shown by heavy centerlines in the sketches) for curved and miter bends, and to the intersection point for tees.

(3) A single intensification factor equal to $0.9/h^{2/3}$ may be used for both i_i and i_o if desired.

(4) The values of k and i can be read directly from Chart A by entering with the characteristic h computed from the formulas given above. Nomenclature is as follows:

D_b = outside diameter of branch
R_1 = bend radius of welding elbow or pipe bend
r_x = see definition in para. 304.3.4(c) and Table 10-48
r_2 = mean radius of matching pipe
s = miter spacing at centerline
$\overline{T}$ = for elbows and miter bends, the nominal wall thickness of the fitting,
= for tees, the nominal wall thickness of the matching pipe
T_c = crotch thickness of branch connections measured at the center of the crotch where shown in the sketches
$\overline{T}_r$ = pad or saddle thickness
θ = one-half angle between adjacent miter axes

(5) Where flanges are attached to one or both ends, the values of k and i in this table shall be corrected by the factors C_1, which can be read directly from Chart B, entering with the computed h.

(6) The designer is cautioned that cast buttwelded fittings may have considerably heavier walls than that of the pipe with which they are used. Large errors may be introduced unless the effect of these greater thicknesses is considered.

(7) In large diameter thin-wall elbows and bends, pressure can significantly affect the magnitudes of k and i. To correct values from the table, divide k by

$$1 + 6\left(\frac{P}{E}\right)\left(\frac{r_2}{\overline{T}}\right)^{7/3}\left(\frac{R_1}{r_2}\right)^{1/3}$$

divide i by

$$1 + 3.25\left(\frac{P}{E}\right)\left(\frac{r_2}{\overline{T}}\right)^{5/2}\left(\frac{R_1}{r_2}\right)^{2/3}$$

For consistency, use kPa and mm for SI metric, and psi and in, for U.S. customary notation.

(8) When $\overline{T}_r$ is > 1½ $\overline{T}$, use $h = 4\,\overline{T}/r_2$.

(9) The designer must be satisfied that this fabrication has a pressure rating equivalent to straight pipe.

(10) Factors shown apply to bending. Flexibility factor for torsion equals 0.9.

(11) If $r_x \geq \frac{1}{8} D_b$ and $T_c \geq 1.5\overline{T}$ a flexibility characteristic of $4.4\,\overline{T}/r_2$ may be used.

(12) The out-of-plane stress intensification factor (SIF) for a reducing branch connection with branch-to-run diameter ratio of $0.5 < d/D < 1.0$ may be nonconservative. A smooth concave weld contour has been shown to reduce the SIF. Selection of the appropriate SIF is the designer's responsibility.

(13) Stress intensification factors for branch connections are based on tests with at least two diameters of straight run pipe on each side of the branch centerline. More closely loaded branches may require special consideration.

(14) 2.1 max. or $2.1\,\overline{T}/C_x$, but not less than 1.3. C_x is the fillet weld leg length (see Fig. 328.5.2C). For unequal leg lengths, use the smaller leg for C_x.

$\overline{T}_b$ = thickness of pipe matching branch, in (mm)
i_o = out-plane stress-intensification factor (Table 10-48)
i_i = in-plane stress-intensification factor (Table 10-48)

Allowable stress range S_A and permissible additive stresses shall be computed in accordance with Eqs. (10-93) and (10-94).

Required Weld Quality Assurance Any weld at which S_E exceeds $0.8S_A$ for any portion of a piping system and the equivalent number of cycles N exceeds 7000 shall be fully examined in accordance with the requirements for severe cyclic service (presented later in this section).

Reactions: Metallic Piping Reaction forces and moments to be used in the design of restraints and supports and in evaluating the effects of piping displacements on connected equipment shall be based on the reaction range R for the extreme displacement conditions, considering the range previously defined for reactions and using E_a. The designer shall consider instantaneous maximum values of forces and moments in the original and extreme displacement conditions as well as the reaction range in making these evaluations.

Maximum Reactions for Simple Systems For two-anchor systems without intermediate restraints, the maximum instantaneous values

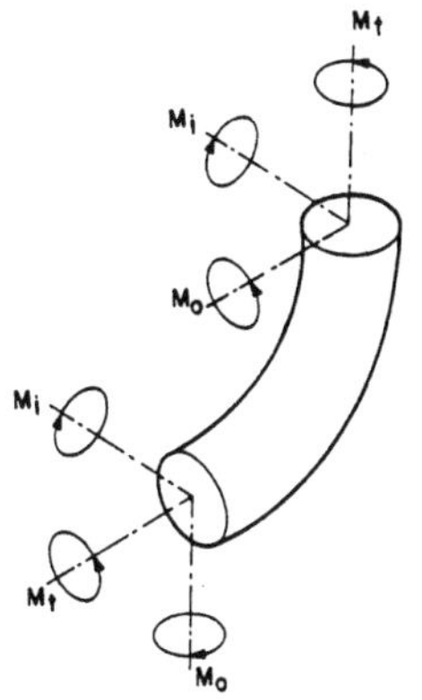

FIG. 10-172 Moments in bends. (*Extracted from the Process Piping Code, B31.3—2004, with permission of the publisher, the American Society of Mechanical Engineers, New York.*)

of reaction forces and moments may be estimated from Eqs. (10-106) and (10-107).

1. *For extreme displacement conditions,* R_m. The temperature for this computation is the design maximum or design minimum temperature as previously defined for reactions, whichever produces the larger reaction:

$$R_m = R\left(1 - \frac{2C}{3}\right)\frac{E_m}{E_a} \qquad (10\text{-}106)$$

where C = cold-spring factor varying from zero for no cold spring to 1.0 for 100 percent cold spring. (The factor ⅔ is based on experience, which shows that specified cold spring cannot be fully assured even with elaborate precautions.)

E_a = modulus of elasticity at installation temperature, lbf/in^2 (MPa)

E_m = modulus of elasticity at design maximum or design minimum temperature, lbf/in^2 (MPa)

R = range of reaction forces or moments (derived from flexibility analysis) corresponding to the full displacement-stress range and based on E_a, lbf or in·lbf (N or N·mm)

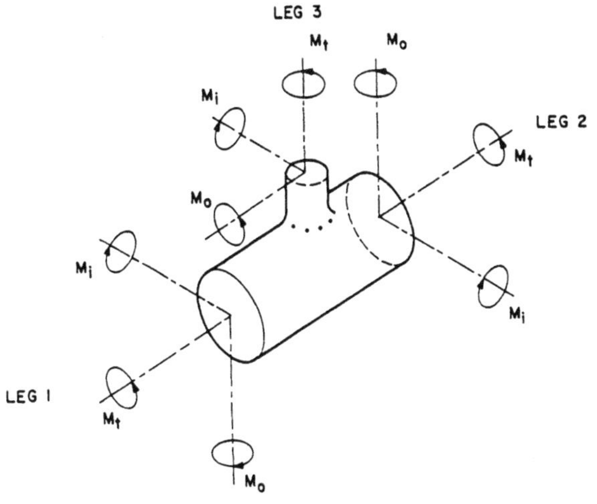

FIG. 10-173 Moments in branch connections. (*Extracted from the Process Piping Code, B31.3—2004, with permission of the publisher, the American Society of Mechanical Engineers, New York.*)

R_m = estimated instantaneous maximum reaction force or moment at design maximum or design minimum temperature, lbf or in·lbf (N or N·mm)

2. *For original condition,* R_a. The temperature for this computation is the expected temperature at which the piping is to be assembled.

$$R_a = CR \quad \text{or} \quad C_1R, \text{ whichever is greater} \qquad (10\text{-}107)$$

where nomenclature is as for Eq. (10-106) and

$C_1 = 1 - (S_hE_a/S_EE_m)$
= estimated self-spring or relaxation factor (use zero if value of C_1 is negative)

R_a = estimated instantaneous reaction force or moment at installation temperature, lbf or in·lbf (N or N·mm)

S_E = computed displacement-stress range, lbf/in^2 (MPa). See Eq. (10-101).

S_h = See Eq. (10-93).

Maximum Reactions for Complex Systems For multianchor systems and for two-anchor systems with intermediate restraints, Eqs. (10-106) and (10-107) are not applicable. Each case must be studied to estimate the location, nature, and extent of local overstrain and its effect on stress distribution and reactions.

Acceptable comprehensive methods of analysis are analytical, model-test, and chart methods, which evaluate for the entire piping system under consideration the forces, moments, and stresses caused by bending and torsion from a simultaneous consideration of terminal and intermediate restraints to thermal expansion and include all external movements transmitted under thermal change to the piping by its terminal and intermediate attachments. Correction factors, as provided by the details of these rules, must be applied for the stress intensification of curved pipe and branch connections and may be applied for the increased flexibility of such component parts.

Expansion Joints All the foregoing applies to "stiff piping systems," i.e., systems without expansion joints (see detail 1 of Fig. 10-174). When space limitations, process requirements, or other considerations result in configurations of insufficient flexibility, capacity for deflection within allowable stress range limits may be increased successively by the use of one or more hinged bellows expansion joints, viz., semirigid (detail 2) and nonrigid (detail 3) systems, and expansion effects essentially eliminated by a free-movement joint (detail 4) system. Expansion joints for semirigid and nonrigid systems are restrained against longitudinal and lateral movement by the hinges with the expansion element under bending movement only and are known as "rotation" or "hinged" joints (see Fig. 10-175). Semirigid systems are limited to one plane; nonrigid systems require a minimum of three joints for two-dimensional and five joints for three-dimensional expansion movement.

Joints similar to that shown in Fig. 10-175, except with two pairs of hinge pins equally spaced around a gimbal ring, achieve similar results with a lesser number of joints.

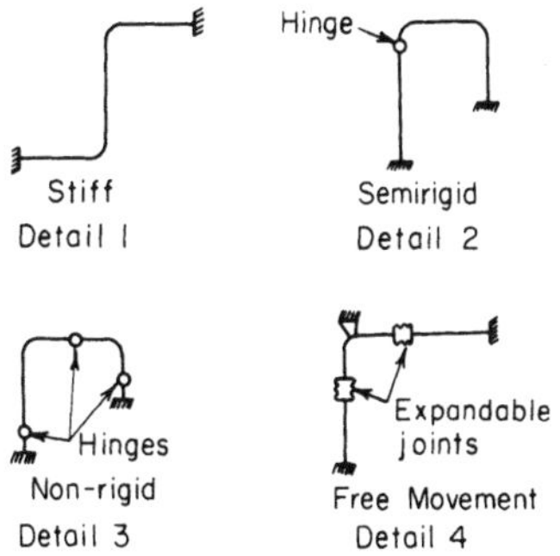

FIG. 10-174 Flexibility classification for piping systems. (*From Kellogg,* Design of Piping Systems, *Wiley, New York, 1965.*)

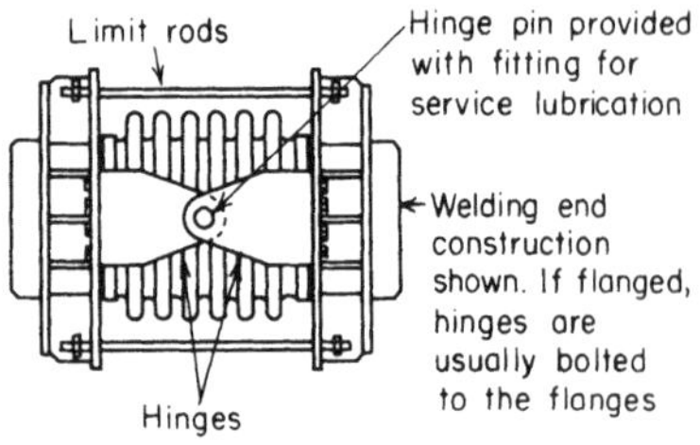

FIG. 10-175 Hinged expansion joint. (*From Kellogg,* Design of Piping Systems, *Wiley, New York, 1965.*)

Expansion joints for free-movement systems can be designed for axial or offset movement alone, or for combined axial and offset movements (see Fig. 10-176). For offset movement alone, the end load due to pressure and weight can be transferred across the joint by tie rods or structural members (see Fig. 10-177). For axial or combined movements, anchors must be provided to absorb the unbalanced pressure load and force bellows to deflect.

Commercial bellows elements are usually light-gauge (of the order of 0.05 to 0.10 in thick and are available in stainless and other alloy steels, copper, and other nonferrous materials. Multi-ply bellows, bellows with external reinforcing rings, and toroidal contour bellows are available for higher pressures. Since bellows elements are ordinarily rated for strain ranges which involve repetitive yielding, predictable performance is assured only by adequate fabrication controls and knowledge of the potential fatigue performance of each design. The attendant cold work can affect corrosion resistance and promote susceptibility to corrosion fatigue or stress corrosion; joints in a horizontal position cannot be drained and have frequently undergone pitting or cracking due to the presence of condensate during operation or offstream. For low-pressure essentially nonhazardous service, nonmetallic bellows of fabric-reinforced rubber or special materials are sometimes used. For corrosive service PTFE bellows may be used.

Because of the inherently greater susceptibility of expansion bellows to failure from unexpected corrosion, failure of guides to control joint movements, etc., it is advisable to examine critically their design choice in comparison with a stiff system.

Slip-type expansion joints (Fig. 10-178) substitute packing (ring or plastic) for bellows. Their performance is sensitive to adequate design

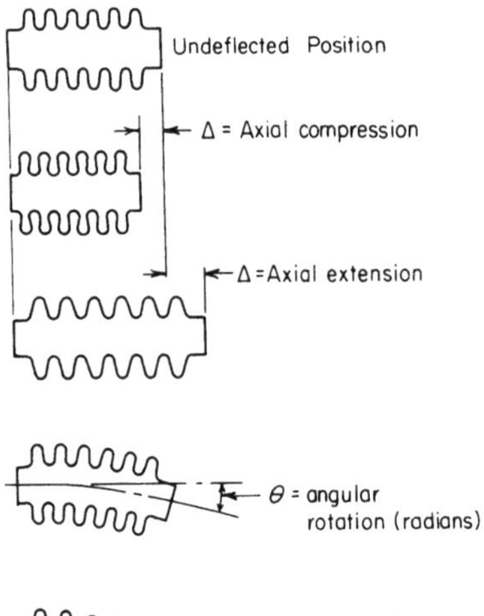

FIG. 10-176 Action of expansion bellows under various movements. (*From Kellogg,* Design of Piping Systems, *Wiley, New York, 1965.*)

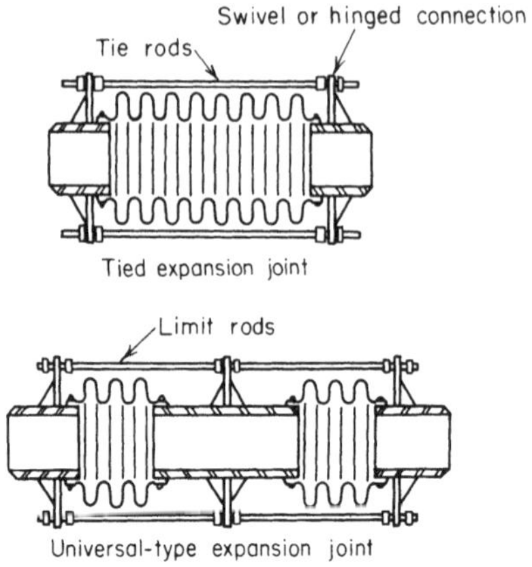

FIG. 10-177 Constrained-bellows expansion joints. (*From Kellogg,* Design of Piping Systems, *Wiley, New York, 1965.*)

with respect to guiding to prevent binding and the adequacy of stuffing boxes and attendant packing, sealant, and lubrication. Anchors must be provided for the unbalanced pressure force and for the friction forces to move the joint. The latter can be much higher than the elastic force required to deflect a bellows joint. Rotary packed joints, ball joints, and other special joints can absorb end load.

Corrugated pipe and corrugated and creased bends are also used to decrease stiffness.

Pipe Supports Loads transmitted by piping to attached equipment and supporting elements include weight, temperature- and pressure-induced effects, vibration, wind, earthquake, shock, and thermal expansion and contraction. The design of supports and restraints is based on concurrently acting loads (if it is assumed that wind and earthquake do not act simultaneously).

Resilient and constant-effort-type supports shall be designed for maximum loading conditions including test unless temporary supports are provided.

Though not specified in the code, supports for discharge piping from relief valves must be adequate to withstand the jet reaction produced by their discharge.

The code states further that pipe-supporting elements shall (1) avoid excessive interference with thermal expansion and contraction of pipe which is otherwise adequately flexible; (2) be such that they do not contribute to leakage at joints or excessive sag in piping requiring drainage; (3) be designed to prevent overstress, resonance, or disengagement due to variation of load with temperature; also, so that combined longitudinal stresses in the piping shall not exceed the code allowable limits; (4) be such that a complete release of the piping load will be prevented in the event of spring failure or misalignment, weight transfer, or added load due to test during erection; (5) be of steel or wrought iron; (6) be of alloy steel or protected from temperature when the temperature limit for carbon steel may be exceeded; (7) not be cast

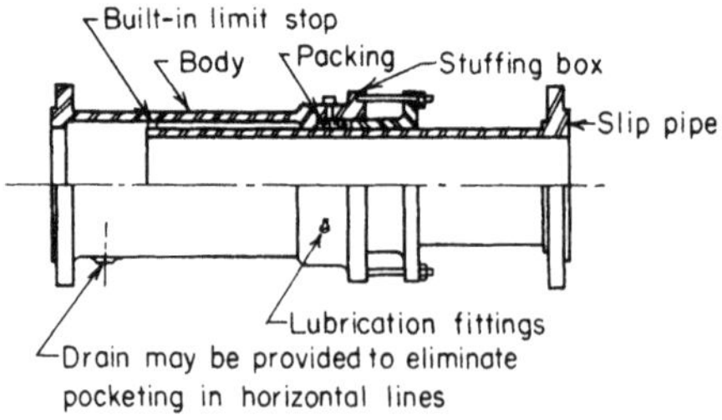

FIG. 10-178 Slip-type expansion joint. (*From Kellogg,* Design of Piping Systems, *Wiley, New York, 1965.*)

iron except for roller bases, rollers, anchor bases, etc., under mainly compression loading; (8) not be malleable or nodular iron except for pipe clamps, beam clamps, hanger flanges, clips, bases, and swivel rings; (9) not be wood except for supports mainly in compression when the pipe temperature is at or below ambient; and (10) have threads for screw adjustment which shall conform to ASME B1.1.

A supporting element used as an anchor shall be designed to maintain an essentially fixed position.

To protect terminal equipment or other (weaker) portions of the system, restraints (such as anchors and guides) shall be provided where necessary to control movement or to direct expansion into those portions of the system that are adequate to absorb them. The design, arrangement, and location of restraints shall ensure that expansion-joint movements occur in the directions for which the joint is designed. In addition to the other thermal forces and moments, the effects of friction in other supports of the system shall be considered in the design of such anchors and guides.

Anchors for Expansion Joints Anchors (such as those of the corrugated, omega, disk, or slip type) shall be designed to withstand the algebraic sum of the forces at the maximum pressure and temperature at which the joint is to be used. These forces are:

1. Pressure thrust, which is the product of the effective thrust area times the maximum pressure to which the joint will be subjected during normal operation. (For slip joints the effective thrust area shall be computed by using the outside diameter of the pipe. For corrugated, omega, or disk-type joints, the effective thrust area shall be that area recommended by the joint manufacturer. If this information is unobtainable, the effective area shall be computed by using the maximum inside diameter of the expansion-joint bellows.)
2. The force required to compress or extend the joint in an amount equal to the calculated expansion movement.
3. The force required to overcome the static friction of the pipe in expanding or contracting on its supports, from installed to operating position. The length of pipe considered should be that located between the anchor and the expansion joint.

Support Fixtures Hanger rods may be pipe straps, chains, bars, or threaded rods which permit free movement for thermal expansion or contraction. Sliding supports shall be designed for friction and bearing loads. Brackets shall be designed to withstand movements due to friction in addition to other loads. Spring-type supports shall be designed for weight load at the point of attachment and to prevent misalignment, buckling, or eccentric loading of springs, and provided with stops to prevent spring overtravel. Compensating-type spring hangers are recommended for high-temperature and critical-service piping to make the supporting force uniform with appreciable movement. Counterweight supports shall have stops to limit travel. Hydraulic supports shall be provided with safety devices and stops to support load in the event of loss of pressure. Vibration dampers or sway braces may be used to limit vibration amplitude.

The code requires that the safe load for threaded hanger rods be based on the root area of the threads. This, however, assumes concentric loading. When hanger rods move to a nonvertical position so that the load is transferred from the rod to the supporting structure via the edge of one flat of the nut on the rod, it is necessary to consider the root area to be reduced by one-third. If a clamp is connected to a vertical line to support its weight, it is recommended that shear lugs be welded to the pipe, or that the clamp be located below a fitting or flange, to prevent slippage. Consideration shall be given to the localized stresses induced in the piping by the integral attachment. Typical pipe supports are shown in Fig. 10-179.

Much piping is supported from structures installed for other purposes. It is common practice to use beam formulas for tubular sections to determine stress, maximum deflection, and maximum slope of piping in **spans between supports.** When piping is supported from structures installed for that sole purpose and those structures rest on driven piles, detailed calculations are usually made to determine maximum permissible spans. Limits imposed on maximum slope to make the contents of the line drain to the lower end require calculations made on the weight per foot of the empty line. To avoid interference with other components, maximum deflection should be limited to 25.4 mm (1 in).

Pipe hangers are essentially frictionless but require taller pipe-support structures which cost more than structures on which pipe is laid. Devices that reduce friction between laid pipe subject to thermal movement and its supports are used to accomplish the following:

1. Reduce loads on anchors or on equipment acting as anchors.
2. Reduce the tendency of pipe acting as a column loaded by friction at supports to buckle sideways off supports.
3. Reduce nonvertical loads imposed by piping on its supports so as to minimize cost of support foundations.
4. Reduce longitudinal stress in pipe.

Linear bearing surfaces made of fluorinated hydrocarbons or of graphite and also rollers are used for this purpose.

Design Criteria: Nonmetallic Pipe In using a nonmetallic material, designers must satisfy themselves as to the adequacy of the material and its manufacture, considering such factors as strength at design temperature, impact- and thermal-shock properties, toxicity, methods of making connections, and possible deterioration in service. Rating information, based usually on ASTM standards or specifications, is generally available from the manufacturers of these materials. Particular attention should be given to provisions for the thermal expansion of nonmetallic piping materials, which may be as much as 5 to 10 times that of steel (Table 10-49). Special consideration should be given to the strength of small pipe connections to piping and equipment and to the need for extra flexibility at the junction of metallic and nonmetallic systems.

Table 10-50 gives values for the modulus of elasticity for nonmetals; however, no specific stress-limiting criteria or methods of stress analysis are presented. Stress-strain behavior of most nonmetals differs considerably from that of metals and is less well-defined for mathematic analysis. The piping system should be designed and laid out so that flexural stresses resulting from displacement due to expansion, contraction, and other movement are minimized. This concept requires special attention to supports, terminals, and other restraints.

Displacement Strains The concepts of strain imposed by restraint of thermal expansion or contraction and by external movement described for metallic piping apply in principle to nonmetals. Nevertheless, the assumption that stresses throughout the piping system can be predicted from these strains because of fully elastic behavior of the piping materials is not generally valid for nonmetals.

In thermoplastics and some thermosetting resins, displacement strains are not likely to produce immediate failure of the piping but may result in detrimental distortion. Especially in thermoplastics, progressive deformation may occur upon repeated thermal cycling or on prolonged exposure to elevated temperature.

In brittle nonmetallics (such as porcelain, glass, impregnated graphite, etc.) and some thermosetting resins, the materials show rigid behavior and develop high displacement stresses up to the point of sudden breakage due to overstrain.

Elastic Behavior The assumption that displacement strains will produce proportional stress over a sufficiently wide range to justify an elastic-stress analysis often is not valid for nonmetals. In brittle nonmetallic piping, strains initially will produce relatively large elastic stresses. The total displacement strain must be kept small, however, since overstrain results in failure rather than plastic deformation. In plastic and resin nonmetallic piping strains generally will produce stresses of the overstrained (plastic) type even at relatively low values of total displacement strain.

FABRICATION, ASSEMBLY, AND ERECTION

Welding, Brazing, or Soldering Code requirements dealing with fabrication are more detailed for welding than for other methods of joining, since welding is the predominant method of construction and the method used for the most demanding applications. The code requirements for welding processes and operators are essentially the same as covered in the subsection on pressure vessels (i.e., qualification to Sec. IX of the ASME Boiler and Pressure Vessel Code) except that welding processes are not restricted, the material grouping (P number) must be in accordance with ASME B31.3 Appendix A-1, and welding positions are related to pipe position. The code also permits one fabricator to accept welders or welding operators qualified by another employer

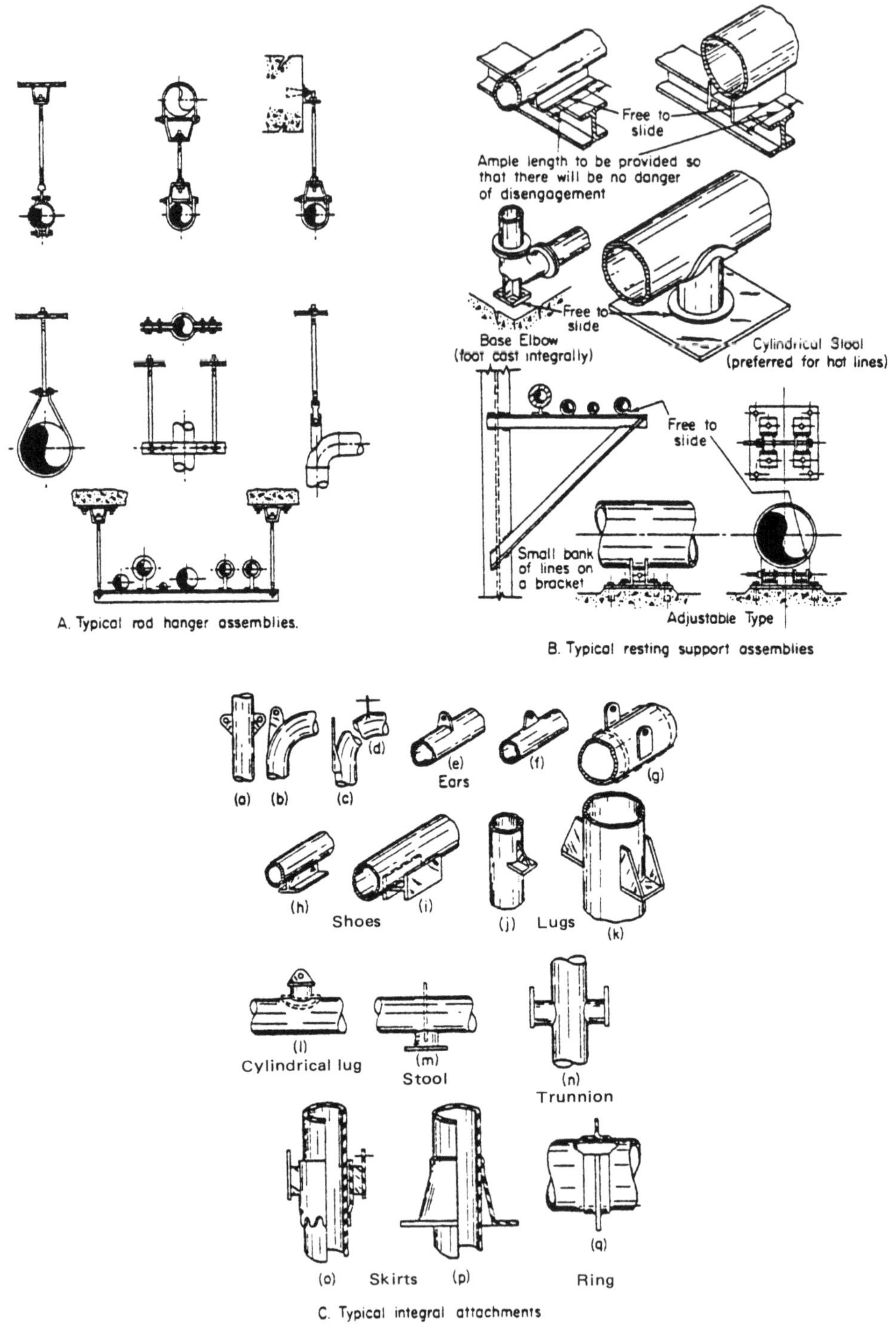

FIG. 10-179 Typical pipe supports and attachments. (*From Kellogg,* Design of Piping Systems, *Wiley, New York, 1965.*)

without requalification when welding pipe by the same or equivalent procedure. The code may require that the welding procedure qualification include low-temperature toughness testing (see. Table 10-51).

Filler metal is required to conform with the requirements of Sec. IX. Backing rings (of ferrous material), when used, shall be of weldable quality with sulfur limited to 0.05 percent. Backing rings of nonferrous and nonmetallic materials may be used provided they are proved satisfactory by procedure-qualification tests and provided their use has been approved by the designer.

The code requires internal alignment within the dimensional limits specified in the welding procedure and the engineering design without specific dimensional limitations. Internal trimming is permitted for correcting internal misalignment provided such trimming does not result in a finished wall thickness before welding of less than required minimum wall thickness t_m. When necessary, weld metal may be deposited on the inside or outside of the component to provide alignment or sufficient material for trimming.

Table 10-53 is a summary of the code acceptance criteria (limits on imperfections) for welds. The defects referred to are illustrated in Fig. 10-180.

Brazing procedures, brazers, and brazing operators must be qualified in accordance with the requirements of Part QB, Sec. IX, ASME Code. At the owner's option, qualification is not required for Category D fluid service not exceeding 93°C (200°F). The clearance between

TABLE 10-49 Thermal Expansion Coefficients, Nonmetals

	Mean coefficients (divide table values by 10^6)			
Material description	in/in, °F	Range, °F	mm/mm, °C	Range, °C
Thermoplastics				
Acetal AP2012	2	...	3.6	...
Acrylonitrile-butadiene-styrene				
ABS 1208	60	...	108	...
ABS 1210	55	45–55	99	7–13
ABS 1316	40	...	72	...
ABS 2112	40	...	72	...
Cellulose acetate butyrate				
CAB MH08	80	...	144	...
CAB S004	95	...	171	...
Chlorinated poly(vinyl chloride)				
CPVC 4120	35	...	63	...
Polybutylene PB 2110	72	...	130	...
Polyether, chlorinated	45	...	81	...
Polyethylene				
PE 1404	100	46–100	180	8–38
PE 2305	90	46–100	162	8–38
PE 2306	80	46–100	144	8–38
PE 3306	70	46–100	126	8–38
PE 3406	60	46–100	108	8–38
Polyphenylene POP 2125	30	...	54	...
Polypropylene				
PP1110	48	33–67	86	1–19
PP1208	43	...	77	...
PP2105	40	...	72	...
Poly(vinyl chloride)				
PVC 1120	30	23–37	54	−5 to +3
PVC 1220	35	34–40	63	1–4
PVC 2110	50	...	90	...
PVC 2112	45	...	81	...
PVC 2116	40	37–45	72	3–7
PVC 2120	30	...	54	...
Poly(vinylidene fluoride)	79	...	142	...
Poly(vinylidene chloride)	100	...	180	...
Polytetrafluoroethylene	55	73–140	99	23–60
Poly(fluorinated ethylenepropylene)	46–58	73–140	83–104	23–60
Poly(perfluoroalkoxy alkane)	67	70–212	121	21–100
Poly(perfluoroalkoxy alkane)	94	212–300	169	100–149
Poly(perfluoroalkoxy alkane)	111	300–408	200	149–209
Reinforced Thermosetting Resins and Reinforced Plastic Mortars				
Glass-epoxy, centrifugally cast	9–13	...	16–23.5	...
Glass-polyester, centrifugally cast	9–15	...	16–27	...
Glass-polyester, filament-wound	9–11	...	16–20	...
Glass-polyester, hand lay-up	12–15	...	21.5–27	...
Glass-epoxy, filament-wound	9–13	...	16–23.5	...
Other Nonmetallic Materials				
Borosilicate glass	1.8	...	3.25	...

GENERAL NOTES:

(a) For Code references to this table, see para. A319.3.1, ASME B31.3—2004. These data are for use in the absence of more applicable data. It is the designer's responsibility to verify that materials are suitable for the intended service at the temperatures shown.

(b) Individual compounds may vary from the values shown. Consult manufacturer for specific values for products.

TABLE 10-50 Modulus of Elasticity, Nonmetals

Material description	E, ksi (73.4°F)	E, MPa (23°C)
Thermoplastics [Note (1)]		
Acetal	410	2,830
ABS, Type 1210	250	1,725
ABS, Type 1316	340	2,345
CAB	120	825
PVC, Type 1120	420	2,895
PVC, Type 1220	410	2,825
PVC, Type 2110	340	2,345
PVC, Type 2116	380	2,620
Chlorinated PVC	420	2,895
Chlorinated polyether	160	1,105
PE, Type 2306	90	620
PE, Type 3306	130	895
PE, Type 3406	150	1,035
Polypropylene	120	825
Poly(vinylidene chloride)	100	690
Poly(vinylidene fluoride)	194	1,340
Poly(tetrafluorethylene)	57	395
Poly(fluorinated ethylenepropylene)	67	460
Poly(perfluoroalkoxy alkane)	100	690
Thermosetting Resins, Axially Reinforced		
Epoxy-glass, centrifugally cast	1,200–1,900	8,275–13,100
Epoxy-glass, filament-wound	1,100–2,000	7,585–13,790
Polyester-glass, centrifugally cast	1,200–1,900	8,275–13,100
Polyester-glass, hand lay-up	800–1,000	5,515–6,895
Other		
Borosilicate glass	9,800	67,570

GENERAL NOTE: For Code references to this table, see para. A319.3.2, ASME B31.3—2004. These data are for use in the absence of more applicable data. It is the designer's responsibility to verify that materials are suitable for the intended service at the temperatures shown.

NOTE:

(1) The modulus of elasticity data shown for thermoplastics are based on short-term tests. The manufacturer should be consulted to obtain values for use under long-term loading.

surfaces to be joined by brazing or soldering shall be no larger than is necessary to allow complete capillary distribution of the filler metal.

The only requirement for solderers is that they follow the procedure in the *Copper Tube Handbook* of the Copper Development Association.

Bending and Forming Pipe may be bent to any radius for which the bend-arc surface will be free of cracks and substantially free of buckles. The use of qualified bends which are creased or corrugated is permitted. Bending may be done by any hot or cold method that produces a product meeting code and service requirements, and that does not have an adverse effect on the essential characteristics of the material. Hot bending and hot forming must be done within a temperature range consistent with material characteristics, end use, and postoperation heat treatment. Postbend heat treatment may be required for bends in some materials; its necessity is dependent upon the type of bending operation and the severity of the bend. Postbend heat treatment requirements are defined in the code.

Piping components may be formed by any suitable hot or cold pressing, rolling, forging, hammering, spinning, drawing, or other method. Thickness after forming shall not be less than required by design. Special rules cover the forming and pressure design verification of flared laps.

The development of fabrication facilities for bending pipe to the radius of commercial butt-welding long-radius elbows and forming flared metallic (Van Stone) laps on pipe are important techniques in reducing welded-piping costs. These techniques save both the cost of the ell or stub end and the welding operation required to attach the fitting to the pipe.

Preheating and Heat Treatment Preheating and postoperation heat treatment are used to avert or relieve the detrimental effects of the high temperature and severe thermal gradients inherent in the welding of metals. In addition, heat treatment may be needed to relieve residual stresses created during the bending or forming of metals. The code provisions shown in Tables 10-54 and 10-55 represent basic practices which are acceptable for most applications of welding, bending, and forming, but they are not necessarily suitable for all service conditions. The specification of more or less stringent preheating and heat-treating requirements is a function of those responsible for the engineering design.

Refer to the code for rules establishing the thickness to be used in determining PWHT requirements for configurations other than butt welds (e.g., fabricated branch connections and socket welds).

Joining Nonmetallic Pipe All joints should be made in accordance with procedures complying with the manufacturer's recommendations and code requirements. General welding and heat fusion procedures are described in ASTM D-2657. ASTM D2855 describes general solvent-cementing procedures.

Depending on size, thermoplastic piping can be joined with mechanical joints, solvent-cemented joints, hot-gas welding, or heat fusion procedures. Mechanical joints are frequently bell-and-spigot joints which employ an elastomer O-ring gasket. Joints of this type are generally not self-restrained against internal pressure thrust.

Thermosetting resin pipe can be joined with mechanical joints or adhesive-bonded joints. Mechanical joints are generally a variation of gasketed bell-and-spigot joints and may be either nonrestrained or self-restrained. Adhesive-bonded joints are typically bell-and-spigot or butt-and-strap. Butt-and-strap joints join piping components with multiple layers of resin-saturated glass reinforcement.

Assembly and Erection Flanged-joint faces shall be aligned to the design plane to within 1⁄10 in/ft (0.5 percent) maximum measured across any diameter, and flange bolt holes shall be aligned to within 3.2-mm (1⁄8-in) maximum offset. Flanged joints involving flanges with widely differing mechanical properties shall be assembled with extra care, and tightening to a predetermined torque is recommended.

The use of flat washers under bolt heads and nuts is a code requirement when assembling nonmetallic flanges. It is preferred that the bolts extend completely through their nuts; however, a lack of complete thread engagement not exceeding one thread is permitted by the code. The assembly of cast-iron bell-and-spigot piping is covered in AWWA Standard C600.

Screwed joints which are intended to be seal-welded shall be made up without any thread compound.

When installing conductive nonmetallic piping and metallic pipe with nonmetallic linings, consideration should be given to the need to provide electrical continuity throughout the system and to grounding requirements. This is particularly critical in areas with potentially explosive atmospheres.

EXAMINATION, INSPECTION, AND TESTING

This subsection provides a general synopsis of code requirements. It should not be viewed as comprehensive.

Examination and Inspection The code differentiates between examination and inspection. "Examination" applies to quality-control functions performed by personnel of the piping manufacturer, fabricator, or erector. "Inspection" applies to functions performed for the owner by the authorized inspector.

The authorized inspector shall be designated by the owner and shall be the owner, an employee of the owner, an employee of an engineering or scientific organization, or an employee of a recognized insurance or inspection company acting as the owner's agent. The inspector shall not represent or be an employee of the piping erector, the manufacturer, or the fabricator unless the owner is also the erector, the manufacturer, or the fabricator.

The authorized inspector shall have a minimum of 10 years' experience in the design, fabrication, or inspection of industrial pressure piping. Each 20 percent of satisfactory work toward an engineering degree accredited by the Engineers' Council for Professional Development shall be considered equivalent to 1 year's experience, up to 5 years total.

TABLE 10-51 Requirements for Low-Temperature Toughness Tests for Metals*

These toughness requirements are in addition to tests required by the material specification.

	Type of material	Column A Design minimum temperature at or above min. temp. in Table A-1 of ASME B31.3—2004 or Table 10-52		Column B Design minimum temperature below min. temp. in Table A-1 of ASME B31.3—2004 or Table 10-52
Listed Materials	1 Gray cast iron	A-1 No additional requirements		B-1 No additional requirements
	2 Malleable and ductile cast iron: carbon steel per Note (1)	A-2 No additional requirements		B-2 Materials designated in Box 2 shall not be used.
		(a) Base metal	(b) Weld metal and heat affected zone (HAD) [Note (2)]	
	3 Other carbon steels, low and intermediate alloy steels, high alloy ferritic steels, duplex stainless steels	A-3 (a) No additional requirements	A-3 (b) Weld metal deposits shall be impact tested per para. 323.3.[a] If design min. temp. < −29°C (−20°F), except as provided in Notes (3) and (5), and except as follows: for materials listed for Curves C and D of Table 10-52, where corresponding welding consumables are qualified by impact testing at the design minimum temperature or lower in accordance with the applicable AWS specification, additional testing is not required.	B-3 Except as provided in Notes (3) and (5), heat treat base metal per applicable ASTM specification listed in para. 323.3.2[a]; then impact test base metal, weld deposits, and HAZ per para. 323.3[a] [see Note (2)]. When materials are used at design min. temp. below the assigned curve as permitted by Notes (2) and (3) of Table 10-52, weld deposits and HAZ shall be impact trested [see Note (2)].
	4 Austenitic stainless steels	A-4 (a) If: (1) carbon content by analysis >0.1%; or (2) material is not in solution heat treated condition; then, impact test per para. 323.3[a] for design min. temp. < −29°C (−20°F) except as provided in Notes (3) and (6)	A-4 (b) Weld metal deposits shall be impact tested per para 323.3.[a] If design min. temp. < −29°C (−20°F) except as provided in para. 323.2.2[b] and in Notes (3) and (6)	B-4 Base metal and weld metal deposits shall be impact tested per para. 323.3[b]. See Notes (2),(3), and (6).
	5 Austenitic ductile iron, ASTM A 571	A-5 (a) No additional requirements	A-5 (b) Welding is not permitted	B-5 Base metal shall be impact tested per para 323.3.[a] Do not use < −196°C (−320°F). Welding is not permitted.
	6 Aluminum, copper, nickel, and their alloys; unalloyed titanium	A-6 (a) No additional requirements	A-6 (b) No additional requirements unless filler metal composition is outside the range for base metal composition; then test per column B-6	B-6 Designer shall be assured by suitable tests [see Note (4)] that base metal, weld deposits, and HAZ are suitable at the design min. temp.
Unlisted Materials	7 An unlisted material shall conform to a published specification. Where composition, heat treatment, and product form are comparable to those of a listed material, requirements for the corresponding listed material shall be met. Other unlisted materials shall be qualified as required in the applicable section of column B.			

NOTES:

(1) Carbon steels conforming to the following are subject to the limitations in Box B-2; plates per ASTM A 36, A 283, and A 570; pipe per ASTM A 134 when made from these plates; and pipe per ASTM A 53 Type F and API 5L Gr. A25 butt weld.

(2) Impact tests that meet the requirements of Table 323.3.1,[a] which are performed as part of the weld procedure qualification, will satisfy all requirements of para. 323.2.2,[b] and need not be repeated for production welds.

(3) Impact testing is not required if the design minimum temperature is below −29°C (−20°F) but at or above −104°C (−155°F) and the Stress Ratio defined in Fig. 323.2.2B[b] does not exceed 0.3.

(4) Tests may include tensile elongation, sharp-notch tensile strength (to be compared with unnotched tensile strength), and/or other tests, conducted at or below design minimum temperature. See also para. 323.3.4.[b]

(5) Impact tests are not required when the maximum obtainable Charpy specimen has a width along the notch of less than 2.5 mm (0.098 in). Under these conditions, the design minimum temperature shall not be less than the lower of −48°C (−55°F) or the minimum temperature for the material in Table A-1.

(6) Impact tests are not required when the maximum obtainable Charpy specimen has a width along the notch of less than 2.5 mm (0.098 in).

[a]Refer to the referenced code paragraph for impact testing methods and acceptance.

[b]Refer to the referenced code paragraph for comments regarding circumstances under which impact testing can be excluded.

*Table 10-51 and notes have been extracted (with minor modifications) from *Process Piping*, ASME B31.3—2004, with permission of the publisher, the American Society of Mechanical Engineers, New York.

TABLE 10-52 Tabular Values for Minimum Temperatures without Impact Testing for Carbon Steel Materials*

Nominal thickness [Note (6)]		Design minimum temperature							
		Curve A [Note (2)]		Curve B [Note (3)]		Curve C [Note (3)]		Curve D	
mm	in	°C	°F	°C	°F	°C	°F	°C	°F
6.4	0.25	–9.4	15	–28.9	–20	–48.3	–55	–48.3	–55
7.9	0.3125	–9.4	15	–28.9	–20	–48.3	–55	–48.3	–55
9.5	0.375	–9.4	15	–28.9	–20	–48.3	–55	–48.3	–55
10.0	0.394	–9.4	15	–28.9	–20	–48.3	–55	–48.3	–55
11.1	0.4375	–6.7	20	–28.9	–20	–41.7	–43	–48.3	–55
12.7	0.5	–1.1	30	–28.9	–20	–37.8	–36	–48.3	–55
14.3	0.5625	2.8	37	–21.7	–7	–35.0	–31	–45.6	–50
15.9	0.625	6.1	43	–16.7	2	–32.2	–26	–43.9	–47
17.5	0.6875	8.9	48	–12.8	9	–29.4	–21	–41.7	–43
19.1	0.75	11.7	53	–9.4	15	–27.2	–17	–40.0	–40
20.6	0.8125	14.4	58	–6.7	20	–25.0	–13	–38.3	–37
22.2	0.875	16.7	62	–3.9	25	–23.3	–10	–36.7	–34
23.8	0.9375	18.3	65	–1.7	29	–21.7	–7	–35.6	–32
25.4	1.0	20.0	68	0.6	33	–19.4	–3	–34.4	–30
27.0	1.0625	22.2	72	2.2	36	–18.3	–1	–33.3	–28
28.6	1.125	23.9	75	3.9	39	–16.7	2	–32.2	–26
30.2	1.1875	25.0	77	5.6	42	–15.6	4	–30.6	–23
31.8	1.25	26.7	80	6.7	44	–14.4	6	–29.4	–21
33.3	1.3125	27.8	82	7.8	46	–13.3	8	–28.3	–19
34.9	1.375	28.9	84	8.9	48	–12.2	10	–27.8	–18
36.5	1.4375	30.0	86	9.4	49	–11.1	12	–26.7	–16
38.1	1.5	31.1	88	10.6	51	–10.0	14	–25.6	–14
39.7	1.5625	32.2	90	11.7	53	–8.9	16	–25.0	–13
41.3	1.625	33.3	92	12.8	55	–8.3	17	–23.9	–11
42.9	1.6875	33.9	93	13.9	57	–7.2	19	–23.3	–10
44.5	1.75	34.4	94	14.4	58	–6.7	20	–22.2	–8
46.0	1.8125	35.6	96	15.0	59	–5.6	22	–21.7	–7
47.6	1.875	36.1	97	16.1	61	–5.0	23	–21.1	–6
49.2	1.9375	36.7	98	16.7	62	–4.4	24	–20.6	–5
50.8	2.0	37.2	99	17.2	63	–3.3	26	–20.0	–4
51.6	2.0325	37.8	100	17.8	64	–2.8	27	–19.4	–3
54.0	2.125	38.3	101	18.3	65	–2.2	28	–18.9	–2
55.6	2.1875	38.9	102	18.9	66	–1.7	29	–18.3	–1
57.2	2.25	38.9	102	19.4	67	–1.1	30	–17.8	0
58.7	2.3125	39.4	103	20.0	68	–0.6	31	–17.2	1
60.3	2.375	40.0	104	20.6	69	0.0	32	–16.7	2
61.9	2.4375	40.6	105	21.1	70	0.6	33	–16.1	3
63.5	2.5	40.6	105	21.7	71	1.1	34	–15.6	4
65.1	2.5625	41.1	106	21.7	71	1.7	35	–15.0	5
66.7	2.625	41.7	107	22.8	73	2.2	36	–14.4	6
68.3	2.6875	41.7	107	22.8	73	2.8	37	–13.9	7
69.9	2.75	42.2	108	23.3	74	3.3	38	–13.3	8
71.4	2.8125	42.2	108	23.9	75	3.9	39	–13.3	8
73.0	2.875	42.8	109	24.4	76	4.4	40	–12.8	9
74.6	2.9375	42.8	109	25.0	77	4.4	40	–12.2	10
76.2	3.0	43.3	110	25.0	77	5.0	41	–11.7	11

It is the owner's responsibility, exercised through the authorized inspector, to verify that all required examinations and testing have been completed and to inspect the piping to the extent necessary to be satisfied that it conforms to all applicable requirements of the code and the engineering design. This verification may include certifications and records pertaining to materials, components, heat treatment, examination and testing, and qualifications of operators and procedures. The authorized inspector may delegate the performance of inspection to a qualified person.

Inspection does not relieve the manufacturer, the fabricator, or the erector of responsibility for providing materials, components, and skill in accordance with requirements of the code and the engineering design, performing all required examinations, and preparing records of examinations and tests for the inspector's use.

Examination Methods The code establishes the types of examinations for evaluating various types of imperfections (see Table 10-56).

Personnel performing examinations other than visual shall be qualified in accordance with applicable portions of SNT TC-1A, *Recommended Practice for Nondestructive Testing Personnel Qualification and Certification*. Procedures shall be qualified as required in Par. T-150, Art. 1, Sec. V of the ASME Code. Limitations on imperfections shall be in accordance with the engineering design but shall at least meet the requirements of the code (see Tables 10-54 and 10-55) for the specific type of examination. Repairs shall be made as applicable.

Visual Examination This consists of observation of the portion of components, joints, and other piping elements that are or can be exposed to view before, during, or after manufacture, fabrication, assembly, erection, inspection, or testing.

TABLE 10-52 Tabular Values for Minimum Temperatures without Impact Testing for Carbon Steel Materials* *(Concluded)*

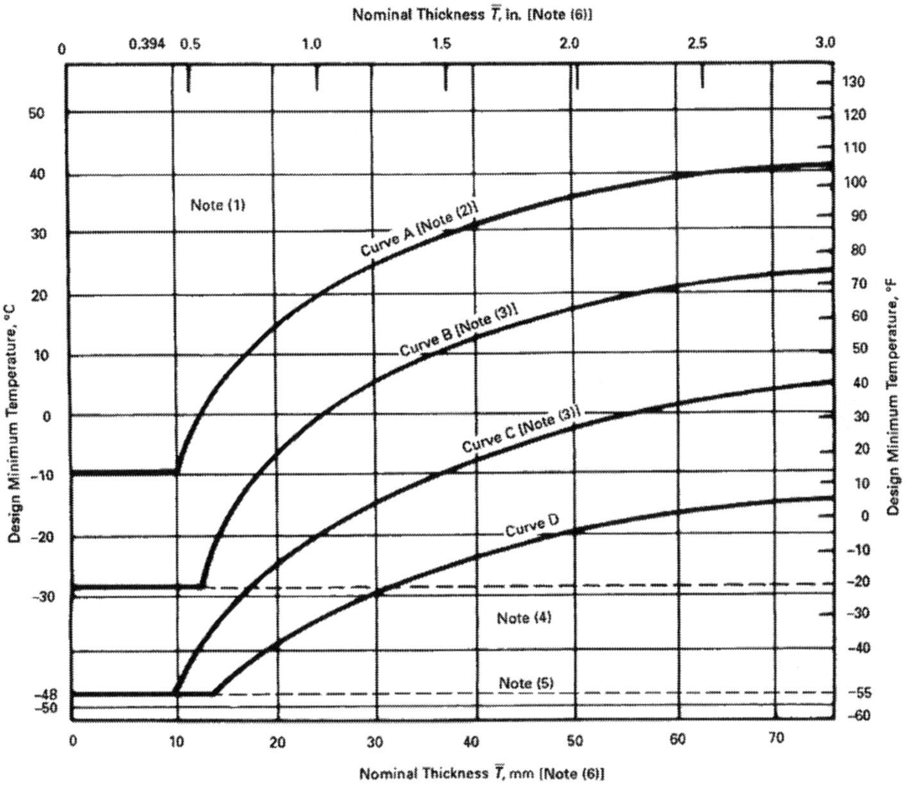

NOTES:

(1) Any carbon steel material may be used to a minimum temperature of −29°C (−20°F) for Category D Fluid Service.

(2) X Grades of API 5L, and ASTM A 381 materials, may be used in accordance with Curve B if normalized or quenched and tempered.

(3) The following materials may be used in accordance with Curve D if normalized:

(*a*) ASTM A 516 Plate, all grades

(*b*) ASTM A 671 Pipe, Grades CE55, CE60, and all grades made with A 516 plate

(*c*) ASTM A 672 Pipe, Grades E55, E60, and all grades made with A 516 plate

(4) A welding procedure for the manufacture of pipe or components shall include impact testing of welds and HAZ for any design minimum temperature below −29°C (−20°F), except as provided in Table 10-51, A-3(b).

(5) Impact testing in accordance with para. 323.3[b] is required for any design minimum temperature below −48°C (−55°F), except as permitted by Note (3) in Table 10-51.

(6) For blind flanges and blanks, $\bar{T}$ shall be $^1/_4$ of the flange thickness.

[a]Refer to the referenced code paragraph for impact testing methods and acceptance.

[b]Refer to the referenced code paragraph for comments regarding circumstances under which impact testing can be excluded.

*Table 10-52 and notes have been extracted (with minor modifications) from *Process Piping,* ASME B31.3—2004, with permission of the publisher, the American Society of Mechanical Engineers, New York.

The examination includes verification of code and engineering design requirements for materials and components, dimensions, joint preparation, alignment, welding, bonding, brazing, bolting, threading and other joining methods, supports, assembly, and erection.

Visual examination shall be performed in accordance with Art. 9, Sec. V of the ASME Code.

Magnetic-Particle Examination This examination shall be performed in accordance with Art. 7, Sec. V of the ASME Code.

Liquid-Penetrant Examination This examination shall be performed in accordance with Art. 6, Sec. V of the ASME Code.

Radiographic Examination The following definitions apply to radiography required by the code or by the engineering design:

1. "Random radiography" applies only to girth and miter groove welds. It is radiographic examination of the complete circumference of a specified percentage of the girth butt welds in a designated lot of piping.

2. Unless otherwise specified in engineering design, *100 percent radiography* applies only to girth welds, miter groove welds, and fabricated branch connections that utilize butt-type welds to join the header and the branch. The design engineer may, however, elect to designate other types of welds as requiring 100 percent radiography. By definition, 100 percent radiography requires radiographic examination of the complete length of all such welds in a designated lot of piping.

TABLE 10-53 Acceptance Criteria for Welds and Examination Methods for Evaluating Weld Imperfections*

Criteria (A to M) for types of welds and for service conditions [Note (1)]											Examination methods			
Normal and Category M fluid service			Severe cyclic conditions			Category D fluid service								
Type of weld			Type of weld			Type of weld								
Girth, miter groove & branch connection [Note (4)]	Longitudinal groove [Note (2)]	Fillet [Note (3)]	Girth, miter groove & branch connection [Note (4)]	Longitudinal groove [Note (2)]	Fillet [Note (3)]	Girth and miter groove	Longitudinal groove [Note (2)]	Fillet [Note (3)]	Branch connection [Note (4)]	Weld imperfection	Visual	Radiography	Magnetic particle	Liquid penetrant
A	A	A	A	A	A	A	A	A	A	Crack	✓	✓	✓	✓
A	A	A	A	A	A	C	A	N/A	A	Lack of fusion	✓	✓	. . .	. . .
B	A	N/A	A	A	N/A	C	A	N/A	B	Incomplete penetration	✓	✓	. . .	. . .
E	E	N/A	D	D	N/A	N/A	N/A	N/A	N/A	Internal porosity	. . .	✓	. . .	. . .
G	G	N/A	F	F	N/A	N/A	N/A	N/A	N/A	Internal slag inclusion, tungsten inclusion, or elongated indication	. . .	✓	. . .	. . .
H	A	H	A	A	A	I	A	H	H	Undercutting	. . .	✓	. . .	. . .
A	A	A	A	A	A	A	A	A	A	Surface porosity or exposed slag inclusion [Note (5)]	✓	. . .	. . .	. . .
N/A	N/A	N/A	J	J	J	N/A	N/A	N/A	N/A	Surface finish	✓	. . .	. . .	. . .
K	K	N/A	K	K	N/A	K	K	N/A	K	Concave root surface (suck up)	✓	✓	. . .	. . .
L	L	L	L	L	L	M	M	M	M	Weld reinforcement or internal protrusion	✓	. . .	. . .	. . .

Criterion value notes

Criterion			
Symbol	Measure	Acceptable value limits [Note (6)]	
A	Extent of imperfection	Zero (no evident imperfection)	
B	Depth of incomplete penetration Cumulative length of incomplete penetration	≤1 mm (1/32 in) and $\leq 0.2\overline{T}_w$ ≤38 mm (1.5 in) in any 150 mm (6 in) weld length	
C	Depth of lack of fusion and incomplete penetration Cumulative length of lack of fusion and incomplete penetration [Note (7)]	$\leq 0.2\overline{T}_w$ ≤38 mm (1.5 in) in any 150 mm (6 in) weld length	
D	Size and distribution of internal porosity	See BPV Code, Section VIII, Division 1, Appendix 4	
E	Size and distribution of internal porosity	For $\overline{T}_w \leq$ 6 mm (1/4 in), limit is same as D For $\overline{T}_w >$ 6 mm (1/4 in), limit is 1.5 × D	
F	Slag inclusion, tungsten inclusion, or elongated indication Individual length Individual width Cumulative length	 $\leq \overline{T}_w/3$ ≤2.5 mm (3/32 in) and $\leq \overline{T}_w/3$ $\leq \overline{T}_w$ in any $12\overline{T}_w$ weld length	
G	Slag inclusion, tungsten inclusion, or elongated indication Individual length Individual width Cumulative length	 $\leq 2\overline{T}_w$ ≤3 mm (1/8 in) and $\leq \overline{T}_w/2$ $\leq 4\overline{T}_w$ in any 150 mm (6 in) weld length	
H	Depth of undercut	≤1 mm (1/32 in) and $\leq \overline{T}_w/4$	
I	Depth of undercut	≤1.5 mm (1/16 in) and ≤[$\overline{T}_w/4$ or 1 mm (1/32 in)]	
J	Surface roughness	≤500 min. *Ra* per ASME B46.1	
K	Depth of root surface concavity	Total joint thickness, incl. weld reinf., $\geq \overline{T}_w$	
L	Height of reinforcement or internal protrusion [Note (8)] in any plane through the weld shall be within limits of the applicable height value in the tabulation at right, except as provided in Note (9). Weld metal shall merge smoothly into the component surfaces.	For $\overline{T}_w$ mm (in) ≤6 (1/4) >6 (1/4), ≤13 (1/2) >13 (1/2), ≤25 (1) >25 (1)	Height, mm (in) ≤1.5 (1/16) ≤3 (1/8) ≤4 (5/32) ≤5 (3/16)
M	Height of reinforcement or internal protrusion [Note (8)] as described in L. Note (9) does not apply.	Limit is twice the value applicable for L above	

TABLE 10-53 Acceptance Criteria for Welds and Examination Methods for Evaluating Weld Imperfections* *(Concluded)*

GENERAL NOTES:

(a) Weld imperfections are evaluated by one or more of the types of examination methods. The types and extent of examination shall comply with code requirements and those specified by engineering design.

(b) "N/A" indicates the code does not establish acceptance criteria or does not require evaluation of this kind of imperfection for this type of weld.

(c) Check (✓) indicates examination method generally used for evaluating this kind of weld imperfection.

(d) Ellipsis (…) indicates examination method not generally used for evaluating this kind of weld imperfection.

NOTES:

(1) Criteria given are for required examination. More stringent criteria may be specified in the engineering design. Other methods of examination may be specified by engineering design to supplement the examination required by the code. The extent of supplementary examination and any acceptance criteria differing from those specified by the code shall be specified by engineering design. Any examination method recognized by the code may be used to resolve doubtful indications. Acceptance criteria shall be those established by the code for the required examination.

(2) Longitudinal groove weld includes straight and spiral seam. Criteria are not intended to apply to welds made in accordance with component standards recognized by the code (ref. ASME.B31.3 Table A1 and Table 326.1); however, alternative leak test requirements dictate that all component welds be examined (see code for specific requirements).

(3) Fillet weld includes socket and seal welds, and attachment welds for slip-on flanges, branch reinforcement, and supports.

(4) Branch connection weld includes pressure containing welds in branches and fabricated laps.

(5) These imperfections are evaluated only for welds ≤ 5 mm (3/16 in) in nominal thickness.

(6) Where two limiting values are separated by *and* the lesser of the values determines acceptance. Where two sets of values are separated by *or* the larger value is acceptable. $\bar{T}_W$ is the nominal wall thickness of the thinner of two components joined by a butt weld.

(7) Tightly butted unfused root faces are unacceptable.

(8) For groove welds, height is the lesser of the measurements made from the surfaces of the adjacent components; both reinforcement and internal protrusion are permitted in a weld. For fillet welds, height is measured from the theoretical throat defined by the code; internal protrusion does not apply.

(9) For welds in aluminum alloy only, internal protrusion shall not exceed the following values:

(a) for thickness ≤2 mm (5/64 in): 1.5 mm (1/16 in)

(b) for thickness >2 mm and ≤6 mm (1/4 in): 2.5 mm (3/32 in)

For external reinforcement and for greater thicknesses, see the tabulation for Symbol L.

*Table 10-53 and notes have been extracted (with minor modifications) from Section 341 of *Process Piping*, ASME B31.3–2004, with permission of the publisher, the American Society of Mechanical Engineers, New York.

3. "Spot radiography" is the practice of making a single-exposure radiograph at a point within a specified extent of welding. Required coverage for a single spot radiograph is as follows:

- For longitudinal welds, at least 150 mm (6 in) of weld length.
- For girth, miter, and branch welds in piping 2½ in NPS and smaller, a single elliptical exposure which encompasses the entire weld circumference, and in piping larger than 2½ in NPS, at least 25 percent of the inside circumference or 150 mm (6 in), whichever is less.

Radiography of components other than castings and of welds shall be in accordance with Art. 2, Sec. V of the ASME Code. Limitations on imperfections in components other than castings and welds shall be as stated in Table 10-53 for the degree of radiography involved.

Ultrasonic Examination Ultrasonic examination of welds shall be in accordance with Art. 5, Sec. V of the ASME Code, except that the modifications stated in Par. 336.4.6 of the code shall be substituted for T542.2.1 and T542.8.1.1. Refer to the code for additional requirements.

Type and Extent of Required Examination The intent of examinations is to provide the examiner and the inspector with reasonable assurance that the requirements of the code and the engineering design have been met. For P-number 3, 4, and 5 materials, examination shall be performed after any heat treatment has been completed.

Examination Normally Required* Piping in normal fluid service shall be examined to the extent specified herein or to any greater extent specified in the engineering design. Acceptance criteria are as stated in the code for Normal Fluid Service unless more stringent requirements are specified.

1. *Visual examination.* At least the following shall be examined in accordance with code requirements:

a. Sufficient materials and components, selected at random, to satisfy the examiner that they conform to specifications and are free from defects.

b. At least 5 percent of fabrication. For welds, each welder's and welding operator's work shall be represented.

c. 100 percent of fabrication for longitudinal welds, except those in components made in accordance with a listed specification. Longitudinal welds required to have a joint efficiency of 0.9 must be spot-radiographed to the extent of 300 mm (12 in) in each 30 m (100 ft) of weld for each welder or welding operator. Acceptance criteria shall comply with code radiography acceptance criteria for Normal Fluid Service.

d. Random examination of the assembly of threaded, bolted, and other joints to satisfy the examiner that they conform to the applicable code requirements for erection and assembly. When pneumatic testing is to be performed, all threaded, bolted, and other mechanical joints shall be examined.

e. Random examination during erection of piping, including checking of alignment, supports, and cold spring.

f. Examination of erected piping for evidence of defects that would require repair or replacement, and for other evident deviations from the intent of the design.

2. *Other examination*

a. Not less than 5 percent of circumferential butt and miter groove welds shall be examined fully by random radiography or random ultrasonic examination in accordance with code requirements established for these methods. The welds to be examined shall be selected to ensure that the work product of each welder or welding operator doing the production welding is included. They shall also be selected to maximize coverage of intersections with longitudinal joints. When a circumferential weld with intersecting longitudinal weld(s) is examined, at least the adjacent 38 mm (1½ in) of each intersecting weld shall be examined. In-process examination in accordance with code requirements may be substituted for all or part of the radiographic or ultrasonic examination on a weld-for-weld basis if specified in the engineering design or specifically authorized by the Inspector.

b. Not less than 5 percent of all brazed joints shall be examined by in-process examination in accordance with the code definition of in-process examination, the joints to be examined being selected to ensure that the work of each brazer making the production joints is included.

3. *Certifications and records.* The examiner shall be assured, by examination of certifications, records, and other evidence, that the materials and components are of the specified grades and that they have received required heat treatment, examination, and testing. The examiner shall provide the Inspector with a certification that all the quality control requirements of the code and of the engineering design have been carried out.

*Extracted (with minor editing) from *Process Piping*, ASME B31.3—2004, paragraph 341, with permission of the publisher, the American Society of Mechanical Engineers, New York.

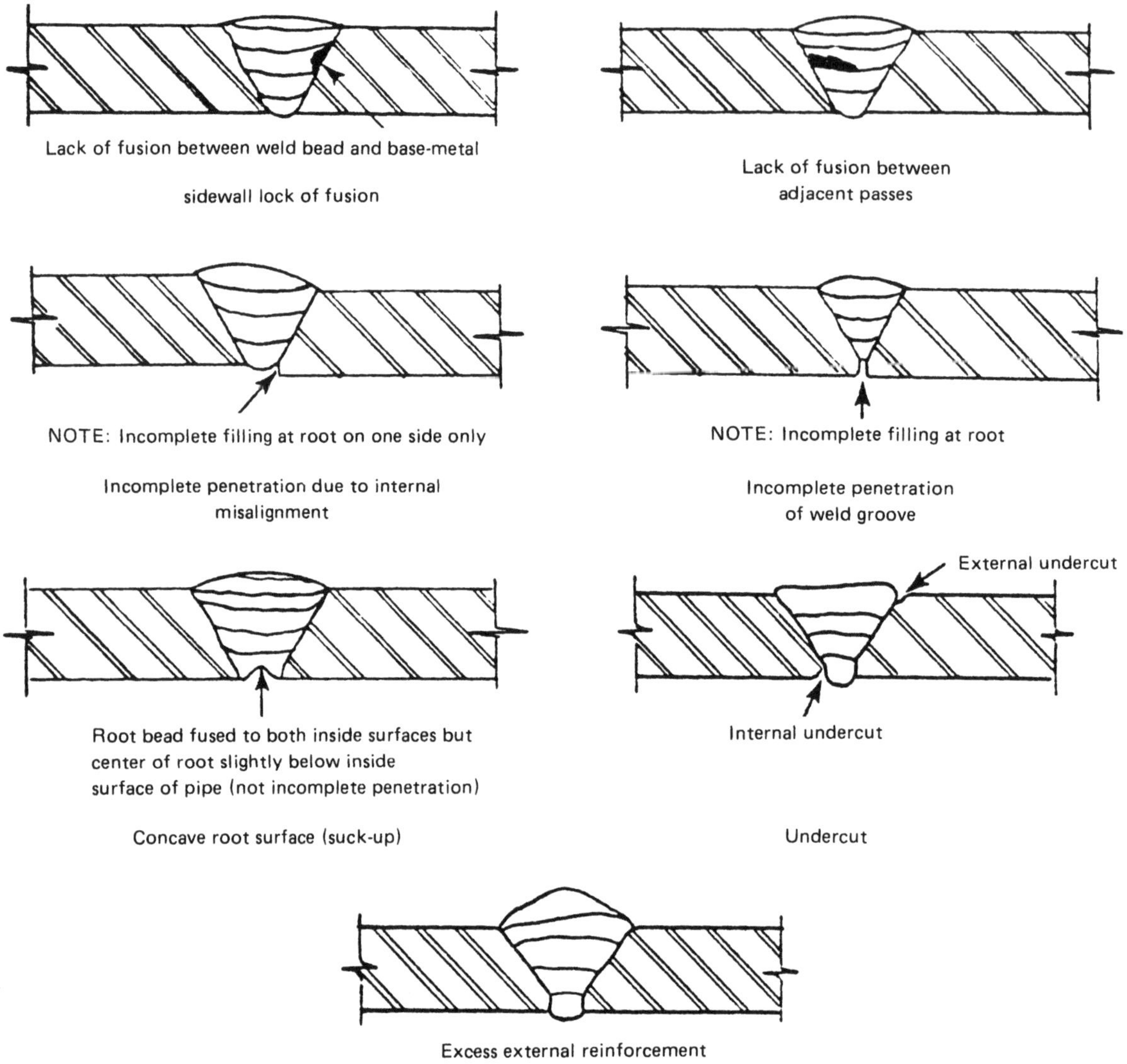

FIG. 10-180 Typical weld imperfections. (*Extracted from Process Piping, ASME B31.3—2004, with permission of the publisher, the American Society of Mechanical Engineers, New York.*)

Examination—Category D Fluid Service° Piping and piping elements for Category D fluid service as designated in the engineering design shall be visually examined in accordance with code requirements for visual examination to the extent necessary to satisfy the examiner that components, materials, and workmanship conform to the requirements of this code and the engineering design. Acceptance criteria shall be in accordance with code requirements and criteria in Table 10-53, for Category D fluid service, unless otherwise specified.

Examination—Severe Cyclic Conditions° Piping to be used under severe cyclic conditions shall be examined to the extent specified herein or to any greater extent specified in the engineering design. Acceptance criteria shall be in accordance with code requirements and criteria in Table 10-53, for severe cyclic conditions, unless otherwise specified.

°Extracted (with minor editing) from *Process Piping*, ASME B31.3—2004, paragraph 341, with permission of the publisher, the American Society of Mechanical Engineers, New York.

1. *Visual examination*. The requirements for Normal Fluid Service apply with the following exceptions.

a. All fabrication shall be examined.

b. All threaded, bolted, and other joints shall be examined.

c. All piping erection shall be examined to verify dimensions and alignment. Supports, guides, and points of cold spring shall be checked to ensure that movement of the piping under all conditions of startup, operation, and shutdown will be accommodated without undue binding or unanticipated constraint.

2. *Other examination*. All circumferential butt and miter groove welds and all fabricated branch connection welds comparable to those recognized by the code (ref. Fig. 10-130) shall be examined by 100% radiography or 100% ultrasonic (if specified in engineering design) in accordance with code requirements. Socket welds and branch connection welds which are not radiographed shall be examined by magnetic particle or liquid penetrant methods in accordance with code requirements.

3. In-process examination in accordance with the code definition, supplemented by appropriate nondestructive examination, may be

TABLE 10-54 Preheat Temperatures*

Base metal P-no. or S-no. [Note (1)]	Weld metal analysis A-no. [Note (2)]	Base metal group	Nominal wall thickness mm	in.	Specified min. tensile strength, base metal MPa	ksi	Min. temperature Required °C	°F	Recommended °C	°F
1	1	Carbon steel	< 25	<1	≤490	≤71	...	...	10	50
			≥ 25	≥1	All	All	...	...	79	175
			All	All	>490	>71	...	...	79	175
3	2,11	Alloy steels, Cr ≤½%	<13	<½	≤490	≤71	...	...	10	50
			≥13	≥½	All	All	...	...	79	175
			All	All	>490	>71	...	...	79	175
4	3	Alloy steels, ½% ≤ Cr ≤ 2%	All	All	All	All	149	300	...	...
5A, 5B, 5C	4, 5	Alloy steels, 2¼% ≤ Cr ≤ 10%	All	All	All	All	177	350	...	...
6	6	High alloy steels martensitic	All	All	All	All	...	...	149[4]	300[4]
7	7	High alloy steels ferritic	All	All	All	All	...	...	10	50
8	8, 9	High alloy steels austenitic	All	All	All	All	...	...	10	50
9A, 9B	10	Nickel alloy steels	All	All	All	All	...	...	93	200
10	...	Cr–Cu steel	All	All	All	All	149–204	300–400	...	...
101	...	27Cr steel	All	All	All	All	149[3]	300[3]	...	...
11A SG 1	...	8Ni, 9Ni steel	All	All	All	All	...	...	10	50
11A SG 2	...	5Ni steel	All	All	All	All	10	50	...	...
21–52	...	...	All	All	All	All	...	...	10	50

*Extracted from *Process Piping*, ASME B31.3—2004, with permission of the publisher, the American Society of Mechanical Engineers, New York.

NOTES:
(1) P-number or S-number from BPV Code, Section IX, QW/QB-442.
(2) A-number from Section IX, QW-442.
(3) Maintain interpass temperature between 177°C and 232°C (350°F and 450°F).
(4) Maximum interpass temperature 316°C (600°F).

substituted for the examination required in 2 above on a weld-for-weld basis if specified in the engineering design or specifically authorized by the Inspector.

4. *Certification and records.* The requirements established by the code for Normal Fluid Service apply.

Impact Testing In specifying materials, it is critical that the low-temperature limits of materials and impact testing requirements of the applicable code edition be clearly understood. In the recent past, code criteria governing low-temperature limits and requirements for impact testing have undergone extensive revision. The code contains extensive criteria detailing when impact testing is required and describing how it is to be performed. Because of the potentially changing requirements and the complexity of the code requirements, this handbook does not attempt to provide a comprehensive treatment of this subject or a comprehensive presentation of the requirements of the current code edition. Some of the general guidelines are provided here; however, the designer should consult the code to clearly understand additional requirements and special circumstances under which impact testing may be omitted. These exclusions permitted by the code may be significant in selecting materials or establishing material requirements.

In general, materials conforming to ASTM specifications listed by the code may be used at temperatures down to the lowest temperature listed for that material in ASME B31.3 Table A-10. When welding or other operations are performed on these materials, additional low-temperature toughness tests (impact testing) may be required. Refer to Table 10-51 for a general summary of these requirements.

Pressure Testing Prior to initial operation, installed piping shall be pressure-tested to assure tightness except as permitted for Category D fluid service described later. The pressure test shall be maintained for a sufficient time to determine the presence of any leaks but not less than 10 min.

If repairs or additions are made following the pressure tests, the affected piping shall be retested except that, in the case of minor repairs or additions, the owner may waive retest requirements when precautionary measures are taken to assure sound construction.

When pressure tests are conducted at metal temperatures near the ductile-to-brittle transition temperature of the material, the possibility of brittle fracture shall be considered.

The test shall be hydrostatic, using water, with the following exceptions. If there is a possibility of damage due to freezing or if the operating fluid or piping material would be adversely affected by water, any other suitable nontoxic liquid may be used. If a flammable liquid is used, its flash point shall not be less than 50°C (120°F), and consideration shall be given to the test environment.

The hydrostatic-test pressure at any point in the system shall be as follows:

1. Not less than 1½ times the design pressure.
2. For a design temperature above the test temperature, the minimum test pressure shall be as calculated by the following formula:

$$P_T = 1.5PS_T/S \qquad (10\text{-}108)$$

where P_T = test hydrostatic gauge pressure, MPa (lbf/in²)
P = internal design pressure, MPa (lbf/in²)
S_T = allowable stress at test temperature, MPa (lbf/in²)
S = allowable stress at design temperature, MPa (lbf/in²)
S_T/S shall not exceed 6.5

If the test pressure as so defined would produce a stress in excess of the yield strength at test temperature, the test pressure may be reduced

TABLE 10-55 Requirements for Heat Treatment*

Base metal P-no. or S-no. [Note (1)]	Weld-metal analysis A-no. [Note (2)]	Base metal group	Nominal wall thickness		Specified min. tensile strength, base metal		Metal temperature range		Holding time			Brinell hardness, max. [Note (4)]
									Nominal wall [Note (3)]		Min. time,	
			mm	in	MPa	ksi	°C	°F	min/mm	h/in	h	
1	1	Carbon steel	≤ 19	≤ ¾	All	All	None	None	. . .	. . .	. . .	. . .
			> 19	> ¾	All	All	593–649	1100–1200	2.4	1	1	. . .
3	2, 11	Alloy steels	≤ 19	≤ ¾	≤ 490	≤ 71	None	None	. . .	. . .	. . .	. . .
		Cr ½%	> 19	> ¾	All	All	593–718	1100–1325	2.4	1	1	225
			All	All	> 490	> 71	593–718	1100–1325	2.4	1	1	225
4 [Note (5)]	3	Alloy steels,	≤ 13	≤ ½	≤ 490	≤ 71	None	None	. . .	. . .	. . .	. . .
		½% < Cr ≤ 2%	> 13	> ½	All	All	704–746	1300–1375	2.4	1	2	225
			All	All	> 490	> 71	704–746	1300–1375	2.4	1	2	225
5A, 5B, 5C [Note (5)]	4, 5	Alloy steels 2¼% < Cr < 10% ≤ 3%										
		Cr and ≤ 0.15%	≤ 13	≤ ½	All	All	None	None	. . .	. . .	. . .	. . .
		C ≤ 3% Cr and ≤	> 13	> ½	All	All	704–760	1300–1400	2.4	1	2	241
		0.15% C > 3% Cr or ≤ 0.15%C	All	All	All	All	704–760	1300–1400	2.4	1	2	241
6	6	High-alloy steels:										
		martensitic	All	All	All	All	732–788	1350–1450	2.4	1	2	241
		A 240, Gr. 429	All	All	All	All	621–663	1150–1225	2.4	1	2	241
7	7	High-alloy steels: ferritic	All	All	All	All	None	None	. . .	. . .	. . .	. . .
8	8, 9	High-alloy steels: austenitic	All	All	All	All	None	None	. . .	. . .	. . .	. . .
9A, 9B	10	Nickel alloy steels	≤ 19	≤ ¾	All	All	None	None	. . .	. . .	. . .	. . .
			> 19	> ¾	All	All	593–635	1100–1175	1.2	½	1	. . .
10	. . .	Cr–Cu steel	All	All	All	All	760–816 [Note (6)]	1400–1500 [Note (6)]	1.2	½	½	. . .
10H	. . .	Duplex stainless steel	All	All	All	All	Note (7)	Note (7)	1.2	½	½	. . .
10I	. . .	27Cr steel	All	All	All	All	663–704 [Note (8)]	1225–1300 [Note (8)]	2.4	1	1	. . .
11A SG 1	. . .	8Ni, 9Ni steel	≤ 51	≤ 2	All	All	None	None	. . .	. . .	. . .	. . .
			> 51	> 2	All	All	552–585 [Note (9)]	1025–1085 [Note (9)]	2.4	1	1	. . .
11A SG 2	. . .	5Ni steel	> 51	> 2	All	All	552–585 [Note (9)]	1025–1085 [Note (9)]	2.4	1	1	. . .
62	. . .	Zr R60705	All	All	All	All	538–593 [Note (10)]	1000–1100 [Note (10)]	Note (10)	Note (10)	1	. . .

* Extracted from *Process Piping*, ASME B31.3—2004, with permission of the publisher, the American Society of Mechanical Engineers, New York.

NOTES:

(1) P-number or S-number from BPV Code, Section IX, QW/QB-422.
(2) A-number from Section IX, QW-442.
(3) For holding time in SI metric units, use min/mm (minutes per mm thickness). For U.S. units, use h/in thickness.
(4) See para. 331.1.7.[a]
(5) See Appendix F, para. F331.1.[a]
(6) Cool as rapidly as possible after the hold period.
(7) Postweld heat treatment is neither required nor prohibited, but any heat treatment applied shall be as required in the material specification.
(8) Cooling rate to 649°C (1200°F) shall be less than 56°C (100°F)/h; thereafter, the cooling rate shall be fast enough to prevent embrittlement.
(9) Cooling rate shall be > 167°C (300°F)/h to 316°C (600°F).
(10) Heat treat within 14 days after welding. Hold time shall be increased by ½ h for each 25 mm (1 in) over 25 mm thickness. Cool to 427°C (800°F) at a rate ≤ 278°C (500°F)/h, per 25 mm (1 in) nominal thickness, 278°C (500°F)/h max. Cool in still air from 427°C (800°F).

[a] Refer to the referenced code paragraph for details.

to the maximum pressure that will not exceed the yield strength at test temperature. If the test liquid in the system is subject to thermal expansion, precautions shall be taken to avoid excessive pressure.

A preliminary air test at not more than 0.17-MPa (25-lbf/in^2) gauge pressure may be made prior to hydrostatic test in order to locate major leaks.

If hydrostatic testing is not considered practicable by the owner, a pneumatic test in accordance with the following procedure may be substituted, using air or another nonflammable gas.

If the piping is tested pneumatically, the test pressure shall be 110 percent of the design pressure. When pneumatically testing nonmetallic materials, assure that the materials are suitable for compressed gas. Pneumatic testing involves a hazard due to the possible release of energy stored in compressed gas. Therefore, particular care must be taken to minimize the chance of brittle failure of metals and thermoplastics. The test temperature is important in this regard and must be considered when material is chosen in the original design. Any pneumatic test shall include a preliminary check at not more than 0.17-MPa (25-lbf/in^2) gauge pressure. The pressure shall be increased gradually in steps providing sufficient time to allow the piping to equalize strains during test and to check for leaks. Once test pressure has been achieved, the pressure shall be reduced to design pressure prior to examining for leakage.

TABLE 10-56 Types of Examination for Evaluating Imperfections*

Type of imperfection	Type of examination			
	Visual	Liquid-penetrant or magnetic-particle	Ultrasonic or radiographic: Random	Ultrasonic or radiographic: 100%
Crack	X	X	X	X
Incomplete penetration	X		X	X
Lack of fusion	X		X	X
Weld undercutting	X			
Weld reinforcement	X			
Internal porosity			X	X
External porosity	X			
Internal slag inclusions			X	X
External slag inclusions	X			
Concave root surface	X		X	X

*Extracted from the Chemical Plant and Petroleum Refinery Piping Code, ANSI B31.3—1980, with permission of the publisher, the American Society of Mechanical Engineers, New York. For code acceptance criteria (limits on imperfections) for welds see Table 10-53.

At the owner's option, a piping system used only for Category D fluid service as defined in the subsection "Classification of Fluid Service" may be tested at the normal operating conditions of the system during or prior to initial operation by examining for leaks at every joint not previously tested. A preliminary check shall be made at not more than 0.17-MPa (25-lbf/in^2) gauge pressure when the contained fluid is a gas or a vapor. The pressure shall be increased gradually in steps providing sufficient time to allow the piping to equalize strains during testing and to check for leaks.

Tests alternative to those required by these provisions may be applied under certain conditions described in the code.

Piping required to have a sensitive leak test shall be tested by the gas- and bubble-formation testing method specified in Art. 10, Sec. V of the ASME Boiler and Pressure Vessel Code or by another method demonstrated to have equal or greater sensitivity. The sensitivity of the test shall be at least (100 Pa·mL)/s [(10^3 atm·mL)/s] under test conditions.

Records shall be kept of each piping installation during the testing.

COST COMPARISON OF PIPING SYSTEMS

Piping may represent as much as 25 percent of the cost of a chemical-process plant. The installed cost of piping systems varies widely with the materials of construction and the complexity of the system. A study of piping costs shows that the most economical choice of material for a simple straight piping run may not be the most economical for a complex installation made up of many short runs involving numerous fittings and valves. The economics also depends heavily on the pipe size and fabrication techniques employed. Fabrication methods such as bending to standard long-radius-elbow dimensions and machine-flaring lap joints have a large effect on the cost of fabricating pipe from ductile materials suited to these techniques. Cost reductions of as high as 35 percent are quoted by some custom fabricators utilizing advanced techniques; however, the basis for pricing comparisons should be carefully reviewed.

Figure 10-181 is based on data extracted from a comparison of the installed cost of piping systems of various materials published by the Dow Chemical Co. The chart shows the relative cost ratios for systems of various materials based on two installations, one consisting of 152 m (500 ft) of 2-in pipe in a complex piping arrangement and the other of 305 m (1000 ft) of 2-in pipe in a straight-run piping arrangement. Figure 10-181 is based on field-fabrication construction techniques using welding stubs, the method commonly used by contractors. A considerably different ranking would result from using other construction methods such as machine-formed lap joints and bends in place of welding elbows. Piping-cost experience shows that it is difficult to generalize and reflect accurate piping-cost comparisons. Although the prices for many of the metallurgies shown in Fig. 10-181 are very volatile even over short periods, Fig. 10-181 may still be used as a reasonable initial estimate of the relative cost of metallic materials. The cost of nonmetallic materials and lined metallic materials versus solid alloy materials should be carefully reviewed prior to material selection. For an accurate comparison the cost for each type of material must be estimated individually on the basis of the actual fabrication and installation methods that will be used, pipe sizes and the conditions anticipated for the proposed installation.

FORCES OF PIPING ON PROCESS MACHINERY AND PIPING VIBRATION

The reliability of process rotating machinery is affected by the quality of the process piping installation. Excessive external forces and moments upset casing alignment and can reduce clearance between motor and casing. Further, the bearings, seals, and coupling can be adversely affected, resulting in repeated failures that may be correctly diagnosed as misalignment, and may have excessive piping forces as the root causes. Most turbine and compressor manufacturers have prescribed specification or will follow NEMA standards for allowable nozzle loading. For most of the pumps, API or ANSI pump standards will be followed when evaluating the pump nozzle loads. Pipe support restraints need to be placed at the proper locations to protect the machinery nozzles during operation.

Prior to any machinery alignment procedure, it is imperative to check for machine pipe strain. This is accomplished by the placement of dial indicators on the shaft and then loosening the hold-down bolts. Movements of greater than 1 mil are considered indication of a pipe strain condition.

This is an important practical problem area, as piping vibration can cause considerable downtime or even pipe failure.

Pipe vibration is caused by:

1. Internal flow (pulsation)
2. Plant machinery (such as compressors, pumps)

Pulsation can be problematic and difficult to predict. Pulsations are also dependent on acoustic resonance characteristics. For reciprocating equipment, such as reciprocating compressors and pumps, in some cases, an analog (digital) study needs to be performed to identify the deficiency in the piping and pipe support systems as well as to evaluate the performance of the machine during operation. The study will also provide recommendations on how to improve the machine and piping system's performance.

When a pulsation frequency coincides with a mechanical or acoustic resonance, severe vibration can result. A common cause for pulsation is the presence of flow control valves or pressure regulators. These often operate with high pressure drops (i.e., high flow velocities), which can result in the generation of severe pulsation. Flashing and cavitation can also contribute.

Modern-day piping design codes can model the vibration situation, and problems can thus be resolved in the design phases.

HEAT TRACING OF PIPING SYSTEMS

Heat tracing is used to maintain pipes and the material that pipes contain at temperatures above the ambient temperature. Two common uses of heat tracing are preventing water pipes from freezing and maintaining fuel oil pipes at high enough temperatures such that the viscosity of the fuel oil will allow easy pumping. Heat tracing is also used to prevent the condensation of a liquid from a gas and to prevent the solidification of a liquid commodity.

A heat-tracing system is often more expensive on an installed cost basis than the piping system it is protecting, and it will also have significant operating costs. A study on heat-tracing costs by a major chemical company showed installed costs of \$31/ft to \$142/ft and yearly operating costs of \$1.40/ft to \$16.66/ft. In addition to being a major cost, the heat-tracing system is an important component of the reliability of a piping system. A failure in the heat-tracing system will often render the piping system inoperable. For example, with a water freeze protection system, the piping system may be destroyed by the expansion of water as it freezes if the heat-tracing system fails.

The vast majority of heat-traced pipes are insulated to minimize heat loss to the environment. A heat input of 2 to 10 watts per foot is generally required to prevent an insulated pipe from freezing. With high wind speeds, an uninsulated pipe could require well over 100

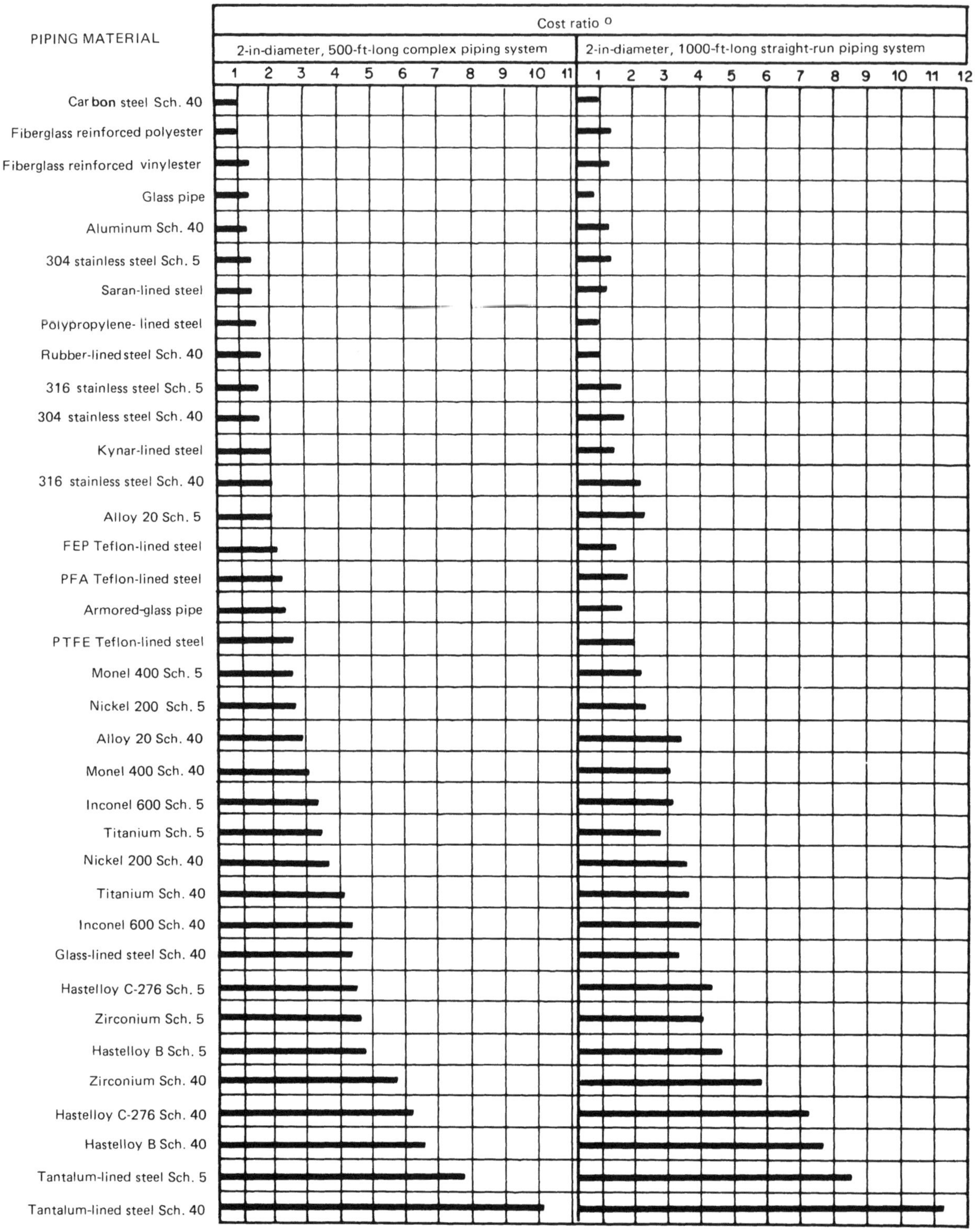

FIG. 10-181 Cost rankings and cost ratios for various piping materials. This figure is based on field-fabrication construction techniques using welding stubs, as this is the method most often employed by contractors. A considerably different ranking would result from using other construction methods, such as machined-formed lap joints, for the alloy pipe. °Cost ratio = (cost of listed item)/(cost of Schedule 40 carbon steel piping system, field-fabricated by using welding stubs). (*Extracted with permission from* Installed Cost of Corrosion Resistant Piping, *copyright 1977, Dow Chemical Co.*)

watts per foot to prevent freezing. Such a high heat input would be very expensive.

Heat tracing for insulated pipes is generally only required for the period when the material in the pipe is not flowing. The heat loss of an insulated pipe is very small compared to the heat capacity of a flowing fluid. Unless the pipe is extremely long (several thousands of feet), the temperature drop of a flowing fluid will not be significant.

The three major methods of avoiding heat tracing are:

1. Changing the ambient temperature around the pipe to a temperature that will avoid low-temperature problems. Burying water pipes below the frost line or running them through a heated building are the two most common examples of this method.

2. Emptying a pipe after it is used. Arranging the piping such that it drains itself when not in use, can be an effective method of avoiding the need for heat tracing. Some infrequently used lines can be pigged or blown out with compressed air. This technique is not recommended for commonly used lines due to the high labor requirement.

3. Arranging a process such that some lines have continuous flow can eliminate the need for tracing these lines. This technique is generally not recommended because a failure that causes a flow stoppage can lead to blocked or broken pipes.

Some combination of these techniques may be used to minimize the quantity of traced pipes. However, the majority of pipes containing fluids that must be kept above the minimum ambient temperature are generally going to require heat tracing.

Types of Heat-Tracing Systems Industrial heat-tracing systems are generally fluid systems or electrical systems. In fluid systems, a pipe or tube called the *tracer* is attached to the pipe being traced, and a warm fluid is put through it. The tracer is placed under the insulation. Steam is by far the most common fluid used in the tracer, although ethylene glycol and more exotic heat-transfer fluids are used. In electrical systems, an electrical heating cable is placed against the pipe under the insulation.

Fluid Tracing Systems Steam tracing is the most common type of industrial pipe tracing. In 1960, over 95 percent of industrial tracing systems were steam traced. By 1995, improvements in electric heating technology increased the electric share to 30 to 40 percent. Fluid systems other than steam are rather uncommon and account for less than 5 percent of tracing systems.

Half-inch copper tubing is commonly used for steam tracing. Three-eighths-inch tubing is also used, but the effective circuit length is then decreased from 150 ft to about 60 ft. In some corrosive environments, stainless steel tubing is used.

In addition to the tracer, a steam tracing system (Fig. 10-182) consists of steam supply lines to transport steam from the existing steam lines to the traced pipe, a steam trap to remove the condensate and hold back the steam, and in most cases a condensate return system to return the condensate to the existing condensate return system. In the past, a significant percentage of condensate from steam tracing was simply dumped to drains, but increased energy costs and environmental rules have caused almost all condensate from new steam tracing systems to be returned. This has significantly increased the initial cost of steam tracing systems.

Applications requiring accurate temperature control are generally limited to electric tracing. For example, chocolate lines cannot be exposed to steam temperatures or the product will degrade and if caustic soda is heated above 65°C (150°F) it becomes extremely corrosive to carbon steel pipes.

For some applications, either steam or electricity is simply not available and this makes the decision. It is rarely economic to install a steam boiler just for tracing. Steam tracing is generally considered only when a boiler already exists or is going to be installed for some other primary purpose. Additional electric capacity can be provided in most situations for reasonable costs. It is considerably more expensive to supply steam from a long distance than it is to provide electricity. Unless steam is available close to the pipes being traced, the automatic choice is usually electric tracing.

For most applications, particularly in processing plants, either steam tracing or electric tracing could be used, and the correct choice is dependent on the installed costs and the operating costs of the competing systems.

Economics of Steam Tracing versus Electric Tracing The question of the economics of various tracing systems has been examined thoroughly. All of these papers have concluded that electric tracing is generally less expensive to install and significantly less expensive to operate. Electric tracing has significant cost advantages in terms of installation because less labor is required than steam tracing. However, it is clear that there are some special cases where steam tracing is more economical.

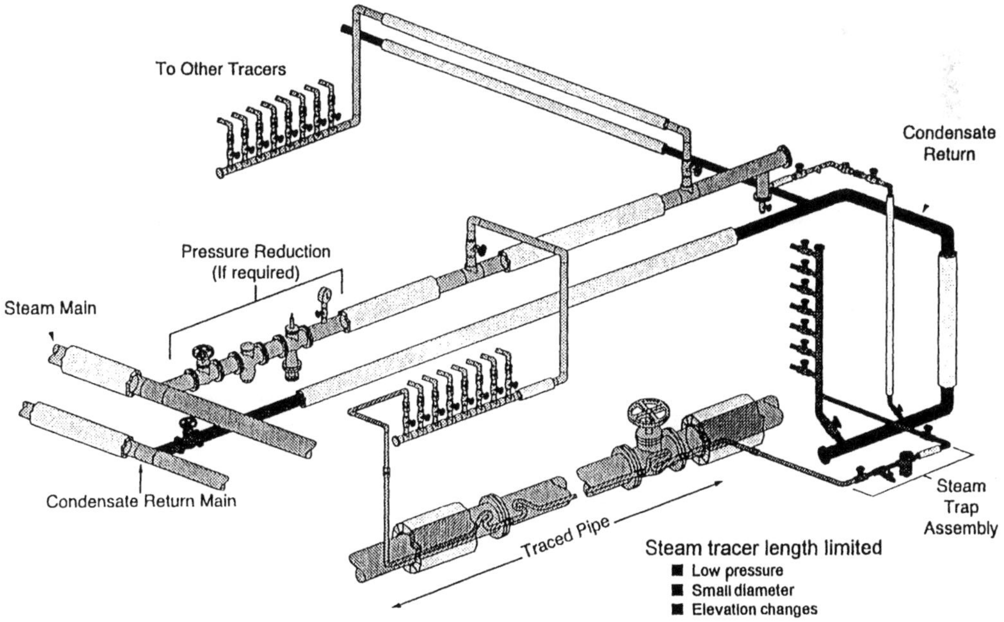

FIG. 10-182 Steam tracing system.

TABLE 10-57 Steam versus Electric Tracing*

Temperature maintained	TIC			TOC		
	Steam	Electric	Ratio S/E	Steam	Electric	Ratio S/E
50°F	22,265	7,733	2.88	1,671	334	5.00
150°F	22,265	13,113	1.70	4,356	1,892	2.30
250°F	22,807	17,624	1.29	5,348	2,114	2.53
400°F	26,924	14,056	1.92	6,724	3,942	1.71

*Specifications: 400 feet of four-inch pipe, $25/hr labor, $0.07/kWh, $4.00/1,000# steam, 100-foot supply lines. TIC = total installed cost; TOC = total operating costs.

The two key variables in the decision to use steam tracing or electric tracing are the temperature at which the pipe must be maintained and the distance to the supply of steam and a source of electric power.

Table 10-57 shows the installed costs and operating costs for 400 ft of 4-in pipe, maintained at four different temperatures, with supply lengths of 100 ft for both electricity and steam and $25/h labor.

The major advantages of a steam tracing system are:

1. *High heat output.* Due to its high temperature, a steam tracing system provides a large amount of heat to the pipe. There is a very high heat transfer rate between the metallic tracer and a metallic pipe. Even with damage to the insulation system, there is very little chance of a low temperature failure with a steam-tracing system.

2. *High reliability.* Many things can go wrong with a steam-tracing system but very few of the potential problems lead to a heat-tracing failure. Steam traps fail, but they usually fail in the open position, allowing for a continuous flow of steam to the tracer. Other problems such as steam leaks that can cause wet insulation are generally prevented from becoming heat-tracing failures by the extremely high heat output of a steam tracer. Also, a tracing tube is capable of withstanding a large amount of mechanical abuse without failure.

3. *Safety.* While steam burns are fairly common, there are generally fewer safety concerns than with electric tracing.

4. *Common usage.* Steam tracing has been around for many years and many operators are familiar with the system. Because of this familiarity, failures due to operator error are not very common.

The weaknesses of a steam-tracing system are:

1. *High installed costs.* The incremental piping required for the steam supply system and the condensate return system must be installed, insulated, and, in the case of the supply system, additional steam traps are often required. The tracer itself is not expensive, but the labor required for installation is relatively high. Studies have shown that steam tracing systems typically cost from 50 to 150 percent more than a comparable electric-tracing system.

2. *Energy inefficiency.* A steam-tracing system's total energy use is often more than 20 times the actual energy requirement to keep the pipe at the desired temperature. The steam tracer itself puts out significantly more energy than required. The steam traps use energy even when they are properly operating and waste large amounts of energy when they fail in the open position, which is the common failure mode. Steam leaks waste large amounts of energy, and both the steam supply system and the condensate return system use significant amounts of energy.

3. *Poor temperature control.* A steam-tracing system offers very little temperature control capability. The steam is at a constant temperature (50 psig steam is 300°F) usually well above that desired for the pipe. The pipe will reach an equilibrium temperature somewhere between the steam temperature and the ambient temperature. However, the section of pipe against the steam tracer will effectively be at the steam temperature. This is a serious problem for temperature-sensitive fluids such as food products. It also represents a problem with fluids such as bases and acids, which are not damaged by high temperatures but often become extremely corrosive to piping systems at higher temperatures.

4. *High maintenance costs.* Leaks must be repaired and steam traps must be checked and replaced if they have failed. Numerous studies have shown that, due to the energy lost through leaks and failed steam traps, an extensive maintenance program is an excellent investment. Steam maintenance costs are so high that for low-temperature maintenance applications, total steam operating costs are sometimes greater than electric operating costs, even if no value is placed on the steam.

Electric Tracing An electric-tracing system (see Fig. 10-183) consists of an electric heater placed against the pipe under the thermal insulation, the supply of electricity to the tracer, and any control or monitoring system that may be used (optional). The supply of electricity to the tracer usually consists of an electrical panel and electrical conduit or cable trays. Depending on the size of the tracing system and the capacity of the existing electrical system, an additional transformer may be required.

Advantages of Electric Tracing

1. *Lower installed and operating costs.* Most studies have shown that electric tracing is less expensive to install and less expensive to

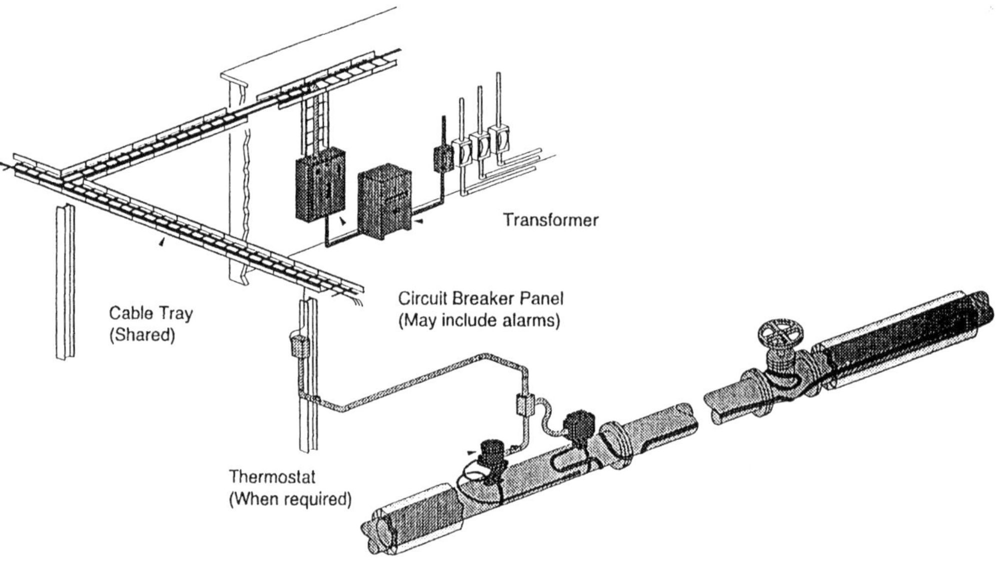

FIG. 10-183 Electrical heat tracing system.

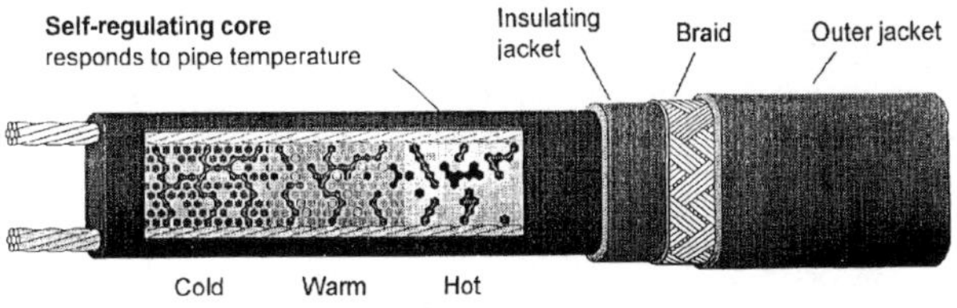

FIG. 10-184 Self-regulating heating cable.

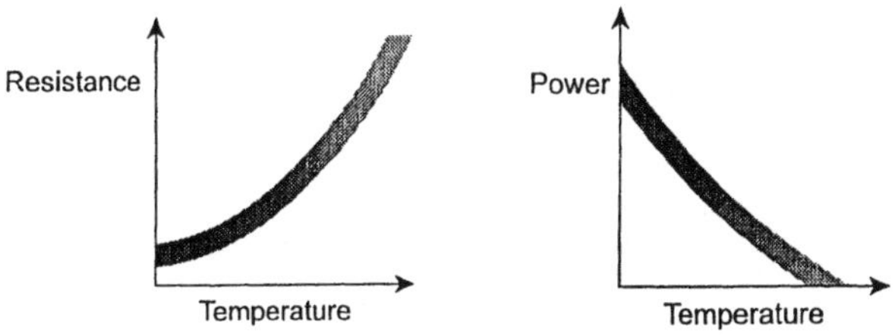

FIG. 10-185 Self-regulation.

operate. This is true for most applications. However, for some applications, the installed costs of steam tracing are equal to or less than electric tracing.

2. *Reliability.* In the past, electric heat tracing had a well-deserved reputation for poor reliability. However, since the introduction of self-regulating heaters in 1971, the reliability of electric heat tracing has improved dramatically. Self-regulating heaters cannot destroy themselves with their own heat output. This eliminates the most common failure mode of polymer-insulated constant wattage heaters. Also, the technology used to manufacture mineral-insulated cables, high-temperature electric heat tracing, has improved significantly, and this has improved their reliability.

3. *Temperature control.* Even without a thermostat or any control system, an electric tracing system usually provides better temperature control than a steam tracing system. With thermostatic or electronic control, very accurate temperature control can be achieved.

4. *Safety.* The use of self-regulating heaters and ground leakage circuit breakers has answered the safety concerns of most engineers considering electric tracing. Self-regulating heaters eliminate the problems from high-temperature failures, and ground leakage circuit breakers minimize the danger of an electrical fault to ground, causing injury or death.

5. *Monitoring capability.* One question often asked about any heat-tracing system is, "How do I know it is working?" Electric tracing now has available almost any level of monitoring desired. The temperature at any point can be monitored with both high and low alarm capability. This capability has allowed many users to switch to electric tracing with a high degree of confidence.

6. *Energy efficiency.* Electric heat tracing can accurately provide the energy required for each application without the large additional energy use of a steam system. Unlike steam tracing systems, other parts of the system do not use significant amounts of energy.

Disadvantages of Electric Tracing

1. *Poor reputation.* In the past, electric tracing has been less than reliable. Due to past failures, some operating personnel are unwilling to take a chance on any electric tracing.

2. *Design requirements.* A slightly higher level of design expertise is required for electric tracing than for steam tracing.

3. *Lower power output.* Since electric tracing does not provide a large multiple of the required energy, it is less forgiving to problems such as damaged insulation or below design ambient temperatures. Most designers include a 10 to 20 percent safety factor in the heat loss calculation to cover these potential problems. Also, a somewhat higher than required design temperature is often specified to provide an additional safety margin. For example, many water systems are designed to maintain 50°F to prevent freezing.

Types of Electric Tracing **Self-regulating electric tracing** (see Fig. 10-184) is by far the most popular type of electric tracing. The heating element in a self-regulating heater is a conductive polymer between the bus wires. This conductive polymer increases its resistance as its temperature increases. The increase in resistance with temperature causes the heater to lower its heat output at any point where its temperature increases (Fig. 10-185). This self-regulating effect eliminates the most common failure mode of constant wattage electric heaters, which is destruction of the heater by its own heat output.

Because self-regulating heaters are parallel heaters, they may be cut to length at any point without changing their power output per unit of length. This makes them much easier to deal with in the field. They may be terminated, teed, or spliced in the field with hazardous-area-approved components.

MI Cables Mineral insulated cables (Fig. 10-186) are the electric heat tracers of choice for high-temperature applications. High-temperature applications are generally considered to maintain temperatures above 250°F or exposure temperatures above 420°F where self-regulating heaters cannot be used. MI cable consists of one or two heating wires, magnesium oxide insulation (from whence it gets its name), and an outer metal sheath. Today the metal sheath is generally inconel. This eliminates both the corrosion problems with copper sheaths and the stress cracking problems with stainless steel.

MI cables can maintain temperatures up to 1200°F and withstand exposure to up to 1500°F. The major disadvantage of MI cable is that it must be factory-fabricated to length. It is very difficult to terminate

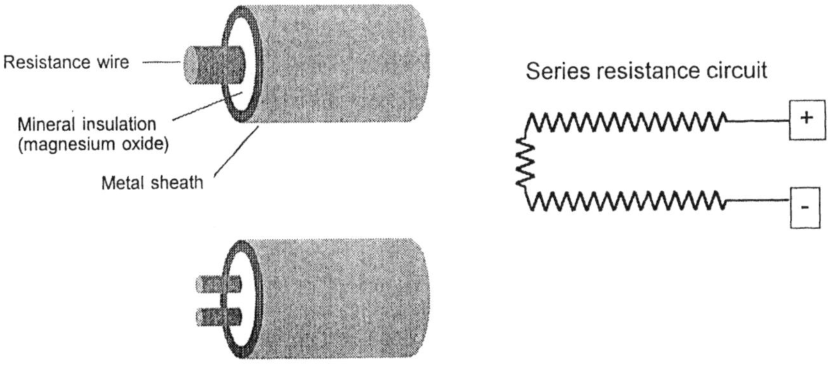

FIG. 10-186 Mineral insulated cable (MI cable).

or splice the heater in the field. This means pipe measurements are necessary before the heaters are ordered. Also, any damage to an MI cable generally requires a complete new heater. It's not as easy to splice in a good section as with self-regulating heaters.

Polymer-Insulated Constant Wattage Electric Heaters These are slightly cheaper than self-regulating heaters, but they are generally being replaced with self-regulating heaters due to inferior reliability. These heaters tend to destroy themselves with their own heat output when they are overlapped at valves or flanges. Since overlapping self-regulating heaters is the standard installation technique, it is difficult to prevent this technique from being used on the similar-looking constant-wattage heaters.

SECT (Skin-Effect Current Tracing) This is a special type of electric tracing employing a tracing pipe, usually welded to the pipe being traced, that is used for extremely long lines. With SECT tracing circuits, up to 10 miles can be powered from one power point. All SECT systems are specially designed by heat-tracing vendors

Impedance Tracing This uses the pipe being traced to carry the current and generate the heat. Less than 1 percent of electric heat-tracing systems use this method. Low voltages and special electrical isolation techniques are used. Impedance heating is useful when extremely high heat densities are required, like when a pipe containing aluminum metal must be melted from room temperature on a regular basis. Most impedance systems are specially designed by heat tracing vendors.

Choosing the Best Tracing System Some applications require either steam tracing or electric tracing regardless of the relative economics. For example, a large line that is regularly allowed to cool and needs to be quickly heated would require steam tracing because of its much higher heat output capability. In most heat-up applications, steam tracing is used with heat-transfer cement, and the heat output is increased by a factor of up to 10. This is much more heat than would be practical to provide with electric tracing. For example, a half-inch copper tube containing 50 psig steam with heat transfer cement would provide over 1100 Btu/hr/ft to a pipe at 50°F. This is over 300 watts per foot or more than 15 times the output of a high-powered electric tracer.

Table 10-57 shows that electric tracing has a large advantage in terms of cost at low temperatures and smaller but still significant advantages at higher temperatures. Steam tracing does relatively better at higher temperatures because steam tracing supplies significantly more power than is necessary to maintain a pipe at low temperatures. Table 10-57 indicates that there is very little difference between the steam tracing system at 50°F and the system at 250°F. However, the electric system more than doubles in cost between these two temperatures because more heaters, higher powered heaters, and higher temperature heaters are required.

The effect of supply lengths on a 150°F system can be seen from Table 10-58. Steam supply pipe is much more expensive to run than electrical conduit. With each system having relatively short supply lines (40 ft each) the electric system has only a small cost advantage (10 percent, or a ratio of 1.1). This ratio is 2.1 at 50°F and 0.8 at 250°F. However, as the supply lengths increase, electric tracing has a large cost advantage.

TABLE 10-58 Effect of Supply Lengths

Ratio of steam TIC to electric TIC maintained at 150°F

	Electric supply length		
Steam supply length	40 feet	100 feet	300 feet
40 feet	1.1	1.0	0.7
100 feet	1.9	1.7	1.1
300 feet	4.9	4.2	2.9

STORAGE AND PROCESS VESSELS

STORAGE OF LIQUIDS

Atmospheric Tanks The term *atmospheric tank* as used here applies to any tank that is designed to operate at pressures from atmospheric through 3.45 kPag (0.5 psig). It may be either open to the atmosphere or enclosed. Atmospheric tanks may be either shop-fabricated or field-erected. The most common application is storage of fuel for transportation and power generation. See the subsections "Pressure Tanks" and "Pressure Vessels" later in this section.

Shop-Fabricated Storage Tanks A shop-fabricated storage tank is a storage tank constructed at a tank manufacturer's plant and transported to a facility for installation. In general, tanks 190 m^3 (50,000 gal) and under can be shop-fabricated and shipped in one piece to an installation site. Shop-fabricated storage tanks may be either underground storage tanks (USTs) or aboveground storage tanks (ASTs).

USTs versus ASTs For decades, USTs were the standard means of storing petroleum and chemicals in quantities of 190 m^3 (50,000 gal) or less. However, during the 1990s, many industrial and commercial facilities shifted to ASTs for hazardous product storage. Reasons include the ability to visually monitor the storage tank as well as to avoid perceived risks and the myriad regulations surrounding underground storage tanks. Nonetheless, AST installations are also subject to certain regulations and codes, particularly fire codes.

The choice of utilizing USTs or ASTs is driven by numerous factors. Local authorities having jurisdiction may allow only underground storage. Total required storage capacity factored against available real estate may also preclude the use of ASTs. In addition, ASTs are subject to minimum distance separations from one another and from buildings, property lines, and public ways. ASTs are visually monitorable, yet that same attribute may be aesthetically undesirable. Other considerations are adequate protection against spills, vandalism, and vehicular damage. Additionally, central design elements regarding product transfer and system functionality must be taken into account.

USTs are subject to myriad EPA regulations as well as fire codes. Further, a commonly cited drawback is the potential for unseen leaks and subsequent environmental damage and cleanup. However, advances in technology have addressed these concerns. Corrosion protection and leak detection are now standard in all UST systems. Sophisticated tank and pipe secondary containment systems have been developed to meet the EPA's secondary containment mandate for underground storage of nonpetroleum chemicals. Computerized in-tank monitoring devices for tracking inventory, evolutionary tank integrity testing equipment, statistical inventory reconciliation analysis, leak-free dry-disconnect pipe and hose joints for loading/unloading, and in-tank fill shutoff valves are just a few of the many pieces of equipment which have surfaced as marketplace solutions.

Properly designed and installed in accordance with industry standards, regulations, and codes, both UST and AST systems are reliable and safe. Because of space limitations and the prevalence of ASTs at plant sites, only ASTs will be discussed further.

Aboveground Storage Tanks Aboveground storage tanks are classified as either field-erected or shop-fabricated. The latter is typically 190-m^3 (50,000-gal) capacity or less and is able to be shipped over the highway, while larger tanks are more economically erected in the field. Whereas field-erected tanks likely constitute the majority of total AST storage *capacity*, shop-fabricated ASTs constitute the clear majority of the total *number* of ASTs in existence today. Most of these shop-fabricated ASTs store flammable or combustible liquids at atmospheric pressure and have 45-m^3 (12,000-gal) capacity or smaller.

Shop-fabricated ASTs can be designed and fabricated as pressure vessels, but are more typically vented to atmosphere. They are oriented for either horizontal or vertical installation and are made in either cylindrical or rectangular form. Tanks are often secondarily contained and may also include insulation for fire safety or temperature control. Compartmented ASTs are also available. Over 90 percent of the atmospheric tank applications store some sort of hydrocarbon. Within that, a majority are used to store fuels for motor vehicle dispensing.

Fire Codes Many chemicals are classified as either flammable or combustible or subject to fire codes. All hydrocarbon tanks are classified as flammable or combustible. Notably, with the tremendous increase in the use of ASTs at private fleet fueling facilities in the 1990s, fire codes were rapidly modified to address the understandable safety concerns. In the United States, two principal organizations publish fire codes pertinent to underground storage tanks, with each state adopting all or part of the respective codes.

The **National Fire Protection Association (NFPA)** has developed several principal codes pertaining to the storage of flammable and combustible liquids:

NFPA 30, Flammable and Combustible Liquids Code (2003)
NFPA 30A, Code for Motor Fuel Dispensing Facilities and Repair Garages (2003)
NFPA 31, Standard for the Installation of Oil Burning Equipment

The **International Code Council (ICC)** was formed by the consolidation of three formerly separate fire code organizations: International Conference of Building Officials (ICBO), which had published the Uniform Fire Code under its fire service arm, the International Fire Code Institute (IFCI); Building Officials and Code Administrators (BOCA), which had published the National Fire Prevention Code; and Southern Building Congress Code International (SBCCI), which had published the Standard Fire Prevention Code. When the three groups merged in 2000, in part to develop a common fire code, the individual codes became obsolete; however, they are noted above since references to them may periodically surface. The consolidated code is **IFC-2006, International Fire Code.**

The **Canadian Commission on Building and Fire Codes (CCBFC)** developed a recommended model code to permit adoption by various regional authorities. The National Research Council of Canada publishes the model code document **National Fire Code of Canada (1995).**

Standards Third-party standards for AST fabrication have evolved significantly over the past two decades, as have various recommended practice guidelines for the installation and operations of AST systems. Standards developed by Underwriters Laboratories (UL) have been perhaps the most predominant guidelines—in fact, ASTs are often categorized according to the UL standard that they meet, such as "a UL 142 tank."

Underwriters Laboratories Inc. standards for steel ASTs storing flammable and combustible liquids include the following:

UL 142, Steel Aboveground Tanks for Flammable and Combustible Liquids, 9th ed., covers steel atmospheric tanks intended for aboveground storage of noncorrosive, stable, flammable, and combustible liquids that have a specific gravity not exceeding that of water. UL 142 applies to single-wall and double-wall horizontal carbon-steel and stainless steel tanks up to 190 m^3 (50,000 gal), with a maximum diameter of 3.66 m (12 ft) and a maximum length-to-diameter ratio of 8 : 1. A formula from *Roark's Formulas for Stress and Strain* has been incorporated within UL 58 to calculate minimum steel wall thicknesses. UL 142 also applies to vertical tanks up to 10.7 m (35 ft) in height. UL 142 has been the primary AST standard since its development in 1922.

Tanks covered by these requirements can be fabricated in cylindrical or rectangular configurations. The standard covers secondary contained tanks, either of dual-wall construction or a tank in a steel dike or bund. It also provides listings for special AST constructions, such as those used under generators for backup power. These tanks are fabricated, inspected, and tested for leakage before shipment from the factory as completely assembled vessels.

UL 142 provides details for steel type, wall thickness, compartments and bulkheads, manways, and other fittings and appurtenances. Issues relating to leakage, venting, and the ability of the tank to withstand the development of internal pressures encountered during operation and production leak testing are also addressed.

These requirements do not apply to large field-erected storage tanks covered by the Standard for Welded Steel Tanks for Oil Storage, ANSI/API 650, or the Specification for Field-Welded Tanks for Storage of Production Liquids, API 12D; or to the Specification for Shop-Welded Tanks for Storage of Production Liquids, API 12F.

UL 2085, Protected Aboveground Tanks for Flammable and Combustible Liquids, covers shop-fabricated, aboveground atmospheric protected tanks intended for storage of stable, flammable or combustible liquids that have a specific gravity not greater than 1.0 and that are compatible with the material and construction of the tank.

The UL 2085 tank construction is intended to limit the heat transferred to the primary tank when the AST is exposed to a 2-h hydrocarbon pool fire of 1093°C (2000°F). The tank must be insulated to withstand the test without leakage and with an average maximum temperature rise on the primary tank not exceeding 127°C (260°F). Temperatures on the inside surface of the primary tank cannot exceed 204°C (400°F).

UL 2085 also provides criteria for resistance against vehicle impact, ballistic impact, and fire hose impact. These tanks are also provided with integral secondary containment intended to prevent any leakage from the primary tank from entering the environment.

UL 2080, Fire Resistant Tanks for Flammable and Combustible Liquids, is similar to UL 2085 tanks, except with an average maximum temperature rise on the primary tank limited to 427°C (800°F) during a 2-h pool fire. Temperatures on the inside surface of the primary tank cannot exceed 538°C (1000°F).

UL 2244, Standard for Aboveground Flammable Liquid Tank Systems, covers factory-fabricated, preengineered aboveground atmospheric tank systems intended for dispensing flammable or combustible liquids, such as gasoline or diesel fuel, into motor vehicles, generators, or aircraft.

UL 80, Steel Tanks for Oil-Burner Fuel, covers the design and construction of welded steel tanks of the atmospheric type with a maximum capacity of 0.23 to 2.5 m^3 (60 to 660 gal) intended for unenclosed installation inside of buildings or for outside aboveground applications as permitted by the Standard for Installation of Oil-Burning Equipment, NFPA 31, primarily for the storage and supply of fuel oil for oil burners.

UL 2245, Below-Grade Vaults for Flammable Liquid Storage Tanks, covers below-grade vaults intended for the storage of flammable or combustible liquids in an aboveground atmospheric tank. Below-grade vaults, constructed of a minimum of 150 mm (6 in) of reinforced concrete or other equivalent noncombustible material, are designed to contain one aboveground tank, which can be a compartment tank. Adjacent vaults may share a common wall. The lid of the vault may be at grade or below. Vaults provide a safe means to install hazardous tanks so that the system is accessible to the operator without unduly exposing the public.

Southwest Research Institute (SwRI) standards for steel ASTs storing flammable and combustible liquids include the following:

SwRI 93-01, Testing Requirements for Protected Aboveground Flammable Liquid/Fuel Storage Tanks, includes tests to evaluate the performance of ASTs under fire, hose stream, ballistics, heavy vehicular impact, and different environments. This standard requires pool-fire resistance similar to that of UL 2085.

SwRI 97-04, Standard for Fire Resistant Tanks, includes tests to evaluate the performance of ASTs under fire and hose stream. This standard is similar to UL 2080 in that the construction is exposed to a 2-h hydrocarbon pool fire of 1093°C (2000°F). However, SwRI 97-04 is concerned only with the integrity of the tank after the 2-h test and not concerned with the temperature inside the tank due to the heat transfer. As a result, UL 142 tanks have been tested to the SwRI standard and passed. Secondary containment with insulation is not necessarily an integral component of the system.

Underwriters Laboratories of Canada (ULC) publishes a number of standards for aboveground tanks and accessories. All the following pertain to the aboveground storage of flammable liquids and combustible liquids such as gasoline, fuel oil, or similar products with a relative density not greater than 1.0:

ULC S601, Shop Fabricated Steel Aboveground Horizontal Tanks for Flammable and Combustible Liquids, covers single- and double-wall cylindrical horizontal steel nonpressure tanks. These requirements do not cover tanks of capacities greater than 200 m^3.

ULC S630, Shop Fabricated Steel Aboveground Vertical Tanks for Flammable and Combustible Liquids, covers single- and double-wall cylindrical vertical steel nonpressure tanks.

ULC S643, Shop Fabricated Steel Utility Tanks for Flammable and Combustible Liquids, covers single- and double-wall cylindrical horizontal steel nonpressure tanks.

ULC S653-94, Aboveground Steel Contained Tank Assemblies for Flammable and Combustible Liquids, covers steel contained tank assemblies.

ULC S655, Aboveground Protected Tank Assemblies for Flammable and Combustible Liquids, covers shop-fabricated primary tanks that provided with secondary containment and protective encasement and are intended for stationary installation and use in accordance with

1. National Fire Code of Canada, Part 4
2. CAN/CSA-B139, Installation Code for Oil Burning Equipment
3. The Environmental Code of Practice for Aboveground Storage Tank Systems Containing Petroleum Products
4. The requirements of the authority having jurisdiction

ULC/ORD C142.20, Secondary Containment for Aboveground Flammable and Combustible Liquid Storage Tanks, covers secondary containments for aboveground primary tanks.

ULC S602, Aboveground Steel Tanks for the Storage of Combustible Liquids Intended to Be Used as Heating and/or Generator Fuels, covers the design and construction of tanks of the nonpressure type, intended for installation inside or outside buildings. This standard covers single-wall tanks and tanks with secondary containment, having a maximum capacity of 2.5 m^3.

The **Petroleum Equipment Institute (PEI)** has developed a recommended practice for AST system installation: **PEI-RP200, RP 200, Recommended Practices for Installation of Aboveground Storage Systems for Motor Vehicle Fueling.**

The **American Petroleum Institute (API)** has developed a series of standards and specifications involving ASTs:

API 12F, Shop Welded Tanks for Storage of Production Liquids
RP 12R1, Setting, Maintenance, Inspection, Operation, and Repair of Tanks in Production Service
RP 575, Inspection of Atmospheric and Low Pressure Storage Tanks
RP 579, Fitness-For-Service
API 650, Welded Steel Tanks for Oil Storage
API 652, Lining of Aboveground Storage Tank Bottoms
API 653, Tank Inspection, Repair, Alteration, and Reconstruction
API 2350, Overfill Protection for Storage Tanks in Petroleum Facilities (overfill is the primary cause of AST releases)

The **American Society of Mechanical Engineers (ASME)** has developed a standard for welded aluminum-alloy storage tanks: **B96.1, Welded Aluminum-Alloy Storage Tanks** (1999), covers the design, materials, fabrication, erection, inspection, and testing requirements for welded aluminum alloy (field-erected or shop-fabricated) aboveground, vertical, cylindrical, flat bottom, open or closed-top tanks storing liquids under pressures approximating atmospheric pressure at ambient temperatures.

The **American Water Works Association (AWWA)** has many standards dealing with water handling and storage. A list of its publications is given in the *AWWA Handbook* (annually). AWWA D100, Standard for Steel Tanks—Standpipes, Reservoirs, and Elevated Tanks for Water Storage, contains rules for design and fabrication. Although AWWA tanks are intended for water, they could be used for the storage of other liquids.

The **Steel Tank Institute (STI)** publishes construction standards and recommended installation practices pertaining to ASTs fabricated to STI technologies. STI's recommended installation practices are notable for their applicability to similar respective technologies:

SP001-06, Standard for Inspection of In-Service Shop Fabricated Aboveground Tanks for Storage of Combustible and Flammable Liquids
R912-00, Installation Instructions for Shop Fabricated Aboveground Storage Tanks for Flammable, Combustible Liquids
F921, Standard for Aboveground Tanks with Integral Secondary Containment
Standard for Fire Resistant Tanks (Flameshield)
Standard for Fireguard Thermally Insulated Aboveground Storage Tanks
F911, Standard for Diked Aboveground Storage Tanks

The **National Association of Corrosion Engineers** (NACE International) has developed the following to protect the soil side of bottoms of on-grade carbon-steel storage tanks: **NACE RP0193-01, Standard Recommended Practice—External Cathodic Protection of On-Grade Metallic Storage Tank Bottoms.**

Environmental Regulations A key **United States Environmental Protection Agency (US EPA)** requirement for certain aboveground storage facilities is the development and submittal of Spill Prevention Control and Countermeasure (SPCC) Plans within 40 CFR 112, the Oil Pollution Prevention regulation, which in turn is part of the Clean Water Act (CWA). SPCC Plans and Facility Response Plans pertain to facilities which may discharge oil into groundwater or storm runoff, which in turn may flow into navigable waters. Enacted in 1973, these requirements were principally used by owners of large, field-fabricated aboveground tanks predominant at that time, although the regulation applied to all bulk containers larger than 2.5 m^3 (660 gal) and included the requirement for a Professional Engineer to certify the spill plan.

In July 2002, the US EPA issued a final rule amending 40 CFR 112 which included differentiation of shop-fabricated from field-fabricated ASTs. The rule also includes new subparts outlining the requirements for various classes of oil, revises the applicability of the regulation, amends the requirements for completing SPCC Plans, and makes other modifications.

The revised rule also states that all bulk storage container installations must provide a secondary means of containment for the entire capacity of the largest single container, with sufficient freeboard to contain precipitation, and that such containment is sufficiently impervious to contain discharged oil. US EPA encourages the use of industry standards to comply with the rules. Many owners of shop-fabricated tanks have opted for double-wall tanks built to STI or UL standards as a means to comply with this requirement.

State and Local Jurisdictions Due to the manner in which aboveground storage tank legislation was promulgated in 1972 for protection of surface waters from oil pollution, state environmental agencies did not receive similar jurisdiction as they did within the underground storage tank rules. Nonetheless, many state environmental agencies, state fire marshals, or Weights and Measures departments—including Minnesota, Florida, Wisconsin, Virginia, Oklahoma, Missouri, Maryland, Delaware, and Michigan—are presently regulating aboveground storage tanks through other means. Other regulations exist for hazardous chemicals and should be consulted for specific requirements.

Aboveground Storage Tank Types and Options Most hydrocarbon storage applications use carbon steel as the most economical and available material that provides suitable strength and compatibility for the specific storage application. For vertical tanks installed on grade, corrosion protection can be given to exterior tank bottoms in contact with soil. The interior of the tank can incorporate special coatings and linings (e.g., polymer, glass, or other metals). Some chemical storage applications will dictate that the storage tank be made from a stainless steel or nickel alloy. Fiberglass-reinforced plastic (FRP), polyethylene, or polypropylene may be used for nonflammable storage in smaller sizes. Suppliers can be contacted to verify the appropriate material to be used.

As stated earlier, shop-fabricated ASTs are often categorized according to the standards to which the tanks are fabricated, e.g., a UL 142 or UL 2085 tank. That said, however, there are defined categories such as diked tanks, protected tanks, fire-resistant tanks, and insulated tanks. It's critical, then, for the tank buyer or specifier to know precisely what is required or desired given the application, code requirements, and/or owner/operator preferences—and to discuss this with the tank contractor and/or manufacturer.

Cylindrical or rectangular tanks storing flammable and combustible liquids (UL 142 ASTs) will normally comply with UL 142. The Seventh Edition published in 1993 was particularly notable, as it incorporated secondary containment designs (diking or steel secondary containment tanks) and rectangular tank designs.

Rectangular tanks became a desirable option for small tanks, typically less than 7.6 m^3 (2000 gal), as operators liked the accessibility of the flat top to perform operations and maintenance without the need for special ladders or catwalks.

Posttensioned concrete is frequently used for tanks to about 57,000 m^3 (15×10^6 gal), usually containing water. Their design is treated in detail by Creasy (*Prestressed Concrete Cylindrical Tanks*, Wiley, New York, 1961). For the most economical design of large open tanks at ground levels, he recommends limiting vertical height to 6 m (20 ft). Seepage can be a problem if unlined concrete is used with some liquids (e.g., gasoline).

Elevated tanks can supply a large flow when required, but pump capacities need be only for average flow. Thus, they may save on pump and piping investment. They also provide flow after pump failure, an important consideration for fire systems.

Open tanks may be used to store materials that will not be harmed by water, weather, or atmospheric pollution. Otherwise, a roof, either fixed or floating, is required. **Fixed roofs** are usually either domed or coned. Large tanks have coned roofs with intermediate supports. Since negligible pressure is involved, snow and wind are the principal design loads. Local building codes often give required values.

Fixed-roof atmospheric tanks require **vents** to prevent pressure changes which would otherwise result from temperature changes and withdrawal or addition of liquid. API Standard 2000, Venting Atmospheric and Low Pressure Storage Tanks, gives practical rules for conservative vent design. The principles of this standard can be applied to fluids other than petroleum products. Excessive losses of volatile liquids, particularly those with flash points below 38°C (100°F), may result from the use of open vents on fixed-roof tanks. Sometimes vents are manifolded and led to a vent tank, or the vapor may be extracted by a recovery system.

An effective way of preventing vent loss is to use one of the many types of **variable-volume tanks.** These are built under API Standard 650. They may have floating roofs of the double-deck or the single-deck type. There are lifter-roof types in which the roof either has a skirt moving up and down in an annular liquid seal or is connected to the tank shell by a flexible membrane. A fabric expansion chamber housed in a compartment on top of the tank roof also permits variation in volume.

Floating roofs must have a seal between the roof and the tank shell. If not protected by a fixed roof, they must have drains for the removal of water, and the tank shell must have a "wind girder" to avoid distortion. An industry has developed to retrofit existing tanks with floating roofs. Much detail on the various types of tank roofs is given in manufacturers' literature. Figure 10-187 shows types. These roofs cause less condensation buildup and are highly recommended.

Fire-Rated or Insulated ASTs: Protected and Fire-Resistant These ASTs have received much attention within the fire regulatory community, particularly for motor fuel storage and dispensing applications and generator base tanks. National model codes have been revised to allow this type of storage aboveground.

An insulated tank can be a protected tank, built to third-party standards UL 2085 and/or SwRI 93-01, or a fire-resistant tank built to UL 2080 or SwRI 97-04. Protected tanks were developed in line with NFPA requirements and terminology, while fire-resistant ASTs were developed in line with Uniform Fire Code (now International Fire Code) requirements and terminology. Both protected tanks and fire-resistant tanks must pass a 1093°C (2000°F), 2-h fire test.

The insulation properties of many fire-rated ASTs marketed today are typically provided by concrete; i.e., the primary steel tank is surrounded by concrete. Due to the weight of concrete, this design is normally limited to small tanks. Another popular AST technology meeting all applicable code requirements for insulated tanks and fabricated to UL 2085 is a tank that utilizes a lightweight monolithic thermal insulation in between two walls of steel to minimize heat transfer from the outer tank to the inner tank and to make tank handling more palatable.

A **secondary containment AST** to prevent contamination of our environment has become a necessity for all hazardous liquid storage, regardless of its chemical nature, in order to minimize liability and protect neighboring property. A number of different regulations exist, but the regulations with the greatest impact are fire codes and the US EPA SPCC rules for oil storage.

In 1991, the Spill Prevention Control and Countermeasure (SPCC) rule proposed a revision to require secondary containment that was impermeable for at least 72 h after a release occurred. The 2003 promulgated EPA SPCC rule no longer mandates a 72-h containment requirement, instead opting to require means to contain releases until they can be detected and removed. Nonetheless, the need for impermeable containment continues to position steel as a material of choice for shop-fabricated tanks. However, release prevention barriers made from plastic or concrete can also meet US EPA requirements when frequently inspected for releases.

Diked ASTs Fire codes dictate flammable and combustible liquid tanks have spill control in the form of dike, remote impounding, or small secondary containment tanks. The dike must contain the content of the largest tank to prevent hazardous liquids from endangering the public and property. Traditional bulk storage systems will include multiple tanks within a concrete or earthen dike wall.

From a shop-fabricated tank manufacturer's perspective, a diked AST generally refers to a steel tank within a factory-fabricated steel box, or dike. An example of a diked AST is the STI F911 standard, providing an open-topped steel rectangular dike and floor as support and secondary containment of a UL 142 steel tank. The dike will contain 110 percent of the tank capacity; as rainwater may already have collected in the dike, the additional 10 percent acts as freeboard should a catastrophic failure dump a full tank's contents into a dike. Many fabricators offer steel dikes with rain shields to prevent precipitation from collecting.

A **double-wall AST** of steel fulfills the same function as a diked AST with rain shield. Double-wall designs consist of a steel wrap over a horizontal or vertical steel storage tank. The steel wrap provides an intimate, secondary containment over the primary tank. One such design is the Steel Tank Institute's F921 standard, based upon

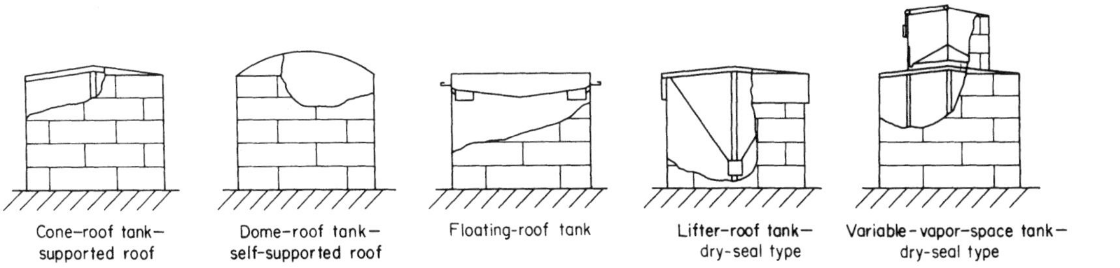

FIG. 10-187 Some types of atmospheric storage tanks.

UL 142-listed construction for the primary tank, outer containment, associated tank supports or skids.

Venting of ASTs is critical, since they are exposed to greater ambient temperature fluctuations than are USTs. Properly designed and sized venting, both normal (atmospheric) and emergency, is required. Normal vents permit the flow of air or inert gas into and out of the tank. The vent line must be large enough to accommodate the maximum filling or withdrawal rates without exceeding the allowable stress for the tank.

Fire codes' recommended installation procedures also detail specifics on pressure/vacuum venting devices and vent line flame arresters. For example, codes mandate different ventilation requirements for Class I-A liquids versus Class I-B or I-C liquids. Tank vent piping is generally not manifolded unless required for special purposes such as air pollution control or vapor recovery. As always, local codes must be consulted.

Emergency venting prevents an explosion during a fire or other emergency. All third-party laboratory standards except UL 80 include emergency relief provisions, since these tanks are designed for atmospheric pressure conditions.

Separation distances are also important. Aboveground storage tanks must be separated from buildings, property lines, fuel dispensers and delivery trucks in accordance with the level of safety the tank design provides, depending upon whether they are constructed of traditional steel or are vault/fire-resistant.

For most chemical storage tanks, codes such as NFPA 30 and the International Fire Code give specific separation distances. For motor vehicle fueling applications, the codes are more stringent on separation requirements due to a greater exposure of the public to the hazards. Hence codes such as NFPA 30A establish variable separation distances depending upon whether the facility is private or public.

Separation distance requirements may dictate whether a tank buyer purchases a traditional steel UL 142 tank, a fire-resistant tank, or a tank in a vault. For example, NFPA 30, NFPA 30A, and the IFC codes allow UL 2085 tanks to be installed closer to buildings and property lines, thereby reducing the real estate necessary to meet fire codes.

Under NFPA 30A, dispensers may be installed directly over vaults or upon fire-resistant tanks at fleet-type installations, whereas a 7.6- to 15.2-m (25- to 50-ft) separation distance is required at retail-type service station installations. The IFC only allows gasoline and diesel to be dispensed from ASTs, designed with a 2-h fire rating. Non-2-h fire-rated UL 142 tanks dispensing diesel can be installed upon approval of the local authority.

Maintenance and Operations Water in any storage system can cause myriad problems from product quality to system degradation due to corrosion from trace contaminants and microbial action. Subsequently, any operations and maintenance program must include a proactive program of monitoring for water—and removing it if found. Other operations and maintenance procedures would include periodic integrity testing and corrosion control for vertical tank bottoms. Additional guidance is available from organizations such as the American Petroleum Institute (API), Petroleum Equipment Institute (PEI), ASTM International, National Oilheat Research Alliance (NORA), and Steel Tank Institute (STI). Also see the Steel Tank Institute document *Keeping Water Out of Your Storage System* (http://www.steeltank.com/library/pubs/waterinfueltanks.pdf).

Integrity testing and visual inspection requirements are discussed in the SPCC requirements, Subpart B, Para. 112.8 c (6). Chemical tanks storing toluene and benzene, e.g., are subject to the rule in addition to traditional fuels. But the need for a good inspection program is recommended regardless of applicable regulations. Both visual inspection and another testing technique are required. Comparison records must be kept, and frequent inspections must be made of the outside of the tank and system components for signs of deterioration, discharges, or accumulation of oil inside diked areas.

For inspection of large field-erected tanks, API 653, *Tank Inspection, Repair, Alteration, and Reconstruction*, is referenced by the US EPA. A certified inspector must inspect tanks. US EPA references the Steel Tank Institute Standard SP001-06, *Standard for Inspection of In-Service Shop Fabricated Aboveground Tanks for Storage of Combustible and Flammable Liquids*, as an industry standard that may assist an owner or operator with the integrity testing and inspection of shop-fabricated tanks. The STI SP001-06 standard includes inspection techniques for all types of shop-fabricated tanks—horizontal cylindrical, vertical, and rectangular. SP001-06 also addresses tanks that rest directly on the ground or on release prevention barriers, tanks that are elevated on supports, and tanks that are either single- or double-wall using a risk-based approach.

Pressure Tanks Vertical cylindrical tanks constructed with domed or coned roofs, which operate at pressures above several hundred pascals (a few pounds per square foot) but which are still relatively close to atmospheric pressure, can be built according to API Standard 650. The pressure force acting against the roof is transmitted to the shell, which may have sufficient weight to resist it. If not, the uplift will act on the tank bottom. The strength of the bottom, however, is limited, and if it is not sufficient, an anchor ring or a heavy foundation must be used. In the larger sizes uplift forces limit this style of tank to very low pressures.

As the size or the pressure goes up, curvature on all surfaces becomes necessary. Tanks in this category, up to and including a pressure of 103.4 kPa (15 lbf/in^2), can be built according to API Standard 620. Shapes used are spheres, ellipsoids, toroidal structures, and circular cylinders with torispherical, ellipsoidal, or hemispherical heads. The ASME Pressure Vessel Code (Sec. VIII of the ASME Boiler and Pressure Vessel Code), although not required below 103.4 kPa (15 lbf/in^2), is also useful for designing such tanks.

Tanks that could be subjected to vacuum should be provided with vacuum-breaking valves or be designed for vacuum (external pressure). The ASME Pressure Vessel Code contains design procedures.

Calculation of Tank Volume A tank may be a single geometrical element, such as a cylinder, a sphere, or an ellipsoid. It may also have a compound form, such as a cylinder with hemispherical ends or a combination of a toroid and a sphere. To determine the volume, each geometrical element usually must be calculated separately. Calculations for a full tank are usually simple, but calculations for partially filled tanks may be complicated.

To calculate the volume of a **partially filled horizontal cylinder** refer to Fig. 10-188. Calculate the angle α in degrees. Any units of length can be used, but they must be the same for H, R, and L. The liquid volume

$$V = LR^2\left(\frac{\alpha}{57.30} - \sin\alpha\,\cos\alpha\right) \tag{10-109}$$

This formula may be used for any depth of liquid between zero and the full tank, provided the algebraic signs are observed. If H is greater than R, $\sin\alpha\cos\alpha$ will be negative and thus will add numerically to $\alpha/57.30$. Table 10-59 gives liquid volume, for a partially filled horizontal cylinder, as a fraction of the total volume, for the dimensionless ratio H/D or $H/2R$.

The **volumes of heads** must be calculated separately and added to the volume of the cylindrical portion of the tank. The four types of heads most frequently used are the standard dished head,* torispherical or ASME head, ellipsoidal head, and hemispherical head. Dimensions and volumes for all four of these types are given in *Lukens Spun*

*The standard dished head does not comply with the ASME Pressure Vessel Code.

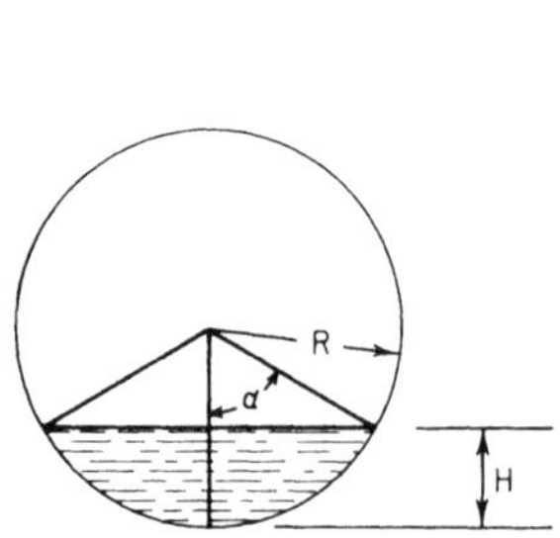

FIG. 10-188 Calculation of partially filled horizontal tanks. H = depth of liquid; R = radius; D = diameter; L = length; α = half of the included angle; and $\cos\alpha = 1 - H/R = 1 - 2H/D$.

TABLE 10-59 Volume of Partially Filled Horizontal Cylinders

H/D	Fraction of volume	H/D	Fraction of volume	H/D	Fraction of volume	H/D	Fraction of volume
0.01	0.00169	0.26	0.20660	0.51	0.51273	0.76	0.81545
.02	.00477	.27	.21784	.52	.52546	.77	.82625
.03	.00874	.28	.22921	.53	.53818	.78	.83688
.04	.01342	.29	.24070	.54	.55088	.79	.84734
.05	.01869	.30	.25231	.55	.56356	.80	.85762
.06	.02450	.31	.26348	.56	.57621	.81	.86771
.07	.03077	.32	.27587	.57	.58884	.82	.87760
.08	.03748	.33	.28779	.58	.60142	.83	.88727
.09	.04458	.34	.29981	.59	.61397	.84	.89673
.10	.05204	.35	.31192	.60	.62647	.85	.90594
.11	.05985	.36	.32410	.61	.63892	.86	.91491
.12	.06797	.37	.33636	.62	.65131	.87	.92361
.13	.07639	.38	.34869	.63	.66364	.88	.93203
.14	.08509	.39	.36108	.64	.67590	.89	.94015
.15	.09406	.40	.37353	.65	.68808	.90	.94796
.16	.10327	.41	.38603	.66	.70019	.91	.95542
.17	.11273	.42	.39858	.67	.71221	.92	.96252
.18	.12240	.43	.41116	.68	.72413	.93	.96923
.19	.13229	.44	.42379	.69	.73652	.94	.97550
.20	.14238	.45	.43644	.70	.74769	.95	.98131
.21	.15266	.46	.44912	.71	.75930	.96	.98658
.22	.16312	.47	.46182	.72	.77079	.97	.99126
.23	.17375	.48	.47454	.73	.78216	.98	.99523
.24	.18455	.49	.48727	.74	.79340	.99	.99831
.25	.19550	.50	.50000	.75	.80450	1.00	1.00000

Heads, Lukens Inc., Coatesville, Pennsylvania. Approximate volumes can also be calculated by the formulas in Table 10-60. Consistent units must be used in these formulas.

A partially filled horizontal tank requires the determination of the partial volume of the heads. The Lukens catalog gives approximate volumes for partially filled (axis horizontal) standard ASME and ellipsoidal heads. A formula for **partially filled heads** (excluding conical), by Doolittle [*Ind. Eng. Chem.,* **21,** 322–323 (1928)], is

$$V = 0.215H^2(3R - H) \qquad (10\text{-}110)$$

where in consistent units V = volume, R = radius, and H = depth of liquid. Doolittle made some simplifying assumptions which affect the volume given by the equation, but the equation is satisfactory for determining the volume as a fraction of the entire head. This fraction, calculated by Doolittle's formula, is given in Table 10-61 as a function of H/D_i (H is the depth of liquid, and D_i is the inside diameter). Table 10-61 can be used for standard dished, torispherical, ellipsoidal, and hemispherical heads with an error of less than 2 percent of the volume of the entire head. The error is zero when H/D_i = 0, 0.5, and 1.0. Table 10-61 cannot be used for conical heads.

When a tank volume cannot be calculated or when greater precision is required, **calibration** may be necessary. This is done by draining (or filling) the tank and measuring the volume of liquid. The measurement may be made by weighing, by a calibrated fluid meter, or by repeatedly filling small measuring tanks which have been calibrated by weight.

Container Materials and Safety Storage tanks are made of almost any structural material. Steel and reinforced concrete are most widely used. Plastics and glass-reinforced plastics are used for tanks up to about 230 m^3 (60,000 gal). Resistance to corrosion, light weight, and lower cost are their advantages. Plastic and glass coatings are also applied to steel tanks. Aluminum and other nonferrous metals are used when their special properties are required. When expensive metals such as tantalum are required, they may be applied as tank linings or as clad metals.

Some grades of steel listed by API and AWWA Standards are of lower quality than is customarily used for pressure vessels. The stresses allowed by these standards are also higher than those allowed by the ASME Pressure Vessel Code. Small tanks containing nontoxic substances are not particularly hazardous and can tolerate a reduced factor of safety. Tanks containing highly toxic substances and very large tanks containing any substance can be hazardous. The designer must consider the magnitude of the hazard. The possibility of brittle behavior of ferrous metal should be taken into account in specifying materials (see subsection "Safety in Design").

Volume 1 of National Fire Codes (National Fire Protection Association, Quincy, Massachusetts) contains recommendations (Code 30) for venting, drainage, and dike construction of tanks for **flammable liquids.**

Container Insulation Tanks containing materials above atmospheric temperature may require insulation to reduce loss of heat. Almost any of the commonly used insulating materials can be employed. Calcium silicate, glass fiber, mineral wool, cellular glass, and plastic foams are among those used. Tanks exposed to weather must have jackets or protective coatings, usually asphalt, to keep water out of the insulation.

Tanks with contents at lower than atmospheric temperature may require insulation to minimize heat absorption. The insulation must have

TABLE 10-60 Volumes of Heads*

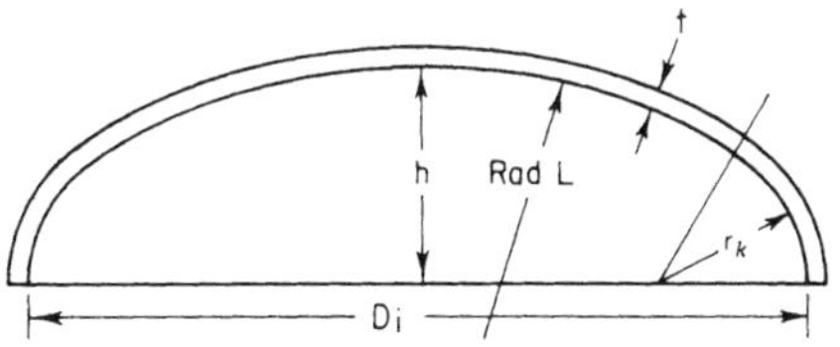

Type of head	Knuckle radius, r_k	h	L	Volume	% Error	Remarks
Standard dished	Approx. $3t$		Approx. D_i	Approx. $0.050D_i^3 + 1.65tD_i^2$	±10	h varies with t
Torispherical or A.S.M.E.	$0.06L$		D_i	$0.0809D_i^3$	±0.1	r_k must be the larger of $0.06L$ and $3t$
Torispherical or A.S.M.E.	$3t$		D_i	Approx. $0.513hD_i^2$	±8	
Ellipsoidal				$\pi D_i^2 h/6$	0	
Ellipsoidal		$D_i/4$		$\pi D_i^3/24$	0	Standard proportions
Hemispherical		$D_i/2$	$D_i/2$	$\pi D_i^3/12$	0	
Conical				$\pi h(D_i^2 + D_i d + d^2)/12$	0	Truncated cone h = height d = diameter at small end

*Use consistent units.

TABLE 10-61 Volume of Partially Filled Heads on Horizontal Tanks*

H/D_i	Fraction of volume	H/D_i	Fraction of volume	H/D_i	Fraction of volume	H/D_i	Fraction of volume
0.02	0.0012	0.28	0.1913	0.52	0.530	0.78	0.8761
.04	.0047	.30	.216	.54	.560	.80	.8960
.06	.0104	.32	.242	.56	.590	.82	.9145
.08	.0182	.34	.268	.58	.619	.84	.9314
.10	.0280	.36	.295	.60	.648	.86	.9467
.12	.0397	.38	.323	.62	.677	.88	.9603
.14	.0533	.40	.352	.64	.705	.90	.9720
.16	.0686	.42	.381	.66	.732	.92	.9818
.18	.0855	.44	.410	.68	.758	.94	.9896
.20	.1040	.46	.440	.70	.784	.96	.9953
.22	.1239	.48	.470	.72	.8087	.98	.9988
.24	.1451	.50	.500	.74	.8324	1.00	1.0000
.26	.1676			.76	.8549		

*Based on Eq. (10-110).

a vapor barrier at the outside to prevent condensation of atmospheric moisture from reducing its effectiveness. An insulation not damaged by moisture is preferable. The insulation techniques presently used for refrigerated systems can be applied (see subsection "Low-Temperature and Cryogenic Storage").

Tank Supports Large vertical atmospheric steel tanks may be built on a base of about 150 cm (6 in) of sand, gravel, or crushed stone if the subsoil has adequate bearing strength. It can be level or slightly coned, depending on the shape of the tank bottom. The porous base provides drainage in case of leaks. A few feet beyond the tank perimeter the surface should drop about 1 m (3 ft) to assure proper drainage of the subsoil. API Standard 650, Appendix B, and API Standard 620, Appendix C, give recommendations for tank foundations.

The bearing pressure of the tank and contents must not exceed the **bearing strength** of the soil. Local building codes usually specify allowable soil loading. Some approximate bearing values are:

	kPa	Tons/ft^2
Soft clay (can be crumbled between fingers)	100	1
Dry fine sand	200	2
Dry fine sand with clay	300	3
Coarse sand	300	3
Dry hard clay (requires a pick to dig it)	350	3.5
Gravel	400	4
Rock	1000–4000	10–40

For high, heavy tanks, a foundation ring may be needed. Prestressed concrete tanks are sufficiently heavy to require foundation rings. Foundations must extend below the frost line. Some tanks that are not flat-bottomed may also be supported by soil if it is suitably graded and drained. When soil does not have adequate bearing strength, it may be excavated and backfilled with a suitable soil, or piles capped with a concrete mat may be required.

Spheres, spheroids, and toroids use steel or concrete saddles or are supported by columns. Some may rest directly on soil. Horizontal cylindrical tanks should have two rather than multiple saddles to avoid indeterminate load distribution. Small horizontal tanks are sometimes supported by legs. Most tanks must be designed to resist the reactions of the saddles or legs, and they may require reinforcing. Neglect of this can cause collapse. Tanks without stiffeners usually need to make contact with the saddles on at least 2.1 rad (120°) of their circumference. An elevated steel tank may have either a circle of steel columns or a large central steel standpipe. Concrete tanks usually have concrete columns. Tanks are often supported by buildings.

Pond and Underground Storage Low-cost liquid materials, if they will not be damaged by rain or atmospheric pollution, may be stored in **ponds.** A pond may be excavated or formed by damming a ravine. To prevent loss by seepage, the soil which will be submerged may require treatment to make it sufficiently impervious. This can also be accomplished by lining the pond with concrete, plastic film, or some other barrier. Prevention of seepage is especially necessary if the pond contains material that could contaminate present or future water supplies.

Underground Cavern Storage Large volumes of liquids are often stored below ground in artificial caverns as an economical alternative to aboveground tanks and other modes of storage. The liquid to be stored must tolerate water, brine, and other contaminants that are usually present to some degree in the cavern. The liquids that are most commonly stored are natural gas liquids (NGLs), LPGs, crude oil, and refined petroleum products. If the liquid is suitable for cavern storage, this method may be less expensive, safer, and more secure than other storage modes.

There are two types of caverns used for storing liquids. *Hard rock (mined)* caverns are constructed by mining rock formations such as shale, granite, limestone, and many other types of rock. *Solution-mined* caverns are constructed by dissolution processes, i.e., solution mining or leaching a mineral deposit, most often salt (sodium chloride). The salt deposit may take the form of a massive salt dome or thinner layers of bedded salt that are stratified between layers of rock. Hard rock and solution-mined caverns have been constructed in the United States and many other parts of the world.

Mined Caverns Caverns mined in hard rock are generally situated 100 to 150 m (325 to 500 ft) below ground level. These caverns are constructed by excavating rock with conventional drill-and-blast mining methods. The excavated cavern consists of a grouping of interconnecting tunnels or storage "galleries." Mined caverns have been constructed for volumes ranging from as little as 3200 to 800,000 m^3 [20,000 API barrels* (bbl) to 5 million bbl]. Figure 10-189 illustrates a typical mined cavern for liquid storage.

Hard rock caverns are designed so that the internal storage pressure is at all times less than the hydrostatic head of the water contained in the surrounding rock matrix. Thus, the depth of a cavern determines its maximum allowable operating pressure. Groundwater that continuously seeps into hard rock caverns in permeable formations is periodically pumped out of the cavern. The maximum operating pressure of the cavern is established after a thorough geological and hydrogeological evaluation is made of the rock formation and the completed cavern is pressure-tested.

Salt Caverns Salt caverns are constructed in both *domal salt*, more commonly referred to as "salt domes," and *bedded salt*, which consists of a body of salt sandwiched between layers of rock. The greatest total volume of underground liquid storage in the United States is stored in salt dome caverns. A salt dome is a large body, mostly consisting of sodium chloride, which over geologic time moved upward thousands of feet from extensive halite deposits deep below the earth's crust. There are numerous salt domes in the United States and other parts of the world [see Harben, P. W., and R. L. Bates, "Industrial Minerals Geology and World Deposits," *Metal Bulletin Plc*, UK, pp. 229–234 (1990)]. An individual salt dome may exceed 1 mi in diameter and contain many storage caverns. The depth to the top of a salt dome may range from a few hundred to several thousand feet, although depths to about 1070 m (3500 ft) are commercially viable for cavern development. The extent of many salt domes allows for caverns of many different sizes and depths to be developed. The extensive nature of salt domes has allowed the development of caverns as large as 5.7 × 10^6 m^3 (36 million bbl) (US DOE Bryan Mound Strategic Petroleum Reserve) and larger; however, cavern volumes of 159,000 to 1.59 × 10^6 m^3 (1 to 10 million bbl) are more common for liquid storage.

The benefits of salt are its high compressive strength of 13.8 to 27.6 MPa (2000 to 4000 psi), its impermeability to hydrocarbon liquids and gases, and its non-chemically reactive (inert) nature. Due to the impervious nature of salt, the maximum allowed storage pressure gradient in this type of cavern is greater than that of a mined cavern. A typical storage pressure gradient for liquids is about 18 kPa/m of depth (0.80 psi/ft) to the bottom of the well casing. Actual maximum and minimum allowable operating pressure gradients are determined from geologic evaluations and rock mechanics studies. Typical depths to the top of a salt cavern may range from 500 to 4000 ft (about 150 to 1200 m).

*One API barrel = 42 US gal = 5.615 ft^3 = 0.159 m^2.

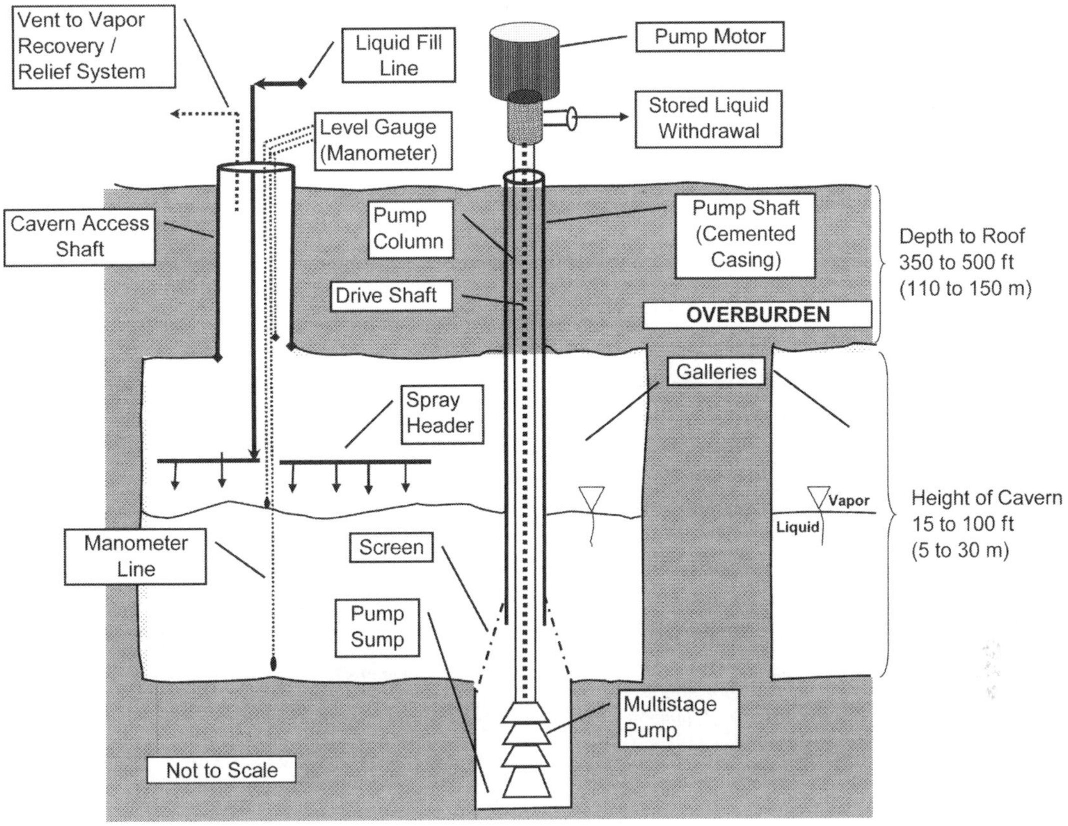

FIG. 10-189 Mined cavern.

Therefore, the maximum storage pressure (2760 to 32,060 kPag, or 400 to 3200 psig) usually exceeds the vapor pressure of all commonly stored hydrocarbon liquids. Higher-vapor-pressure products such as ethylene or ethane cannot be stored in relatively shallow caverns.

Salt caverns are developed by solution mining, a process (leaching) in which water is injected to dissolve the salt. Approximately 7 to 10 units of fresh water are required to leach 1 unit of cavern volume. Figure 10-190 illustrates the leaching process for two caverns. Modern salt dome caverns are shaped as relatively tall, slender cylinders. The leaching process produces nearly saturated brine from the cavern. Brine may be disposed into nearby disposal wells or offshore disposal fields, or it may be supplied to nearby plants as a feedstock for manufacturing of caustic (NaOH) and chlorine (Cl_2). The final portion of the produced brine is retained and stored in artificial surface ponds or tanks to be used to displace the stored liquid from the cavern.

Salt caverns are usually developed using a single well, although some employ two or more wells. The well consists of a series of concentric casings that protect the water table and layers of rock and sediments (overburden) that lie above the salt dome. The innermost well casing is referred to as the last cemented or well "production" casing and is cemented in place, sealing the cavern and protecting the surrounding geology. Once the last cemented casing is in place, a borehole is drilled from the bottom of the well, through the salt to the design cavern depth. For single-well leaching, two concentric tubing strings are then suspended in the well. A liquid, such as diesel, or a gas is then injected through the outer annular space and into the top of the cavern to act as a "blanket" to prevent undesired leaching of the top of the cavern. Water is then injected into one of the suspended tubing strings, and brine is withdrawn from the other. During the leaching process, the flow path for the injected water is alternated between the innermost

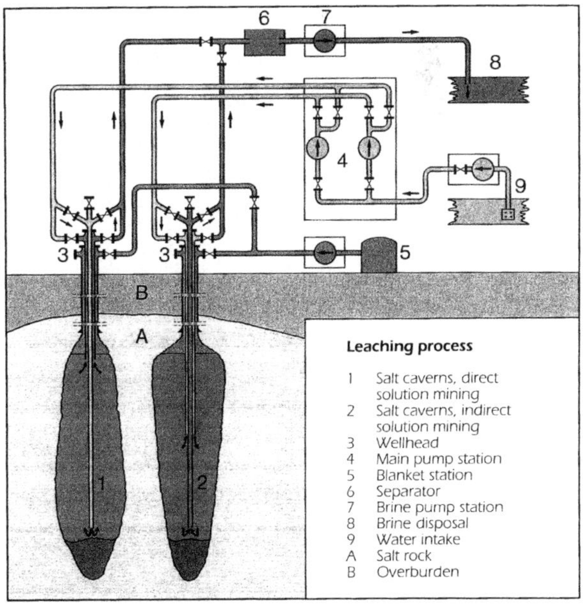

FIG. 10-190 Cavern leaching process.

tubing and the inner annulus, and these strings are periodically raised upward to control the cavern shape. A typical salt dome cavern may require 18 to 30 months of leaching time, whereas smaller, bedded salt caverns may be developed in a shorter time frame.

Brine-Compensated Storage As the stored product is pumped into the cavern, brine is displaced into an aboveground brine storage reservoir. To withdraw the product from the cavern, brine is pumped back into the cavern, displacing the stored liquid. This method of product transfer is termed *brine-compensated*, and caverns that operate in this fashion remain liquid-filled at all times. Figure 10-191 illustrates brine-compensated storage operations.

Uncompensated Storage Hard rock caverns and a few bedded salt caverns do not use brine for product displacement. This type of storage operation is referred to as pumpout or *uncompensated* storage operations. When the cavern is partially empty of liquid, the void space is filled with the vapor that is in equilibrium with the stored liquid. When liquid is introduced into the cavern, it compresses and condenses this saturated vapor phase. In some cases, vapor may be vented to the surface where it may be refrigerated and recycled to the cavern.

Submersible pumps or vertical line shaft pumps are used for withdrawing the stored liquid. Vertical line shaft pumps are suited for depths of no more than several hundred feet. Figure 10-189 illustrates an example of uncompensated storage operations.

Water is also stored underground when suitable formations are available. When an excess of surface water is available part of the time, the excess is treated, if required, and pumped into the ground to be retrieved when needed. Sometimes pumping is unnecessary, and it will seep into the ground.

Underground chambers are also constructed in frozen earth (see subsection "Low-Temperature and Cryogenic Storage"). Underground tunnel or tank storage is often the most practical way of storing hazardous or radioactive materials, such as proposed at Yucca Mountain, Nevada. A cover of 30 m (100 ft) of rock or dense earth can exert a pressure of about 690 kPa (100 lbf/in^2).

STORAGE OF GASES

Gas Holders Gas is sometimes stored in expandable gas holders of either the liquid-seal or dry-seal type. The liquid-seal holder is a familiar sight. It has a cylindrical container, closed at the top, and varies its volume by moving it up and down in an annular water-filled seal tank. The seal tank may be staged in several lifts (as many as five). Seal tanks have been built in sizes up to 280,000 m^3 (10 × 10^6 ft^3). The dry-seal holder has a rigid top attached to the sidewalls by a flexible fabric diaphragm which permits it to move up and down. It does not involve the weight and foundation costs of the liquid-seal holder. Additional information on gas holders can be found in *Gas Engineers Handbook*, Industrial Press, New York, 1966.

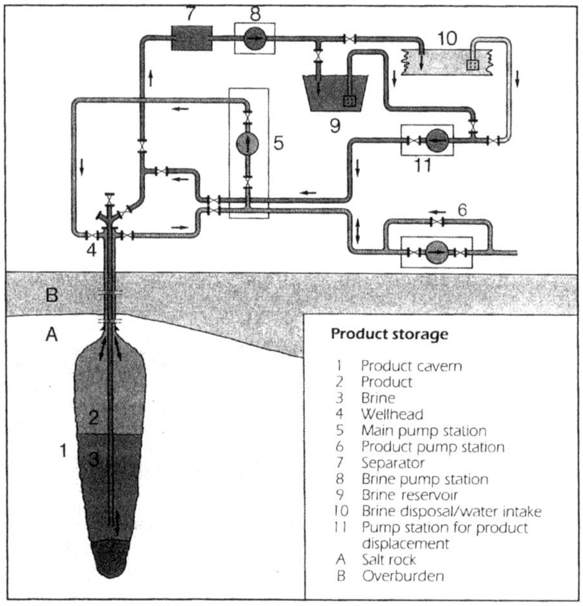

FIG. 10-191 Brine-compensated storage.

Solution of Gases in Liquids Certain gases will dissolve readily in liquids. In some cases in which the quantities are not large, this may be a practical storage procedure. Examples of gases that can be handled in this way are ammonia in water, acetylene in acetone, and hydrogen chloride in water. Whether or not this method is used depends mainly on whether the end use requires the anhydrous or the liquid state. Pressure may be either atmospheric or elevated. The solution of acetylene in acetone is also a safety feature because of the instability of acetylene.

Storage in Pressure Vessels, Bottles, and Pipe Lines The distinction between pressure vessels, bottles, and pipes is arbitrary. They can all be used for storing gases under pressure. A storage pressure vessel is usually a permanent installation. Storing a gas under pressure not only reduces its volume but also in many cases liquefies it at ambient temperature. Some gases in this category are carbon dioxide, several petroleum gases, chlorine, ammonia, sulfur dioxide, and some types of Freon. Pressure tanks are frequently installed underground.

Liquefied petroleum gas (LPG) is the subject of API Standard 2510, The Design and Construction of Liquefied Petroleum Gas Installations at Marine and Pipeline Terminals, Natural Gas Processing Plants, Refineries, and Tank Farms. This standard in turn refers to:

1. National Fire Protection Association (NFPA) Standard 58, Standard for the Storage and Handling of Liquefied Petroleum Gases
2. NFPA Standard 59, Standard for the Storage and Handling of Liquefied Petroleum Gases at Utility Gas Plants
3. NFPA Standard 59A, Standard for the Production, Storage, and Handling of Liquefied Natural Gas (LNG)

The API Standard gives considerable information on the construction and safety features of such installations. It also recommends minimum distances from property lines. The user may wish to obtain added safety by increasing these distances.

The term **bottle** is usually applied to a pressure vessel that is small enough to be conveniently portable. Bottles range from about 57 L (2 ft^3) down to CO_2 capsules of about 16.4 mL (1 in^3). Bottles are convenient for small quantities of many gases, including air, hydrogen, nitrogen, oxygen, argon, acetylene, Freon, and petroleum gas. Some are one-time-use disposable containers.

Pipe Lines A pipe line is not ordinarily a storage device. Pipes, however, have been buried in a series of connected parallel lines and used for storage. This avoids the necessity of providing foundations, and the earth protects the pipe from extremes of temperature. The economics of such an installation would be doubtful if it were designed to the same stresses as a pressure vessel. Storage is also obtained by increasing the pressure in operating pipe lines and thus using the pipe volume as a tank.

Low-Temperature and Cryogenic Storage This type is used for gases that liquefy under pressure at atmospheric temperature. In cryogenic storage the gas is at, or near to, atmospheric pressure and remains liquid because of low temperature. A system may also operate with a combination of pressure and reduced temperature. The term "cryogenic" usually refers to temperatures below −101°C (−150°F). Some gases, however, liquefy between −101°C and ambient temperatures. The principle is the same, but cryogenic temperatures create different problems with insulation and construction materials.

The liquefied gas must be maintained at or below its boiling point. Refrigeration can be used, but the usual practice is to cool by evaporation. The quantity of liquid evaporated is minimized by insulation. The vapor may be vented to the atmosphere (wasteful), it may be compressed and reliquefied, or it may be used.

At very low temperatures with liquid air and similar substances, the tank may have double walls with the interspace evacuated. The well-known Dewar flask is an example. Large tanks and even pipe lines are now built this way. An alternative is to use double walls without vacuum but with an insulating material in the interspace. Perlite and plastic foams are two insulating materials employed in this way. Sometimes both insulation and vacuum are used.

Materials Materials for liquefied-gas containers must be suitable for the temperatures, and they must not be brittle. Some carbon steels can be used down to −59°C (−75°F), and low-alloy steels to −101°C (−150°F) and sometimes −129°C (−200°F). Below these temperatures austenitic stainless steel (AISI 300 series) and aluminum are the principal materials. (See discussion of brittle fracture on p.10-160.)

Low temperatures involve problems of **differential thermal expansion.** With the outer wall at ambient temperature and the inner wall at the liquid boiling point, relative movement must be accommodated. Some systems for accomplishing this are patented. The Gaz Transport of France reduces dimensional change by using a thin inner liner of Invar. Another patented French system accommodates this change by means of the flexibility of thin metal which is creased. The creases run in two directions, and the form of the crossings of the creases is a feature of the system.

Low-temperature tanks may be installed underground to take advantage of the insulating value of the earth. Frozen-earth storage is also used. The frozen earth forms the tank. Some installations using this technique have been unsuccessful because of excessive heat absorption.

Cavern Storage Gases are also stored below ground in salt caverns. The most common type of gas stored in caverns is natural gas, although hydrogen and air have also been stored. Hydrogen storage requires special consideration in selecting metallurgy for the wellhead and the tubular goods. Air is stored for the purpose of providing compressed air energy for peak shaving power plants. Two such plants are in operation, one in the United States (Alabama), the other in Germany. A discussion of the Alabama plant is presented in *History of First U.S. Compressed Air Energy Storage (CAES) Plant (110-MW-26 h)*, vol. 1, *Early CAES Development*, Electric Power Research Institute (EPRI), Palo Alto, Calif. (1992).

Since salt caverns contain brine and other contaminants, the type of gas to be stored should not be sensitive to the presence of contaminants. If the gas is determined suitable for cavern storage, then cavern storage may not offer only economic benefits and enhanced safety and security; salt caverns also offer relatively high rates of deliverability compared to reservoir and aquifer storage fields. Solution-mined gas storage caverns in salt formations operate as *uncompensated* storage—no fluid is injected into the well to displace the compressed gas.

Surface gas handling facilities for storage caverns typically include receipt and delivery measurement stations, gas compressors, and gas processing equipment. When compressors are required for cavern injection and/or withdrawal, banks of positive-displacement-type compressors are used, since this compressor type is well suited for handling the highly variable compression ratios and flow rates associated with cavern injection and withdrawal operations. Cavern withdrawal operations typically involve single or multiple pressure reduction stations and full or partial gas dehydration. Large pressure throttling requirements often require heating the gas upon withdrawal, and injection of methanol to help control hydrate formation is also often necessary.

An in-depth discussion on natural gas storage in underground caverns may be found in *Gas Engineering and Operating Practices, Supply*, Book S-1, Part 1, *Underground Storage of Natural Gas,* and Part 2, Chapter 2, "Leached Caverns," American Gas Association, Arlington, Va. (1990).

Additional References API Recommended Practice 1114, *Design of Solution-Mined Underground Storage Facilities,* 1st ed., Washington, June, 1994. API 1115, *Operation of Solution-Mined Underground Storage Facilities,* 1st ed., Washington, September 1994. LeFond, Stanley J., *Handbook of World Salt Resources, Monographs in Geoscience*, Department of Geology, Columbia University, New York, 1969. *SME Mining Engineering Handbook*, 2d ed., vol. 2, 1992.

COST OF STORAGE FACILITIES

Contractors' bids offer the most reliable information on cost. Order-of-magnitude costs, however, may be required for preliminary studies. One way of estimating them is to obtain cost information from similar facilities and scale it to the proposed installation. Costs of steel storage tanks and vessels have been found to vary approximately as the 0.6 to 0.7 power of their weight [see Happel, *Chemical Process Economics*, Wiley, 1958, p. 267; also Williams, *Chem. Eng.,* **54**(12), 124 (1947)]. All estimates based on the costs of existing equipment must be corrected for changes in the price index from the date when the equipment was built. Considerable uncertainty is involved in adjusting data more than a few years old.

Based on a survey in 1994 for storage tanks, the prices for field-erected tanks are for multiple-tank installations erected by the contractor on foundations provided by the owner. Some cost information on tanks is given in various references cited in Sec. 9. Cost data vary considerably from one reference to another.

Prestressed (posttensioned) concrete tanks cost about 20 percent more than steel tanks of the same capacity. Once installed, however, concrete tanks require very little maintenance. A true comparison with steel would, therefore, require evaluating the maintenance cost of both types.

BULK TRANSPORT OF FLUIDS

Transportation is often an important part of product cost. Bulk transportation may provide significant savings. When there is a choice between two or more forms of transportation, the competition may result in rate reduction. Transportation is subject to considerable regulation, which will be discussed in some detail under specific headings.

Pipe Lines For quantities of fluid which an economic investigation indicates are sufficiently large and continuous to justify the investment, pipe lines are one of the lowest-cost means of transportation. They have been built up to 1.22 m (48 in) or more in diameter and about 3200 km (2000 mi) in length for oil, gas, and other products. Water is usually not transported more than 160 to 320 km (100 to 200 miles), but the conduits may be much greater than 1.22 m (48 in) in diameter. Open canals are also used for water transportation.

Petroleum pipe lines before 1969 were built to ASA (now ANSI) Standard B31.4 for liquids and Standard B31.8 for gas. These standards were seldom mandatory because few states adopted them. The U.S. Department of Transportation (DOT), which now has responsibility for pipe-line regulation, issued Title 49, Part 192—Transportation of Natural Gas and Other Gas by Pipeline: Minimum Safety Standards, and Part 195—Transportation of Liquids by Pipeline. These contain considerable material from B31.4 and B31.8. They allow generally higher stresses than the ASME Pressure Vessel Code would allow for steels of comparable strength. The enforcement of their regulations is presently left to the states and is therefore somewhat uncertain.

Pipe-line pumping stations usually range from 16 to 160 km (10 to 100 miles) apart, with maximum pressures up to 6900 kPa (1000 lbf/in^2) and velocities up to 3 m/s (10 ft/s) for liquid. Gas pipe lines have higher velocities and may have greater spacing of stations.

Tanks Tank cars (single and multiple tank), tank trucks, portable tanks, drums, barrels, carboys, and cans are used to transport fluids (see Figs. 10-192 to 10-194). Interstate transportation is regulated by the DOT. There are other regulating agencies—state, local, and private. Railroads make rules determining what they will accept, some states require compliance with DOT specifications on intrastate movements, and tunnel authorities as well as fire chiefs apply restrictions. Water shipments involve regulations of the U.S. Coast Guard. The American Bureau of Shipping sets rules for design and construction which are recognized by insurance underwriters.

The most pertinent **DOT regulations** (*Code of Federal Regulations,* Title 18, Parts 171–179 and 397) were published by R. M. Graziano (then agent and attorney for carriers and freight forwarders) in his tariff titled *Hazardous Materials Regulations of the Department of Transportation* (1978). New tariffs identified by number are issued at intervals, and interim revisions are sent out. Agents change at intervals.

Graziano's tariff lists many regulated (dangerous) commodities (Part 172, DOT regulations) for transportation. This includes those that are poisonous, flammable, oxidizing, corrosive, explosive, radioactive, and compressed gases. Part 178 covers specifications for

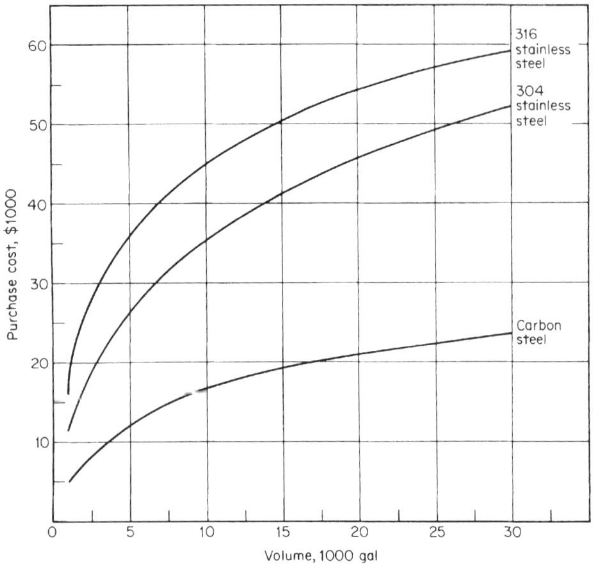

FIG. 10-192 Cost of shop-fabricated tanks in mid-1980 with ¼-in walls. Multiplying factors on carbon steel costs for other materials are: carbon steel, 1.0; rubber-lined carbon steel, 1.5; aluminum, 1.6; glass-lined carbon steel, 4.5; and fiber-reinforced plastic, 0.75 to 1.5. Multiplying factors on type 316 stainless-steel costs for other materials are: 316 stainless steel, 1.0; Monel, 2.0; Inconel, 2.0; nickel, 2.0; titanium, 3.2; and Hastelloy C, 3.8. Multiplying factors for wall thicknesses different from ¼ in are:

Thickness, in	Carbon steel	304 stainless steel	316 stainless steel
½	1.4	1.8	1.8
¾	2.1	2.5	2.6
1	2.7	3.3	3.5

To convert gallons to cubic meters, multiply by 3.785×10^{-3}.

all types of containers from carboys to large portable tanks and tank trucks. Part 179 deals with tank-car construction.

An Association of American Railroads (AAR) publication, *Specifications for Tank Cars*, covers many requirements beyond the DOT regulations.

Some additional details are given later. Because of frequent changes, it is always necessary to check the latest rules. The **shipper,** not the carrier, has the ultimate responsibility for shipping in the correct container.

Tank Cars These range in size from about 7.6 to 182 m³ (2000 to 48,000 gal), and a car may be single or multiunit. The DOT now limits them to 130 m³ (34,500 gal) and 120,000 kg (263,000 lb) gross mass. Large cars usually result in lower investment per cubic meter

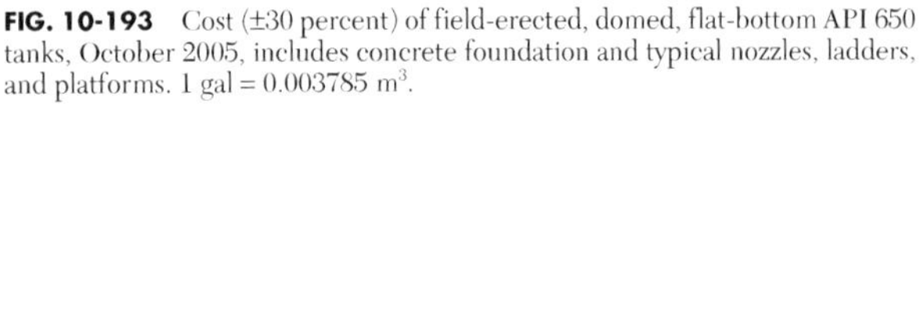

FIG. 10-193 Cost (±30 percent) of field-erected, domed, flat-bottom API 650 tanks, October 2005, includes concrete foundation and typical nozzles, ladders, and platforms. 1 gal = 0.003785 m³.

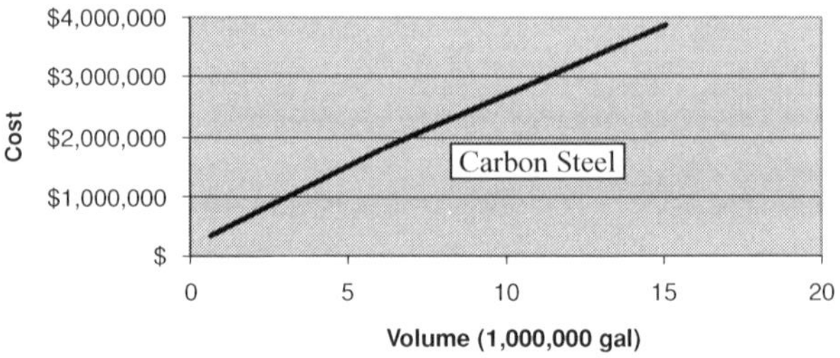

FIG. 10-194 Cost (±30 percent) of field-erected, floating roof tanks, October 2005, includes concrete foundation and typical nozzles, ladders, and platforms. 1 gal = 0.003785 m³.

and take lower shipping rates. Cars may be insulated to reduce heating or cooling of the contents. Certain liquefied gases may be carried in insulated cars; temperatures are maintained by evaporation (see subsection "Low-Temperature and Cryogenic Storage"). Cars may be heated by steam coils or by electricity. Some products are loaded hot, solidify in transport, and are melted for removal. Some low-temperature cargoes must be unloaded within a given time (usually 30 days) to prevent pressure buildup.

Tank cars are classified as pressure or general-purpose. Pressure cars have relief-valve settings of 517 kPa (75 lbf/in²) and above. Those designated as general-purpose cars are, nevertheless, pressure vessels and may have relief valves or rupture disks. The DOT specification code number indicates the type of car. For instance, 105A500W indicates a pressure car with a test pressure of 3447 kPa (500 lbf/in²) and a relief-valve setting of 2585 kPa (375 lbf/in²). In most cases, loading and unloading valves, safety valves, and vent valves must be in a dome or an enclosure.

Companies shipping dangerous materials sometimes build tank cars with metal thicker than required by the specifications in order to reduce the possibility of leakage during a wreck or fire. The punching of couplers or rail ends into heads of tanks is a hazard.

Older tank cars have a center sill or beam running the entire length of the car. Most modern cars have no continuous sill, only short stub sills at each end. Cars with full sills have tanks anchored longitudinally at the center of the sill. The anchor is designed to be weaker than either the tank shell or the doubler plate between anchor and shell. Cars with stub sills have similar safeguards. Anchors and other parts are designed to meet AAR requirements.

The impact forces on car couplers put high stresses in sills, anchors, and doublers. This may start fatigue cracks in the shell, particularly at the corners of welded doubler plates. With brittle steel in cold weather, such cracks sometimes cause complete rupture of the tank. Large end radii on the doublers and tougher steels will reduce this hazard. Inspection of older cars can reveal cracks before failure.

A difference between tank cars and most pressure vessels is that tank cars are designed in terms of the theoretical ultimate or bursting strength of the tank. The test pressure is usually 40 percent of the bursting pressure (sometimes less). The safety valves are set at 75 percent of the test pressure. Thus, the maximum operating pressure is usually 30 percent of the bursting pressure. This gives a nominal factor of safety of 3.3, compared with 3.5 for Division 1 of the ASME Pressure Vessel Code.

The DOT rules require that pressure cars have relief valves designed to limit pressure to 82.5 percent (with certain exceptions) of test pressure (110 percent of maximum operating pressure) when exposed to fire. Appendix A of AAR Specifications deals with the flow capacity of relief devices. The formulas apply to cars in the upright position with the device discharging vapor. They may not protect the car adequately when it is overturned and the device is discharging liquid.

Appendix B of AAR Specifications deals with the certification of facilities. Fabrication, repairing, testing, and specialty work on tank cars must be done in certified facilities. The AAR certifies shops to build cars of certain materials, to do test work on cars, or to make certain repairs and alterations.

Tank Trucks These trucks may have single, compartmented, or multiple tanks. Many of their requirements are similar to those for tank cars, except that thinner shells are permitted in most cases. Trucks for nonhazardous materials are subject to few regulations other than the normal highway laws governing all motor vehicles. But trucks carrying hazardous materials must comply with DOT regulations, Parts 173, 177, 178, and 397. Maximum weight, axle loading, and length are governed by state highway regulations. Many states have limits in the vicinity of 31,750 kg (70,000 lb) total mass, 14,500 kg (32,000 lb) for tandem axles, and 18.3 m (60 ft) or less overall length. Some allow tandem trailers.

Truck cargo tanks (for dangerous materials) are built under Part 173 and Subpart J of Part 178, DOT regulations. This includes Specifications MC-306, MC-307, MC-312, and MC-331. MC-331 is required for compressed gas. Subpart J requires tanks for pressures above 345 kPa (50 lbf/in^2) in one case and 103 kPa (15 lbf/in^2) in another to be built according to the ASME Pressure Vessel Code. A particular issue of the code is specified.

Because of the demands of highway service, the DOT specifications have a number of requirements in addition to the ASME Code. These include design for impact forces and rollover protection for fittings.

Portable tanks, drums, or bottles are shipped by rail, ship, air, or truck. Portable tanks containing hazardous materials must conform to DOT regulations, Parts 173 and 178, Subpart H.

Some tanks are designed to be shipped by trailer and transferred to railcars or ships (see following discussion).

Marine Transportation Seagoing **tankers** are for high tonnage. The traditional tanker uses the ship structure as a tank. It is subdivided into a number of tanks by means of transverse bulkheads and a centerline bulkhead. More than one product can be carried. An elaborate piping system connects the tanks to a pumping plant which can discharge or transfer the cargo. Harbor and docking facilities appear to be the only limit to tanker size. The largest crude oil tanker size to date is about 560,000 deadweight tons. In the United States, tankers are built to specifications of the American Bureau of Shipping and the U.S. Coast Guard.

Low-temperature liquefied gases are shipped in special ships with insulation between the hull and an inner tank. The largest LNG carrier's capacity is about 145,000 m^3. Poisonous materials are shipped in separate tanks built into the ship. This prevents tank leakage from contaminating harbors. Separate tanks are also used to transport pressurized gases.

Barges are used on inland waterways. Popular sizes are up to 16 m (52½) wide by 76 m (250 ft) long, with 2.6 m (8½ ft) to 4.3 m (14 ft) draft. Cargo requirements and waterway limitations determine design. Use of barges of uniform size facilitates rafting them together.

Portable tanks may be stowed in the holds of conventional cargo ships or special container ships, or they may be fastened on deck.

Container ships have guides in the hold and on deck which hold boxlike containers or tanks. The tank is latched to a trailer chassis and hauled to shipside. A movable gantry, sometimes permanently installed on the ship, hoists the tank from the trailer and lowers it into the guides on the ship. This system achieves large savings in labor, but its application is sometimes limited by lack of agreement between ship operators and unions.

Portable tanks for regulated commodities in marine transportation must be designed and built under Coast Guard regulations (see discussion under "Pressure Vessels").

Materials of Construction for Bulk Transport Because of the more severe service, construction materials for transportation usually are more restricted than for storage. Most large pipe lines are constructed of steel conforming to API Specification 5L or 5LX. Most tanks (cars, etc.) are built of pressure-vessel steels or AAR specification steels, with a few of aluminum or stainless steel. Carbon steel tanks may be lined with rubber, plastic, nickel, glass, or other materials. In many cases this is practical and cheaper than using a stainless-steel tank. Other materials for tank construction may be proposed and used if approved by the appropriate authorities (AAR and DOT).

PRESSURE VESSELS

This discussion of pressure vessels is intended as an overview of the codes most frequently used for the design and construction of pressure vessels. Chemical engineers who design or specify pressure vessels should determine the federal and local laws relevant to the problem and then refer to the most recent issue of the pertinent code or standard before proceeding. Laws, codes, and standards are frequently changed.

A pressure vessel is a closed container of limited length (in contrast to the indefinite length of piping). Its smallest dimension is considerably larger than the connecting piping, and it is subject to pressures above 7 or 14 kPa (1 or 2 lbf/in^2). It is distinguished from a boiler, which in most cases is used to generate steam for use external to itself.

Code Administration The American Society of Mechanical Engineers has written the ASME Boiler and Pressure Vessel Code, which contains rules for the design, fabrication, and inspection of boilers and pressure vessels. The ASME Code is an American National Standard. Most states in the United States and all Canadian provinces have passed legislation which makes the ASME Code or certain parts of it their legal requirement. Only a few jurisdictions have adopted the code for all vessels. The others apply it to certain types of vessels or to boilers. States employ inspectors (usually under a chief boiler inspector) to enforce code provisions. The authorities also depend a great deal on insurance company inspectors to see that boilers and pressure vessels are maintained in a safe condition.

The ASME Code is written by a large committee and many subcommittees, composed of engineers appointed by the ASME. The Code Committee meets regularly to review the code and consider requests for its revision, interpretation, or extension. **Interpretation and extension** are accomplished through "code cases." The decisions are published in *Mechanical Engineering*. Code cases are also mailed to those who subscribe to the service. A typical code case might be the approval of the use of a metal which is not presently on the list of approved code materials. Inquiries relative to code cases should be addressed to the secretary of the ASME Boiler and Pressure Vessel Committee, American Society of Mechanical Engineers, New York.

A new edition of the code is issued every 3 years. Between editions, alterations are handled by issuing semiannual addenda, which may be purchased by subscription. The ASME considers any issue of the code to be adequate and safe, but some government authorities specify certain issues of the code as their legal requirement.

Inspection Authority The National Board of Boiler and Pressure Vessel Inspectors is composed of the chief inspectors of states and municipalities in the United States and Canadian provinces which have made any part of the Boiler and Pressure Vessel Code a legal requirement. This board promotes uniform enforcement of boiler and pressure-vessel rules. One of the board's important activities is providing examinations for, and commissioning of, inspectors. Inspectors so qualified and employed by an insurance company, state, municipality, or Canadian province may inspect a pressure vessel and permit it to be stamped ASME—NB (National Board). An inspector employed by a vessel user may authorize the use of only the ASME stamp. The ASME Code Committee authorizes fabricators to use the various ASME stamps. The stamps, however, may be applied to a vessel only with the approval of the inspector.

The ASME Boiler and Pressure Vessel Code consists of eleven sections as follows:

I. Power Boilers
II. Materials
  a. Ferrous
  b. Nonferrous
  c. Welding rods, electrodes, and filler metals
  d. Properties
III. Rules for Construction of Nuclear Power Plant Components
IV. Heating Boilers
V. Nondestructive Examination
VI. Rules for Care and Operation of Heating Boilers
VII. Guidelines for the Care of Power Boilers
VIII. Pressure Vessels
IX. Welding and Brazing Qualifications
X. Fiber-Reinforced Plastic Pressure Vessels
XI. Rules for Inservice Inspection of Nuclear Power Plant Components

Pressure vessels (as distinguished from boilers) are involved with Secs. II, III, V, VIII, IX, X, and XI. Section VIII, Division 1, is the

Pressure Vessel Code as it existed in the past (and will continue). Division 1, was brought out as a means of permitting higher design stresses while ensuring at least as great a degree of safety as in Division 1. These two divisions plus Secs. III and X will be discussed briefly here. They refer to Secs. II and IX.

ASME Code Section VIII, Division 1 Most pressure vessels used in the process industry in the United States are designed and constructed in accordance with Sec. VIII, Division 1 (see Fig. 10-195). This division is divided into three subsections followed by appendixes.

Introduction The Introduction contains the scope of the division and defines the responsibilities of the user, the manufacturer, and the inspector. The scope defines pressure vessels as containers for the containment of pressure. It specifically excludes vessels having an internal pressure not exceeding 103 kPa (15 lbf/in^2) and further states that the rules are applicable for pressures not exceeding 20,670 kPa (3000 lbf/in^2). For higher pressures it is usually necessary to deviate from the rules in this division.

The scope covers many other less basic exclusions, and inasmuch as the scope is occasionally revised, except for the most obvious cases, it is prudent to review the current issue before specifying or designing pressure vessels to this division. Any vessel which meets all the requirements of this division may be stamped with the code *U* symbol even though exempted from such stamping.

Subsection A This subsection contains the general requirements applicable to all materials and methods of construction. Design temperature and pressure are defined here, and the loadings to be considered in design are specified. For stress failure and yielding, this section of the code uses the maximum-stress theory of failure as its criterion.

This subsection refers to the tables elsewhere in the division in which the maximum allowable tensile-stress values are tabulated. The basis for the establishment of these allowable stresses is defined in detail in Appendix P; however, as the safety factors used were very important in establishing the various rules of this division, it is noted that the safety factors for internal-pressure loads are 3.5 on ultimate strength and 1.6 or 1.5 on yield strength, depending on the material. For external-pressure loads on cylindrical shells, the safety factors are 3 for both elastic buckling and plastic collapse. For other shapes subject to external pressure and for longitudinal shell compression, the safety factors are 3.5 for both elastic buckling and plastic collapse. Longitudinal compressive stress in cylindrical elements is limited in this subsection by the lower of either stress failure or buckling failure.

Internal-pressure design rules and formulas are given for cylindrical and spherical shells and for ellipsoidal, torispherical (often called ASME heads), hemispherical, and conical heads. The formulas given assume membrane-stress failure, although the rules for heads include consideration for buckling failure in the transition area from cylinder to head (knuckle area).

Longitudinal joints in cylinders are more highly stressed than circumferential joints, and the code takes this fact into account. When forming heads, there is usually some thinning from the original plate thickness in the knuckle area, and it is prudent to specify the minimum allowable thickness at this point.

Unstayed flat heads and covers can be designed by very specific rules and formulas given in this subsection. The stresses caused by pressure on these members are bending stresses, and the formulas include an allowance for additional edge moments induced when the head, cover, or blind flange is attached by bolts. Rules are provided for quick-opening closures because of the risk of incomplete attachment or opening while the vessel is pressurized. Rules for braced and stayed surfaces are also provided.

External-pressure failure of shells can result from overstress at one extreme or from elastic instability at the other or at some intermediate loading. The code provides the solution for most shells by using a number of charts. One chart is used for cylinders where the shell diameter-to-thickness ratio and the length-to-diameter ratio are the variables. The rest of the charts depict curves relating the geometry of cylinders and spheres to allowable stress by curves which are determined from the modulus of elasticity, tangent modulus, and yield strength at temperatures for various materials or classes of materials. The text of this subsection explains how the allowable stress is determined from the charts for cylinders, spheres, and hemispherical, ellipsoidal, torispherical, and conical heads.

Frequently cost savings for cylindrical shells can result from reducing the effective length-to-diameter ratio and thereby reducing shell thickness. This can be accomplished by adding circumferential stiffeners to the shell. Rules are included for designing and locating the stiffeners.

Openings are always required in pressure-vessel shells and heads. Stress intensification is created by the existence of a hole in an otherwise symmetrical section. The code compensates for this by an area-replacement method. It takes a cross section through the opening, and it measures the area of the metal of the required shell that is removed and replaces it in the cross section by additional material (shell wall, nozzle wall, reinforcing plate, or weld) within certain distances of the opening centerline. These rules and formulas for calculation are included in Subsec. A.

When a cylindrical shell is drilled for the insertion of multiple tubes, the shell is significantly weakened and the code provides rules for tube-hole patterns and the reduction in strength that must be accommodated.

Fabrication tolerances are covered in this subsection. The tolerances permitted for shells for external pressure are much closer than those for internal pressure because the stability of the structure is dependent on the symmetry. Other paragraphs cover repair of defects during fabrication, material identification, heat treatment, and impact testing.

Inspection and testing requirements are covered in detail. Most vessels are required to be hydrostatic-tested (generally with water) at 1.3 times the maximum allowable working pressure. Some enameled (glass-lined) vessels are permitted to be hydrostatic-tested at lower pressures. Pneumatic tests are permitted and are carried to at least $1\frac{1}{4}$ times the maximum allowable working pressure, and there is provision for proof testing when the strength of the vessel or any of its parts cannot be computed with satisfactory assurance of accuracy. Pneumatic or proof tests are rarely conducted.

Pressure-relief-device requirements are defined in Subsec. A. Set point and maximum pressure during relief are defined according to the service, the cause of overpressure, and the number of relief devices. Safety, safety relief, relief valves, rupture disk, breaking pin, and rules on tolerances for the relieving point are given.

Testing, certification, and installation rules for relieving devices are extensive. Every chemical engineer responsible for the design or operation of process units should become very familiar with these rules. The pressure-relief-device paragraphs are the only parts of Sec. VIII, Division 1, that are concerned with the installation and ongoing operation of the facility; all other rules apply only to the design and manufacture of the vessel.

Subsection B This subsection contains rules pertaining to the methods of fabrication of pressure vessels. Part UW is applicable to welded vessels. Service restrictions are defined. Lethal service is for "lethal substances," which are defined as poisonous gases or liquids of such a nature that a very small amount of the gas or the vapor of the liquid mixed or unmixed with air is dangerous to life when inhaled. It is stated that it is the user's responsibility to advise the designer or manufacturer if the service is lethal. All vessels in lethal service shall have all butt-welded joints fully radiographed, and when practical, joints shall be butt-welded. All vessels fabricated of carbon or low-alloy steel shall be postweld-heat-treated.

Low-temperature service is defined as being below −29°C (−20°F), and impact testing of many materials is required. The code is restrictive in the type of welding permitted.

Unfired steam boilers with design pressures exceeding 345 kPa (50 lbf/in^2) have restrictive rules on welded-joint design, and all butt joints require full radiography.

Pressure vessels subject to direct firing have special requirements relative to welded-joint design and postweld heat treatment.

This subsection includes rules governing welded-joint designs and the degree of radiography, with efficiencies for welded joints specified as functions of the quality of joint. These efficiencies are used in the formulas in Subsec. A for determining vessel thicknesses.

Details are provided for head-to-shell welds, tube sheet-to-shell welds, and nozzle-to-shell welds. Acceptable forms of welded staybolts and plug and slot welds for staying plates are given here.

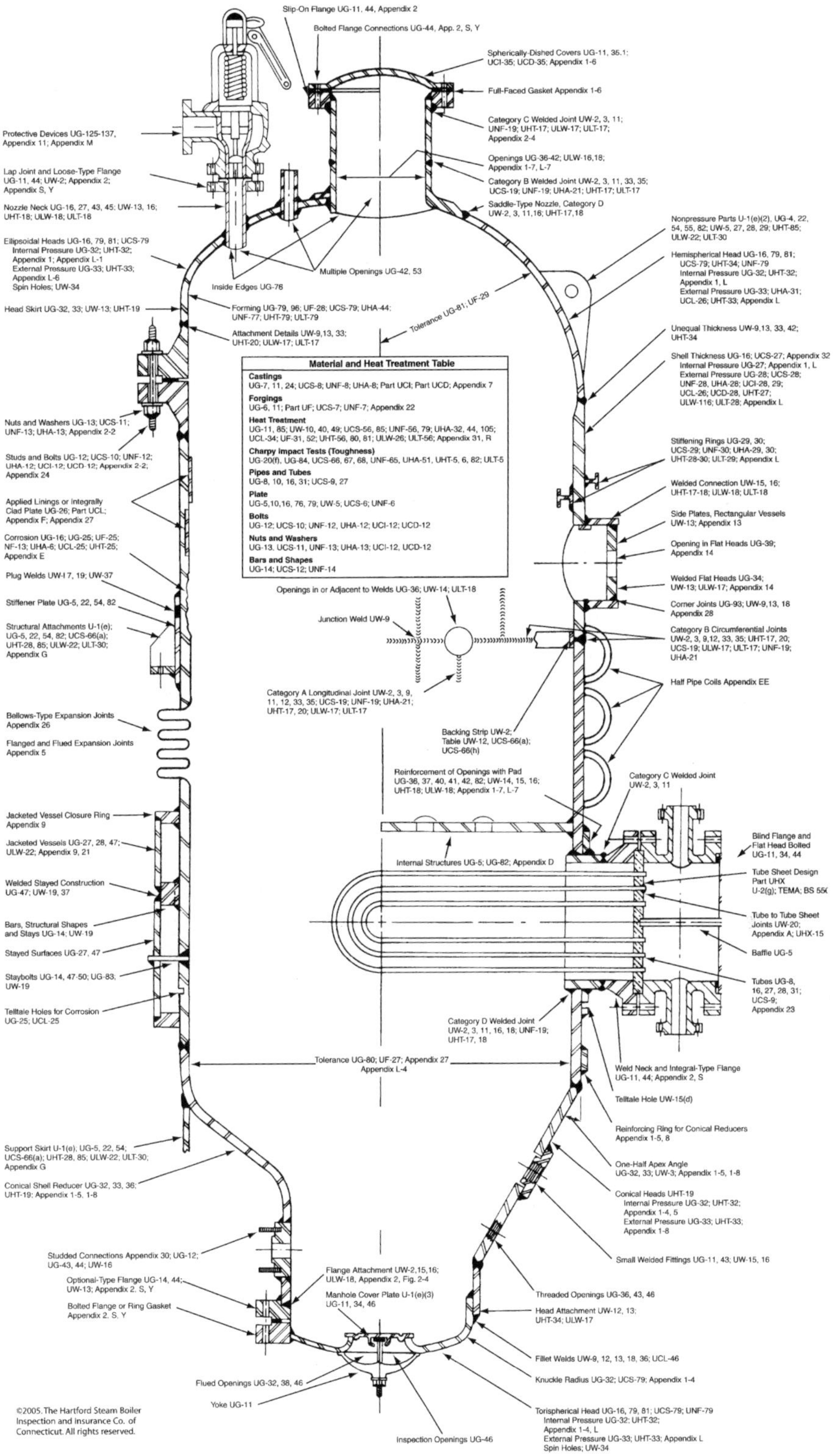

FIG. 10-195 Quick reference guide to ASME Boiler and Pressure Vessel Code Section VIII, Division 1 (2004 edition). (*Reprinted with permission of publisher, HSB Global Standards, Hartford, Conn.*)

ORGANIZATION

Introduction	Scope and Applicability
Subsection A	Part UG-General requirements for all construction and all materials
Subsection B	Requirements for methods of fabrication
Part UW	Welding
Part UF	Forging
Part UB	Brazing
Subsection C	Requirements for classes of material
Part UCS	Carbon and low alloy steels
Part UNF	Nonferrous materials
Part UHA	High alloy steels
Part UCI	Cast iron
Part UCL	Clad plate and corrosion resistant liners
Part UCD	Cast ductile iron
Part UHT	Ferritic steels with tensile properties enhanced by heat treatment
Part ULW	Layered Construction
Part ULT	Low Temperature Materials
Part UHX	Rules for Shell and Tube Heat Exchangers
Mandatory Appendices	1-33
Nonmandatory Appendices	A-Y, DD, EE, FF, GG

GENERAL NOTES

Code Jurisdiction for Piping...U-1
Design Pressure...UG-21; UG-98; UCD-3
Design Temperature...UG-20; UCL-24; UCD-3
Dimpled or Embossed Assemblies...Appendix 17
Inspector's Responsibility...U-2; UG-90
Loadings...UG-22; Appendix G, H
Low Temperature Service...UW-2; UCS-66-68; UHA-51; Part ULT
Manufacturer's Responsibility...U-2; UG-90
Material, General...UG-4, 10, 11, 15
- a. Bolts and Studs...UG-12
- b. Castings...UG-7
- c. Forgings...UG-6
- d. Nuts and Washers...UG-13
- e. Pipes and Tubes...UG-8
- f. Plates...UG-5
- g. Rod and Bar...UG-14
- h. Standard Parts...UG-11, 44
- i. Welding...UG-9

Material Identification, Marking and Certification...UG-77, 93, 94
Material Tolerances...UG-16
Nameplates, Stamping and Reports...UG-115-120; UHT-115; ULW-115; ULT-115; App. W & 18
Nondestructive Examination
- a. Liquid Penetrant...Appendix 8
- b. Magnetic Particle...Appendix 6
- c. Radiography...UW-51, 52
- d. Ultrasonic...Appendix 12

Porosity Charts...Appendix 4
Pressure Tests...UG-99, 100, 101; UW-50; UCI-99; UCD-99; ULT-99
Quality Control System...U-2; Appendix 10
Quick Actuating Closures...U-1; UG-35; App. FF
Service Restrictions...UW-2; UB-3; UCL-3; UCD-2
Stress, Maximum Allowable...UG-23; App. P; UCS-23; UNF-23; UHA-23; UCD-23; UHT-23; ULT-23; UCL-23; UCI-23;
User's Responsibility...UG-125; U-2
Welding Preheat Recommendations...Appendix R
Welding Qualifications...UW-26 thru 29; ULT-82; UHT-82; UHA-52; UNF-95

PARA.	REQUIREMENTS/REMARKS	PT	MT	UT	RT
UG-24	General requirements for castings	X	X	X	X
UG-93	General requirements for the inspection of all materials	X	X		
UW-11	Radiographic and ultrasonic examination required for pressure vessels and vessel parts (also see UW-51, UW-52)			X	X
UW-42	Surface weld metal buildup	X	X		X
UW-50	Welds on pneumatically tested vessels	X	X		
UF-31	Vessels fabricated from SA-372 forging material to be liquid quenched and tempered	X	X		
UF-32	Finished welds after postweld heat treatment	X	X		
UF-37	Repair welds in forgings	X	X		X
UF-55	Vessels constructed of SA-372 Class VIII material			X	
UCS-57	Examination in addition to UW-11 for butt welded joints on carbon and low alloy steel pressure vessels and vessel parts				X
UNF-57	Examination in addition to UW-11 for pressure vessels and vessel parts constructed of nonferrous materials				X
UNF-58	All groove and fillet welds in vessels constructed of certain nonferrous materials	X			
UHA-33	Exceptions for radiographic examinations of high alloy steel vessels				X
UHA-34	Austenitic chromium-nickel alloy steel butt and fillet welds	X			
UCI-78	Repair of defects in cast iron pressure vessels and vessel parts	X	X		X

PARA.	REQUIREMENTS/REMARKS	PT	MT	UT
UCL-35	Vessels or parts of vessels constructed of clad plate and those having applied corrosion resistant linings			
UCL-36	Chromium stainless steel cladding or lining			
UCD-78	Repair of defects in cast ductile iron pressure vessels and vessel parts			
UHT-57	Pressure vessels or vessel parts constructed of ferritic steels having tensile properties enhanced by heat treatment	X	X	
UHT-83	Metal removal accomplished by methods involving melting on pressure vessels and vessel parts constructed of ferritic steels having tensile properties enhanced by heat treatment	X	X	
UHT-85	Removal of temporary welds on pressure vessels and vessel parts constructed of ferritic steels having tensile properties enhanced by heat treatment	X	X	
ULW-51	Inner shells and inner heads of layered pressure vessels			
ULW-52	Welded joints in the layers of layered pressure vessels		X	X
ULW-53	Step welded girth joints in the layers of layered pressure vessels		X	X
ULW-54	Butt welded joints in layered pressure vessels			
ULW-55	Flat head and tube sheet welded joints in layered pressure vessels		X	X
ULW-56	Nozzle and communicating chamber welded joints in layered pressure vessels	X	X	
ULW-57	Random spot examinations and repairs of welds in layered pressure vessels		X	X
ULT-57	Welds in pressure vessels and vessel parts constructed of materials having increased design stress values due to low temperature applications	X	X	X

Qualifications of Personnel Performing Section VIII, Division 1 Nondestructive Examinations

Examination	Reference	Examination	Reference
PT	Appendix 8	UT	Appendix 12
MT	Appendix 6	RT	UW-51

FIG. 10-195 *(Continued)*

Rules for the welded fabrication of pressure vessels cover welding processes, manufacturer's record keeping on welding procedures, welder qualification, cleaning, fit-up alignment tolerances, and repair of weld defects. Procedures for postweld heat treatment are detailed. Checking the procedures and welders and radiographic and ultrasonic examination of welded joints are covered.

Requirements for vessels fabricated by forging in Part UF include unique design requirements with particular concern for stress risers, fabrication, heat treatment, repair of defects, and inspection. Vessels fabricated by brazing are covered in Part UB. Brazed vessels cannot be used in lethal service, for unfired steam boilers, or for direct firing. Permitted brazing processes as well as testing of brazed joints for strength are covered. Fabrication and inspection rules are also included.

Subsection C This subsection contains requirements pertaining to classes of materials. Carbon and low-alloy steels are governed by Part UCS, nonferrous materials by Part UNF, high-alloy steels by Part UHA, and steels with tensile properties enhanced by heat treatment by Part UHT. Each of these parts includes tables of maximum allowable stress values for all code materials for a range of metal temperatures. These stress values include appropriate safety factors. Rules governing the application, fabrication, and heat treatment of the vessels are included in each part.

Part UHT also contains more stringent details for nozzle welding that are required for some of these high-strength materials. Part UCI has rules for cast-iron construction, Part UCL has rules for welded vessels of clad plate as lined vessels, and Part UCD has rules for ductile-iron pressure vessels.

A relatively recent addition to the code is Part ULW, which contains requirements for vessels fabricated by layered construction. This type of construction is most frequently used for high pressures, usually in excess of 13,800 kPa (2000 lbf/in^2).

There are several methods of layering in common use: (1) thick layers shrunk together; (2) thin layers, each wrapped over the other and the longitudinal seam welded by using the prior layer as backup; and (3) thin layers spirally wrapped. The code rules are written for either thick or thin layers. Rules and details are provided for all the usual welded joints and nozzle reinforcement. Supports for layered vessels require special consideration, in that only the outer layer could contribute to the support. For lethal service only the inner shell and inner heads need comply with the requirements in Subsec. B. Inasmuch as radiography would not be practical for inspection of many of the welds, extensive use is made of magnetic-particle and ultrasonic inspection. When radiography is required, the code warns the inspector that indications sufficient for rejection in single-wall vessels may be acceptable. Vent holes are specified through each layer down to the inner shell to prevent buildup of pressure between layers in the event of leakage at the inner shell.

Mandatory Appendixes These include a section on supplementary design formulas for shells not covered in Subsec. A. Formulas are given for thick shells, heads, and dished covers. Another appendix gives very specific rules, formulas, and charts for the design of bolted-flange connections. The nature of these rules is such that they are readily programmable for a digital computer, and most flanges now are designed by using computers. One appendix includes only the charts used for calculating shells for external pressure discussed previously. Jacketed vessels are covered in a separate appendix in which very specific rules are given, particularly for the attachment of the jacket to the inner shell. Other appendixes cover inspection and quality control.

Nonmandatory Appendixes These cover a number of subjects, primarily suggested good practices and other aids in understanding the code and in designing with the code. Several current nonmandatory appendixes will probably become mandatory.

Figure 10-195 illustrates a pressure vessel with the applicable code paragraphs noted for the various elements. Additional important paragraphs are referenced at the bottom of the figure.

ASME Code Section VIII, Division 2 Paragraph AG-100e of Division 2 states: "In relation to the rules of Division 1 of Section VIII, these rules of Division 2 are more restrictive in the choice of materials which may be used but permit higher design stress intensity values to be employed in the range of temperatures over which the design stress intensity value is controlled by the ultimate strength or the yield strength; more precise design procedures are required and some common design details are prohibited; permissible fabrication procedures are specifically delineated and more complete testing and inspection are required." Most Division 2 vessels fabricated to date have been large or intended for high pressure and, therefore, expensive when the material and labor savings resulting from smaller safety factors have been greater than the additional engineering, administrative, and inspection costs.

The organization of Division 2 differs from that of Division 1.

Part AG This part gives the scope of the division, establishes its jurisdiction, and sets forth the responsibilities of the user and the manufacturer. Of particular importance is the fact that no upper limitation in pressure is specified and that a user's design specification is required. The user or the user's agent shall provide requirements for intended operating conditions in such detail as to constitute an adequate basis for selecting materials and designing, fabricating, and inspecting the vessel. The user's design specification shall include the method of supporting the vessel and any requirement for a fatigue analysis. If a fatigue analysis is required, the user must provide information in sufficient detail so that an analysis for cyclic operation can be made.

Part AM This part lists permitted individual construction materials, applicable specifications, special requirements, design stress-intensity values, and other property information. Of particular importance are the ultrasonic-test and toughness requirements. Among the properties for which data are included are thermal conductivity and diffusivity, coefficient of thermal expansion, modulus of elasticity, and yield strength. The design stress-intensity values include a safety factor of 3 on ultimate strength at temperature or 1.5 on yield strength at temperature.

Part AD This part contains requirements for the design of vessels. The rules of Division 2 are based on the maximum-shear theory of failure for stress failure and yielding. Higher stresses are permitted when wind or earthquake loads are considered. Any rules for determining the need for fatigue analysis are given here.

Rules for the design of shells of revolution under internal pressure differ from the Division 1 rules, particularly the rules for formed heads when plastic deformation in the knuckle area is the failure criterion. Shells of revolution for external pressure are determined on the same criterion, including safety factors, as in Division 1. Reinforcement for openings uses the same area-replacement method as Division 1; however, in many cases the reinforcement metal must be closer to the opening centerline.

The rest of the rules in Part AD for flat heads, bolted and studded connections, quick-actuating closures, and layered vessels essentially duplicate Division 1. The rules for support skirts are more definitive in Division 2.

Part AF This part contains requirements governing the fabrication of vessels and vessel parts.

Part AR This part contains rules for pressure-relieving devices.

Part AI This part contains requirements controlling inspection of vessel.

Part AT This part contains testing requirements and procedures.

Part AS This part contains requirements for stamping and certifying the vessel and vessel parts.

Appendixes Appendix 1 defines the basis used for defining stress-intensity values. Appendix 2 contains external-pressure charts, and Appendix 3 has the rules for bolted-flange connections; these two are exact duplicates of the equivalent appendixes in Division 1.

Appendix 4 gives definitions and rules for stress analysis for shells, flat and formed heads, and tube sheets, layered vessels, and nozzles including discontinuity stresses. Of particular importance are Table 4-120.1, "Classification of Stresses for Some Typical Cases," and Fig. 4-130.1, "Stress Categories and Limits of Stress Intensity." These are very useful in that they clarify a number of paragraphs and simplify stress analysis.

Appendix 5 contains rules and data for stress analysis for cyclic operation. Except in short-cycle batch processes, pressure vessels are usually subject to few cycles in their projected lifetime, and the endurance-limit data used in the machinery industries are not applicable. Curves are given for a broad spectrum of materials, covering a range from 10 to 1 million cycles with allowable stress values as high as 650,000 lbf/in^2. This low-cycle fatigue has been developed from strain-fatigue work in which stress values are obtained by multiplying the strains by the modulus of elasticity. Stresses of this magnitude cannot occur, but strains do. The curves given have a factor of safety of 2 on stress or 20 on cycles.

Appendix 6 contains requirements of experimental stress analysis, Appendix 8 has acceptance standards for radiographic examination, Appendix 9 covers nondestructive examination, Appendix 10 gives rules for capacity conversions for safety valves, and Appendix 18 details quality-control-system requirements.

The remaining appendixes are nonmandatory but useful to engineers working with the code.

General Considerations Most pressure vessels for the chemical-process industry will continue to be designed and built to the rules of Sec. VIII, Division 1. While the rules of Sec. VIII, Division 2, will frequently provide thinner elements, the cost of the engineering analysis, stress analysis and higher-quality construction, material control, and inspection required by these rules frequently exceeds the savings from the use of thinner walls.

Additional ASME Code Considerations

ASME Code Sec. III: Nuclear Power Plant Components This section of the code includes vessels, storage tanks, and concrete containment vessels as well as other nonvessel items.

ASME Code Sec. X: Fiberglass–Reinforced-Plastic Pressure Vessels This section is limited to four types of vessels: bag-molded and centrifugally cast, each limited to 1000 kPa (150 lbf/in^2); filament-wound with cut filaments limited to 10,000 kPa (1500 lbf/in^2); and filament-wound with uncut filaments limited to 21,000 kPa (3000 lbf/in^2). Operating temperatures are limited to the range from +66°C (150°F) to −54°C (−65°F). Low modulus of elasticity and other property differences between metal and plastic required that many of the procedures in Sec. X be different from those in the sections governing metal vessels. The requirement that at least one vessel of a particular design and fabrication shall be tested to destruction has prevented this section from being widely used. The results from the combined fatigue and burst test must give the design pressure a safety factor of 6 to the burst pressure.

Safety in Design Designing a pressure vessel in accordance with the code will, under most circumstances, provide adequate safety. In the code's own words, however, the rules "cover minimum construction requirements for the design, fabrication, inspection, and certification of pressure vessels." The significant word is "minimum." The **ultimate responsibility** for safety rests with the user and the designer. They must decide whether anything beyond code requirements is necessary. The code cannot foresee and provide for all the unusual conditions to which a pressure vessel might be exposed. If it tried to do so, the majority of pressure vessels would be unnecessarily restricted. Some of the conditions that a vessel might encounter are unusually low temperatures, unusual thermal stresses, stress ratcheting caused by thermal cycling, vibration of tall vessels excited by von Karman vortices caused by wind, very high pressures, runaway chemical reactions, repeated local overheating, explosions, exposure to fire, exposure to materials that rapidly attack the metal, containment of extremely toxic materials, and very large sizes of vessels. Large vessels, although they may contain nonhazardous materials, could, by their very size, create a serious hazard if they burst. The failure of the Boston molasses tank in 1919 killed 12 people. For pressure vessels which are outside code jurisdiction, there are sometimes special hazards in very-high-strength materials and plastics. There may be many others which the designers should recognize if they encounter them.

Metal fatigue, when it is present, is a serious hazard. Section VIII, Division 1, mentions rapidly fluctuating pressures. Division 2 and Sec. III do require a fatigue analysis. In extreme cases vessel contents may affect the fatigue strength (endurance limit) of the material. This is corrosion fatigue. Although most ASME Code materials are not particularly sensitive to corrosion fatigue, even they may suffer an endurance limit loss of 50 percent in some environments. High-strength heat-treated steels, on the other hand, are very sensitive to corrosion fatigue. It is not unusual to find some of these which lose 75 percent of their endurance in corrosive environments. In fact, in corrosion fatigue many steels do not have an endurance limit. The curve of stress versus cycles to failure (*S/N* curve) continues to slope downward regardless of the number of cycles.

Brittle fracture is probably the most insidious type of pressure-vessel failure. Without brittle fracture, a pressure vessel could be pressurized approximately to its ultimate strength before failure. With brittle behavior some vessels have failed well below their design pressures (which are about 25 percent of the theoretical bursting pressures). In order to reduce the possibility of brittle behavior, Division 2 and Sec. III require impact tests.

The subject of brittle fracture has been understood only since about 1950, and knowledge of some of its aspects is still inadequate. A notched or cracked plate of pressure-vessel steel, stressed at 66°C (150°F), would elongate and absorb considerable energy before breaking. It would have a ductile or plastic fracture. As the temperature is lowered, a point is reached at which the plate would fail in a brittle manner with a flat fracture surface and almost no elongation. The transition from ductile to brittle fracture actually takes place over a temperature range, but a point in this range is selected as the **transition temperature.** One of the ways of determining this temperature is the Charpy impact test (see ASTM Specification E-23). After the transition temperature has been determined by laboratory impact tests, it must be correlated with service experience on full-size plates. The literature on brittle fracture contains information on the relation of impact tests to service experience on some carbon steels.

A more precise but more elaborate method of dealing with the ductile-brittle transition is the **fracture-analysis diagram.** This uses a transition known as the **nil-ductility temperature** (NDT), which is determined by the drop-weight test (ASTM Standard E208) or the drop-weight tear test (ASTM Standard E436). The application of this diagram is explained in two papers by Pellini and Puzak [*Trans. Am. Soc. Mech. Eng.*, 429 (October 1964); Welding Res. Counc. Bull. 88, 1963].

Section VIII, Division 1, is rather lax with respect to brittle fracture. It allows the use of many steels down to −29°C (−20°F) without a check on toughness. Occasional brittle failures show that some vessels are operating below the nil-ductility temperature, i.e., the lower limit of ductility. Division 2 has resolved this problem by requiring impact tests in certain cases. Tougher grades of steel, such as the SA516 steels (in preference to SA515 steel), are available for a small price premium. Stress relief, steel made to fine-grain practice, and normalizing all reduce the hazard of brittle fracture.

Nondestructive testing of both the plate and the finished vessel is important to safety. In the analysis of fracture hazards, it is important to know the size of the flaws that may be present in the completed vessel. The four most widely used methods of examination are radiographic, magnetic-particle, liquid-penetrant, and ultrasonic.

Radiographic examination is either by **x-rays** or by **gamma radiation.** The former has greater penetrating power, but the latter is more portable. Few x-ray machines can penetrate beyond 300-mm (12-in) thickness.

Ultrasonic techniques use vibrations with a frequency between 0.5 and 20 MHz transmitted to the metal by a transducer. The instrument sends out a series of pulses. These show on a cathode-ray screen as they are sent out and again when they return after being reflected from the opposite side of the member. If there is a crack or an inclusion along the way, it will reflect part of the beam. The initial pulse and its reflection from the back of the member are separated on the screen by a distance which represents the thickness. The reflection from a flaw will fall between these signals and indicate its magnitude and position. Ultrasonic examination can be used for almost any thickness of material from a fraction of an inch to several feet. Its use is dependent upon the shape of the body because irregular surfaces may give confusing reflections. Ultrasonic transducers can transmit pulses normal to the surface or at an angle. Transducers transmitting pulses that are oblique to the surface can solve a number of special inspection problems.

Magnetic-particle examination is used only on magnetic materials. Magnetic flux is passed through the part in a path parallel to the surface. Fine magnetic particles, when dusted over the surface, will concentrate near the edges of a crack. The sensitivity of magnetic-particle examination is proportional to the sine of the angle between the direction of the magnetic flux and the direction of the crack. To be sure of picking up all cracks, it is necessary to probe the area in two directions.

Liquid-penetrant examination involves wetting the surface with a fluid which penetrates open cracks. After the excess liquid has been wiped off, the surface is coated with a material which will reveal any liquid that has penetrated the cracks. In some systems a colored dye will seep out of cracks and stain whitewash. Another system uses a penetrant that becomes fluorescent under ultraviolet light.

Each of these four popular methods has its advantages. Frequently, best results are obtained by using more than one method. Magnetic particles or liquid penetrants are effective on surface cracks. Radiography and ultrasonics are necessary for subsurface flaws. *No known method of nondestructive testing can guarantee the absence of flaws.* There are other less widely used methods of examination. Among these are eddy-current, electrical-resistance, acoustics, and thermal testing. *Nondestructive Testing Handbook* [Robert C. McMaster (ed.), Ronald, New York, 1959] gives information on many testing techniques.

The **eddy-current technique** involves an alternating-current coil along and close to the surface being examined. The electrical impedance of the coil is affected by flaws in the structure or changes in composition. Commercially, the principal use of eddy-current testing is for the examination of tubing. It could, however, be used for testing other things.

The **electrical-resistance method** involves passing an electric current through the structure and exploring the surface with voltage probes. Flaws, cracks, or inclusions will cause a disturbance in the voltage gradient on the surface. Railroads have used this method for many years to locate transverse cracks in rails.

The **hydrostatic test** is, in one sense, a method of examination of a vessel. It can reveal gross flaws, inadequate design, and flange leaks. Many believe that a hydrostatic test guarantees the safety of a vessel. This is not necessarily so. A vessel that has passed a hydrostatic test is probably safer than one that has not been tested. It can, however, still fail in service, even on the next application of pressure. Care in material selection, examination, and fabrication do more to guarantee vessel integrity than the hydrostatic test.

The ASME Codes recommend that hydrostatic tests be run at a temperature that is usually above the nil-ductility temperature of the material. This is, in effect, a pressure-temperature treatment of the vessel. When tested in the relatively ductile condition above the nil-ductility temperature, the material will yield at the tips of cracks and flaws and at points of high residual weld stress. This procedure will actually reduce the residual stresses and cause a redistribution at crack tips. The vessel will then be in a safer condition for subsequent operation. This procedure is sometimes referred to as **notch nullification.**

It is possible to design a hydrostatic test in such a way that it probably will be a proof test of the vessel. This usually requires, among other things, that the test be run at a temperature as low as and preferably lower than the minimum operating temperature of the vessel. Proof tests of this type are run on vessels built of ultrahigh-strength steel to operate at cryogenic temperatures.

Other Regulations and Standards Pressure vessels may come under many types of regulation, depending on where they are and what they contain. Although many states have adopted the ASME Boiler and Pressure Vessel Code, either in total or in part, any state or municipality may enact its own requirements. The federal government regulates some pressure vessels through the Department of Transportation, which includes the Coast Guard. If pressure vessels are shipped into foreign countries, they may face additional regulations.

Pressure vessels carried aboard United States–registered ships must conform to rules of the **U.S. Coast Guard.** Subchapter F of Title 46, *Code of Federal Regulations,* covers marine engineering. Of this, Parts 50 through 61 and 98 include pressure vessels. Many of the rules are similar to those in the ASME Code, but there are differences.

The **American Bureau of Shipping** (ABS) has rules that insurance underwriters require for the design and construction of pressure vessels which are a permanent part of a ship. Pressure cargo tanks may be permanently attached and come under these rules. Such tanks supported at several points are independent of the ship's structure and are distinguished from "integral cargo tanks" such as those in a tanker. ABS has pressure vessel rules in two of its publications. Most of them are in *Rules for Building and Classing Steel Vessels.*

Standards of Tubular Exchanger Manufacturers Association (TEMA) give recommendations for the construction of tubular heat exchangers. Although TEMA is not a regulatory body and there is no legal requirement for the use of its standards, they are widely accepted as a good basis for design. By specifying TEMA standards, one can obtain adequate equipment without having to write detailed specifications for each piece. TEMA gives formulas for the thickness of tube sheets. Such formulas are not in ASME Codes. (See further discussion of TEMA in Sec. 11.)

Vessels with Unusual Construction High pressures create design problems. The ASME Code Sec. VIII, Division 1, applies to vessels rated for pressures up to 20,670 kPa (3000 lbf/in^2). Division 2 is unlimited. At high pressures, special designs not necessarily in accordance with the code are sometimes used. At such pressures, a vessel designed for ordinary low-carbon-steel plate, particularly in large diameters, would become too thick for practical fabrication by ordinary methods. The alternatives are to make the vessel of high-strength plate, use a solid forging, or use multilayer construction.

High-strength steels with tensile strengths over 1380 MPa (200,000 lbf/in^2) are limited largely to applications for which weight is very important. Welding procedures are carefully controlled, and preheat is used. These materials are brittle at almost any temperature, and vessels must be designed to prevent brittle fracture. Flat spots and variations in curvature are avoided. Openings and changes in shape require appropriate design. The maximum permissible size of flaws is determined by fracture mechanics, and the method of examination must assure as much as possible that larger flaws are not present. All methods of nondestructive testing may be used. Such vessels require the most sophisticated techniques in design, fabrication, and operation.

Solid forgings are frequently used in construction for pressure vessels above 20,670 kPa (3000 lbf/in^2) and even lower. Almost any shell thickness can be obtained, but most of them range between 50 and 300 mm (2 and 12 in). The ASME Code lists forging materials with tensile strengths from 414 to 930 MPa (from 60,000 to 135,000 lbf/in^2). Brittle fracture is a possibility, and the hazard increases with thickness. Furthermore, some forging alloys have nil-ductility temperatures as high as 121°C (250°F). A forged vessel should have an NDT at least 17°C (30°F) below the design temperature. In operation, it should be slowly and uniformly heated at least to NDT before it is subjected to pressure. During construction, nondestructive testing should be used to detect dangerous cracks or flaws. Section VIII of the ASME Code, particularly Division 2, gives design and testing techniques.

As the size of a forged vessel increases, the sizes of ingot and handling equipment become larger. The cost may increase faster than the weight. The problems of getting sound material and avoiding brittle fracture also become more difficult. Some of these problems are avoided by use of **multilayer construction.** In this type of vessel, the heads and flanges are made of forgings, and the cylindrical portion is built up by a series of layers of thin material. The thickness of these layers may be between 3 and 50 mm (⅛ and 2 in), depending on the type of construction. There is an inner lining which may be different from the outer layers.

Although there are multilayer vessels as small as 380-mm (15-in) inside diameter and 2400 mm (8 ft) long, their principal advantage applies to the larger sizes. When properly made, a multilayer vessel is probably safer than a vessel with a solid wall. The layers of thin material are tougher and less susceptible to brittle fracture, have less probability of defects, and have the statistical advantage of a number of small elements instead of a single large one. The heads, flanges, and welds, of course, have the same hazards as other thick members. Proper attention is necessary to avoid cracks in these members.

There are several assembly techniques. One frequently used is to form successive layers in half cylinders and butt-weld them over the previous layers. In doing this, the welds are staggered so that they do not fall together. This type of construction usually uses plates from 6 to 12 mm (¼ to ½ in) thick. Another method is to weld each layer separately to form a cylinder and then shrink it over the previous layers. Layers up to about 50-mm (2-in) thickness are assembled in this way. A third method of fabrication is to wind the layers as a continuous sheet. This technique is used in Japan. The Wickel construction, fabricated in Germany, uses helical winding of interlocking metal strip. Each method has its advantages and disadvantages, and choice will depend upon circumstances.

Because of the possibility of voids between layers, it is preferable not to use multilayer vessels in applications where they will be subjected to fatigue. Inward thermal gradients (inside temperature lower than outside temperature) are also undesirable.

Articles on these vessels have been written by Fratcher [*Pet. Refiner,* **34**(11), 137 (1954)] and by Strelzoff, Pan, and Miller [*Chem. Eng.,* **75**(21), 143–150 (1968)].

Vessels for high-temperature service may be beyond the temperature limits of the stress tables in the ASME Codes. Section VIII, Division 1, makes provision for construction of pressure vessels up to 650°C (1200°F) for carbon and low-alloy steel and up to 815°C (1500°F) for stainless steels (300 series). If a vessel is required for temperatures above these values and above 103 kPa (15 lbf/in^2), it would be necessary, in a code state, to get permission from the state authorities to build it as a special project. Above 815°C (1500°F), even the 300 series stainless steels are weak, and creep rates increase rapidly. If the metal which resists the pressure operates at these temperatures, the vessel pressure and size will be limited. The vessel must also be expendable because its life will be short. Long exposure to high temperature may cause the metal to deteriorate and become brittle. Sometimes, however, economics favor this type of operation.

One way to circumvent the problem of low metal strength is to use a metal inner liner surrounded by insulating material, which in turn is confined by a pressure vessel. The liner, in some cases, may have perforations which will allow pressure to pass through the insulation and act on the outer shell, which is kept cool to obtain normal strength. The liner has no pressure differential acting on it and, therefore, does not need much strength. Ceramic linings are also useful for high-temperature work.

Lined vessels are used for many applications. Any type of lining can be used in an ASME Code vessel, provided it is compatible with the metal of the vessel and the contents. Glass, rubber, plastics, rare metals, and ceramics are a few types. The lining may be installed separately, or if a metal is used, it may be in the form of clad plate. The cladding on plate can sometimes be considered as a stress-carrying part of the vessel.

A **ceramic lining** when used with high temperature acts as an insulator so that the steel outer shell is at a moderate temperature while the temperature at the inside of the lining may be very high. Ceramic linings may be of unstressed brick, or prestressed brick, or cast in place. Cast ceramic linings or unstressed brick may develop cracks and are used when the contents of the vessel will not damage the outer shell. They are usually designed so that the high temperature at the inside will expand them sufficiently to make them tight in the outer (and cooler) shell. This, however, is not usually sufficient to prevent some penetration by the product.

Prestressed-brick linings can be used to protect the outer shell. In this case, the bricks are installed with a special thermosetting-resin mortar. After lining, the vessel is subjected to internal pressure and heat. This expands the steel vessel shell, and the mortar expands to take up the space. The pressure and temperature must be at least as high as the maximum that will be encountered in service. After the mortar has set, reduction of pressure and temperature will allow the vessel to contract, putting the brick in compression. The upper temperature limit for this construction is about 190°C (375°F). The installation of such linings is highly specialized work done by a few companies. Great care is usually exercised in operation to protect the vessel from exposure to unsymmetrical temperature gradients. Side nozzles and other unsymmetrical designs are avoided insofar as possible.

Concrete pressure vessels may be used in applications that require large sizes. Such vessels, if made of steel, would be too large and heavy to ship. Through the use of posttensioned (prestressed) concrete, the vessel is fabricated on the site. In this construction, the reinforcing steel is placed in tubes or plastic covers, which are cast into the concrete. Tension is applied to the steel after the concrete has acquired most of its strength.

Concrete nuclear reactor vessels, of the order of magnitude of 15-m (50-ft) inside diameter and length, have inner linings of steel which confine the pressure. After fabrication of the liner, the tubes for the cables or wires are put in place and the concrete is poured. High-strength reinforcing steel is used. Because there are thousands of reinforcing tendons in the concrete vessel, there is a statistical factor of safety. The failure of 1 or even 10 tendons would have little effect on the overall structure.

Plastic pressure vessels have the *advantages of chemical resistance* and light weight. Above 103 kPa (15 lbf/in^2), with certain exceptions, they must be designed according to the ASME Code Section X (see "Storage of Gases") and are confined to the three types of approved code construction. Below 103 kPa (15 lbf/in^2), any construction may be used. Even in this pressure range, however, the code should be used for guidance. Solid plastics, because of low strength and creep, can be used only for the lowest pressures and sizes. A stress of a few hundred pounds-force per square inch is the maximum for most plastics. To obtain higher strength, the filled plastics or filament-wound vessels, specified by the code, must be used. Solid-plastic parts, however, are often employed inside a steel shell, particularly for heat exchangers.

Graphite and ceramic vessels are used fully armored; that is, they are enclosed within metal pressure vessels. These materials are also used for boxlike vessels with backing plates on the sides. The plates are drawn together by tie bolts, thus putting the material in compression so that it can withstand low pressure.

ASME Code Developments At the time of this publication, ASME Section VIII is currently being rewritten and reorganized into three classes. Class 1 will be for low-pressure vessels employing spot radiography. Class 2 will be for vessels requiring full radiography. Class 3 will be for vessels experiencing fatigue. This rewriting effort is also using material stress levels similar to those of competing vessel codes from Europe and Asia.

Vessel Codes Other than ASME Different design and construction rules are used in other countries. Chemical engineers concerned with pressure vessels outside the United States must become familiar with local pressure-vessel laws and regulations. *Boilers and Pressure Vessels*, an international survey of design and approval requirements published by the British Standards Institution, Maylands Avenue, Hemel Hempstead, Hertfordshire, England, in 1975, gives pertinent information for 76 political jurisdictions.

The British Code (British Standards) and the German Code (*A. D. Merkblätter*) in addition to the ASME Code are most commonly permitted, although Netherlands, Sweden, and France also have codes. The major difference between the codes lies in factors of safety and in whether or not ultimate strength is considered. ASME Code, Sec. VIII, Division 1, vessels are generally heavier than vessels built to the other codes; however, the differences in allowable stress for a given material are less in the higher temperature (creep) range.

Engineers and metallurgists have developed alloys to comply economically with individual codes. In Germany, where design stress is determined from yield strength and creep-rupture strength and no allowance is made for ultimate strength, steels which have a very high yield-strength-to-ultimate-strength ratio are used.

Other differences between codes include different bases for the design of reinforcement for openings and the design of flanges and heads. Some codes include rules for the design of heat-exchanger tube sheets, while others (ASME Code) do not. The Dutch Code (*Grondslagen*) includes very specific rules for calculation of wind loads, while the ASME Code leaves this entirely to the designer.

There are also significant differences in construction and inspection rules. Unless engineers make a detailed study of the individual codes and keep current, they will be well advised to make use of responsible experts for any of the codes.

Vessel Design and Construction The ASME Code lists a number of loads that must be considered in designing a pressure vessel. Among them are impact, weight of the vessel under operating and test conditions, superimposed loads from other equipment and piping, wind and earthquake loads, temperature-gradient stresses, and localized loadings from internal and external supports. In general, the code gives no values for these loads or methods for determining them, and no formulas are given for determining the stresses from these loads. Engineers must be knowledgeable in mechanics and strength of materials to solve these problems.

Some of the problems are treated by Brownell and Young, *Process Equipment Design*, Wiley, New York, 1959. ASME papers treat others, and a number of books published by the ASME are collections of papers on pressure-vessel design: *Pressure Vessels and Piping Design: Collected Papers, 1927–1959; Pressure Vessels and Piping Design and Analysis*, four volumes; and *International Conference: Pressure Vessel Technology*, published annually.

Throughout the year the Welding Research Council publishes bulletins which are final reports from projects sponsored by the council, important papers presented before engineering societies, and other reports of current interest which are not published in *Welding Research*. A large number of the published bulletins are pertinent for vessel designers.

Care of Pressure Vessels Protection against **excessive pressure** is largely taken care of by code requirements for relief devices. Exposure to fire is also covered by the code. The code, however, does not provide for the possibility of local overheating and weakening of a vessel in a fire. Insulation reduces the required relieving capacity and also reduces the possibility of local overheating.

A pressure-reducing valve in a line leading to a pressure vessel is not adequate protection against overpressure. Its failure will subject the vessel to full line pressure.

Vessels that have an operating cycle which involves the solidification and remelting of solids can develop excessive pressures. A solid plug of material may seal off one end of the vessel. If heat is applied at that end to cause melting, the expansion of the liquid can build up a high pressure and possibly result in yielding or rupture. Solidification in connecting piping can create similar problems.

Some vessels may be exposed to a runaway chemical reaction or even an explosion. This requires relief valves, rupture disks, or, in extreme cases, a frangible roof design or barricade (the vessel is expendable). A vessel with a large rupture disk needs anchors designed for the jet thrust when the disk blows.

Vacuum must be considered. It is nearly always possible that the contents of a vessel might contract or condense sufficiently to subject it to an internal vacuum. If the vessel cannot withstand the vacuum, it must have vacuum-breaking valves.

Improper operation of a process may result in the vessel's **exceeding design temperature.** Proper control is the only solution to this problem. Maintenance procedures can also cause excessive temperatures. Sometimes the contents of a vessel may be burned out with torches. If the flame impinges on the vessel shell, overheating and damage may occur.

Excessively low temperature may involve the hazard of brittle fracture. A vessel that is out of use in cold weather could be at a subzero temperature and well below its nil-ductility temperature. In startup, the vessel should be warmed slowly and uniformly until it is above the NDT. A safe value is 38°C (100°F) for plate if the NDT is unknown. The vessel should not be pressurized until this temperature is exceeded. Even after the NDT has been passed, excessively rapid heating or cooling can cause high thermal stresses.

Corrosion is probably the greatest threat to vessel life. Partially filled vessels frequently have severe pitting at the liquid-vapor interface. Vessels usually do not have a corrosion allowance on the outside. Lack of protection against the weather or against the drip of corrosive chemicals can reduce vessel life. Insulation may contain damaging substances. Chlorides in insulating materials can cause cracking of stainless steels. Water used for hydrotesting should be free of chlorides.

Pressure vessels should be **inspected periodically.** No rule can be given for the frequency of these inspections. Frequency depends on operating conditions. If the early inspections of a vessel indicate a low corrosion rate, intervals between inspections may be lengthened. Some vessels are inspected at 5-year intervals; others, as frequently as once a year. Measurement of corrosion is an important inspection item. One of the most convenient ways of measuring thickness (and corrosion) is to use an ultrasonic gauge. The location of the corrosion and whether it is uniform or localized in deep pits should be observed and reported. Cracks, any type of distortion, and leaks should be observed. Cracks are particularly dangerous because they can lead to sudden failure. Insulation is usually left in place during inspection of insulated vessels. If, however, severe external corrosion is suspected, the insulation should be removed. All forms of nondestructive testing are useful for examinations.

There are many ways in which a pressure vessel can suffer **mechanical damage.** The shells can be dented or even punctured, they can be dropped or have hoisting cables improperly attached, bolts can be broken, flanges are bent by excessive bolt tightening, gasket contact faces can be scratched and dented, rotating paddles can drag against the shell and cause wear, and a flange can be bolted up with a gasket half in the groove and half out. Most of these forms of damage can be prevented by care and common sense. If damage is repaired by straightening, as with a dented shell, it may be necessary to stress-relieve the repaired area. Some steels are susceptible to embrittlement by aging after severe straining. A safer procedure is to cut out the damaged area and replace it.

The National Board Inspection Code, published by the National Board of Boiler and Pressure Vessel Inspectors, Columbus, Ohio, is helpful. Any repair, however, is acceptable if it is made in accordance with the rules of the Pressure Vessel Code.

Care in **reassembling** the vessel is particularly important. Gaskets should be properly located, particularly if they are in grooves. Bolts should be tightened in proper sequence. In some critical cases and with large bolts, it is necessary to control bolt tightening by torque wrenches, micrometers, patented bolt-tightening devices, or heating bolts. After assembly, vessels are sometimes given a hydrostatic test.

Pressure-Vessel Cost and Weight Figure 10-196 can be used for estimating carbon-steel vessel cost when a weight estimate is not available and Fig. 10-197 with a weight estimate. Weight and cost include skirts and other supports. The cost is based on several 2005 pressure-vessels. Costs are for vessels not of unusual design. Complicated vessels could cost considerably more. Guthrie [*Chem. Eng.*, **76**(6), 114–142 (1969)] also gives pressure-vessel cost data.

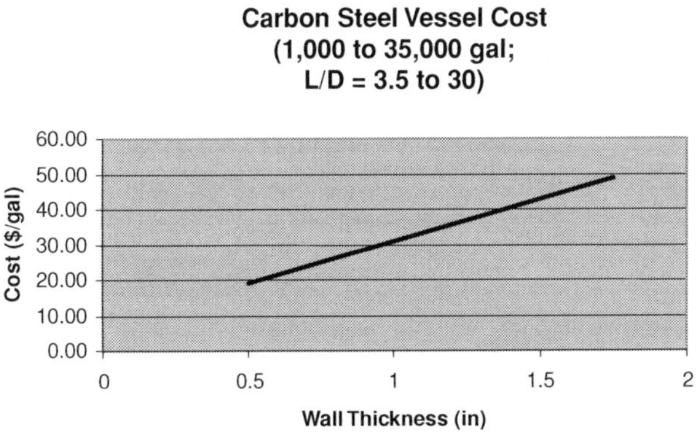

FIG. 10-196 Carbon-steel pressure-vessel cost as a function of wall thickness. 1 gal = 0.003875 cm^3; 1 in = 0.0254 m. (*Courtesy of E. S. Fox, Ltd.*)

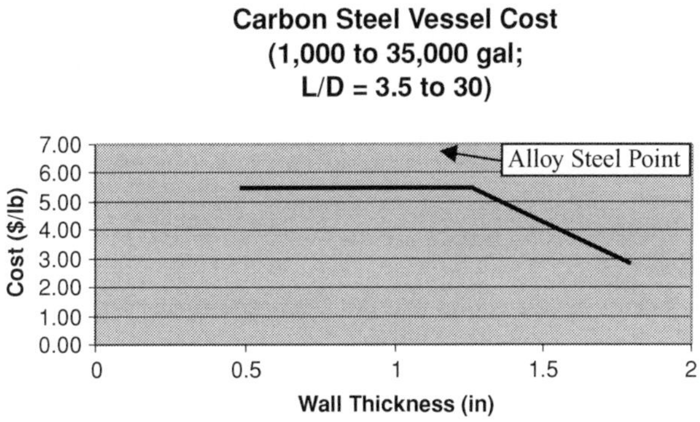

FIG. 10-197 Carbon-steel pressure-vessel cost as a function of wall thickness. 1 gal = 0.003875 cm^3; 1 in = 0.0254 m; 1 lb = 0.4536 kg. (*Courtesy of E. S. Fox, Ltd.*)

When vessels have complicated construction (large, heavy bolted connections, support skirts, etc.), it is preferable to estimate their weight and apply a unit cost in dollars per pound.

Pressure-vessel weights are obtained by calculating the cylindrical shell and heads separately and then adding the weights of nozzles and attachments. Steel weighs 7817 kg/m^3 (488 lb/ft^3). Metal in heads can be approximated by calculating the area of the blank (disk) used for forming the head. The required diameter of blank can be calculated by multiplying the head outside diameter by the approximate factors given in Table 10-62. These factors make no allowance for the straight flange which is a cylindrical extension that is formed on the head. The blank diameter obtained from these factors must be increased by twice the length of straight flange, which is usually 1½ to 2 in but can be up to several inches in length. Manufacturers' catalogs give weights of heads.

Forming a head thins it in certain areas. To obtain the required minimum thickness of a head, it is necessary to use a plate that is initially thicker. Table 10-63 gives allowances for additional thickness.

Nozzles and flanges may add considerably to the weight of a vessel. Their weights can be obtained from manufacturers' catalogs (Taylor Forge Division of Gulf & Western Industries, Inc., Tube Turns Inc., Ladish Co., Lenape Forge, and others). Other parts such as skirts, legs, support brackets, and other details must be calculated.

TABLE 10-62 Factors for Estimating Diameters of Blanks for Formed Heads

	Ratio d/t	Blank diameter factor
ASME head	Over 50	1.09
	30–50	1.11
	20–30	1.15
Ellipsoidal head	Over 20	1.24
	10–20	1.30
Hemispherical head	Over 30	1.60
	18–30	1.65
	10–18	1.70

d = head diameter
t = nominal minimum head thickness

TABLE 10-63 Extra Thickness Allowances for Formed Heads*

Minimum head thickness, in	Extra thickness, in		
	ASME and ellipsoidal		
	Head o.d. up to 150 in incl.	Head o.d. over 150 in	Hemispherical
Up to 0.99	1⁄16	1⁄8	3⁄16
1 to 1.99	1⁄8	1⁄8	3⁄8
2 to 2.99	1⁄4	1⁄4	5⁄8

*Lukens, Inc.

Made in United States
Orlando, FL
04 November 2026

90200802R00090